International Conference "Transport and Storage of Hydrocarbons"

IOP Conference Series: Materials Science and Engineering
Volume 357

Tyumen, Russia
10 – 12 January 2018

ISBN: 978-1-5108-6313-2
ISSN: 1757-8981

Printed from e-media with permission by:

Curran Associates, Inc.
57 Morehouse Lane
Red Hook, NY 12571

Printed by Curran Associates, Inc. (2018)

For permission requests, please contact the Institute of Physics
at the address below.

Institute of Physics
Dirac House, Temple Back
Bristol BS1 6BE UK

Phone: 44 1 17 929 7481
Fax: 44 1 17 920 0979

techtracking@iop.org

Additional copies of this publication are available from:

Curran Associates, Inc.
57 Morehouse Lane
Red Hook, NY 12571 USA
Phone: 845-758-0400
Fax: 845-758-2633
Email: curran@proceedings.com
Web: www.proceedings.com

Table of contents

Volume 357

International Conference "Transport and Storage of Hydrocarbons"

10–12 January 2018, Tyumen, Russian Federation

Accepted papers received: 20 April 2018
Published online: 18 May 2018

Preface

Papers

IOP Conf. Series: Materials Science and Engineering **357** (2018) 011001 doi:10.1088/1757-899X/357/1/011001

Preface

The International Conference "Transport and Storage of Hydrocarbons" was held in Tyumen, Russia, on **January 10-12, 2018**.

The 2018 Conference is aiming to convene undergraduates, post-graduates and researchers from Russia and other countries to present their research works and share their ideas on the most urgent topics in the sphere of transportation and storage of hydrocarbons.

The three-day Conference brought together the specialists from different countries and organizations and provided an excellent opportunity for knowledge exchange, making oral contributions and poster presentations, and initiating discussions on the topics that are of interest to the Conference participants.

The selected papers of the Conference cover a wide range of technical areas and modern aspects of transportation and storage of hydrocarbons.

IOP Conf. Series: Materials Science and Engineering **357** (2018) 011002 doi:10.1088/1757-899X/357/1/011002

Peer review statement

All papers published in this volume of *IOP Conference Series: Materials Science and Engineering* have been peer reviewed through processes administered by the proceedings Editors. Reviews were conducted by expert referees to the professional and scientific standards expected of a proceedings journal published by IOP Publishing.

IOP Conf. Series: Materials Science and Engineering **357** (2018) 012001 doi:10.1088/1757-899X/357/1/012001

Noospheric ecological imperative in culture of technocratic society

V D Shantarin, N A Bezzubtseva and Yu D Zemenkov

Industrial University of Tyumen, 38 Volodarskogo St., Tyumen, 625000, Russia

E-mail: shantarin33@mail.ru, muzemenkova@mail.ru

Abstract. Technogenic, dynamic and mobile civilization has been in a state of systemic ecological and economic crisis for several decades. Numerous studies of its origins and causes show that the culture of a technocratic society associated with a quantitative increase in consumption has exhausted its capabilities and proved incapable of alleviating the sustainable future of humankind. Comprehension of this problem gives grounds to believe that its solvation is largely predetermined by the formation of a new culture, the core of which is the noospheric ecological imperative: only the thing that does not violate the natural balance is allowed. Life in the conditions of imposed restrictions is connected with the change of worldview attitudes of society, understanding the ideas of noospheregenesis and eco-centrism, the need for strict regulation, economy and conscious restriction of consumption.

1. Introduction

The philosophical comprehension of the situation shows that the key to solving the problem is the increase in the society humanistic culture level, the limitation of the desire for material excess, the perfection of human qualities, the development and encouragement of the spiritual, its harmonious combination with the material in each individual person, the logical transition to the ideology of noospherism, the foundations of which were laid by V.I. Vernadsky [5, 6]. This transfers the problem of overcoming the crisis into the sphere of education. It is education that is called upon to nurture a "noospheric person" who will be able to voluntarily and consciously adopt a way of life within the framework of an ecological imperative, to implement such patterns of activity and behavior that would support the stability of natural biogeochemical cycles and social processes. The increasing contradiction between the interaction of Nature and Reason urgently requires both innovative approaches in the education of the younger generation, whose behavior should be adequate to the socio-natural dynamics, and the preservation and augmentation of the traditions of Russian education. The apparent contradictions of the modern era between the growing needs of society and the inability of nature to meet these needs have matured [7, 8].

2. Materials and methods

The aim of modern higher education is the formation of a professionally competent, socially active, creatively independent personality. The increasing role of education in the current socio-economic situation leads to the fact that along with traditional forms of education, more and more progressive innovative ideas are being increasingly used. The meaning of the word *education* is "to withdraw, induce, that is, to activate hereditary inclinations and abilities for transferring to a more developed, manifested state." It must be clearly understood: everything that is given to a person ensures one's

IOP Conf. Series: Materials Science and Engineering 357 (2018) 012001 doi:10.1088/1757-899X/357/1/012001

self-preservation in the world. The task of education is to help a person to move on the ladder of self-improvement to the heights of personal development. The Law of the Russian Federation "On Education" says: "Education means a purposeful process of education and training in the interests of man, society, and state, accompanied by a statement of the citizen's achievement of educational levels established by the state." "... The Russian Federation proclaims the field of education a priority. The content of education should ensure: the general and professional culture of society adequate to the world level; formation of an adequate picture of the world in the student; integration of the individual into national and world culture; formation of a person and a citizen integrated into modern society and aimed at improving this society ".

The main objectives of the social and economic policy of the Government of the Russian Federation determine the priority areas for updating the education system in order to achieve a new quality of education of specialists. In the National Project "Education", in "Support and Development of the Best Samples of Domestic Education" direction, the main block of activities is the stimulation of universities that actively implement innovative educational programs. The priorities of state policy, reflected in the National Project, determine the special degree of attention of the scientific and pedagogical community to the innovative processes in Russian education. Education plays a key role in ensuring the dynamic development of society, contributes to changing the system of values, priorities, ideals of each person for the sake of radical transformation of civilization. The idea of the development of the world as a whole, the idea of a man as a part of nature, about one's intellectual and spiritual power at the present time should become, in the authors' opinion, the world outlook basis of the educational process [9, 10].

The studied and generalized philosophical, scientific, and pedagogical literature shows that the need to create an educational system, focused on sustainable development, survival and development of modern civilization and preservation of the biosphere, makes the problem of noospheric education urgent.

A retrospective analysis of the literature makes it possible to conclude that now the teachers are busy with an intensive search for ways that will ensure a higher quality of education for university graduates. This can be achieved by actively involving students in the learning process.

Meanwhile, the crisis in the education system is determined by many modern thinkers not only as a turning point in the painful, but temporary transition of society from one era to another, but also as a fingerprint of the system itself. Radically changed civilizational realities run counter to the foundations of high school, and a controversial issue, what is better - acquisition and modernization of new or adherence to traditions and stability, remains open.

Today people are participants of a difficult process connected with the modernization of the whole system of Russian education. At the heart of the higher education development, there are still the "Charters" of higher education institutions, which are a reflection of the specific educational environment, prevailing in Russia, but not freed from the burden of problems originally predetermined by the very essence of the higher education system. In this context, the problems of upbringing come first, as the demands of the society for graduates have changed greatly, and not only in the professional sphere, but, first of all, in the social sphere.

Upbringing is a personality-oriented process regulated from within and determined by the inner world of a man. In the center of it, there is a man with his vital existential problems. The goal of education is to introduce a young man into the world of culture of the scientific and technical society. Personality, due to the active efforts of human consciousness, comprehends the values and knowledge offered by culture, develops its own attitude, and acquires independence, uniqueness, subjectness. Upbringing is not so much the appropriation of the meaning of the "I" as the movement of the subject beyond the limits of oneself, into another, more organized sense. Overcoming identification with the other actors in this process is possible through the development of the ability to create individual life mission. In education, the semantic core of upbringing, a constant comprehension (rethinking) of the new is carried out, which transforms the undeveloped into the acquired. Ambiguity of meanings and internal contradictions inherent in cultural reality initiate the process of interpretation. Interpretation is a creative act of changing accumulated meanings, a point of cultural growth. Modeling of upbringing allows us to look at the organization of upbringing itself in a different way, to bring the management theory of educational systems closer to the generally accepted theories of management without loss of content and uniqueness.

Specificity of upbringing in an engineering university is evident, first of all, because of the high consolidation of the internal educational space, oriented not to the non-science, but to the technical component of culture. Teachers in engineering high school are, as a rule, far from the theory of upbringing and in the upbringing process, they are focused primarily on their own life values and meanings, which, however, are not alien to their pupils. At the same time, the need for acquiring pedagogical knowledge, understanding and appropriating the psychological and pedagogical culture, understanding all aspects of the educational process, reaching an understanding of the essence of upbringing cannot but become a "stumbling block" for engineering teachers.

In such circumstances, the role, on the one hand, of a competent curator and mentor, which is a teacher-psychologist by vocation, on the other hand, of student self- and co-management, contributing to the development of socially significant qualities of personality and social competence, is increasingly growing. In this regard, the problem of humanization and humanitarization of education, which under the new methodology acquires a much deeper meaning than simply familiarizing the individual with a humanitarian culture, is quite clearly recognized [12].

It should be emphasized that almost all developed countries have passed the reform of the education system: the reforms of higher education have acquired the status of state policy; they were and are being invested with huge amounts of money. This is evident in view of the fact that the peoples and governments are aware of the depth of the connection between the level of higher education and progress. At the same time, reforms have never started in a successful system: why is it necessary to reform something that works in a proper way? At the same time, the perfect is always the enemy of the good. In Russia, the educational crisis has a dual nature. On the one hand, this is a clear consequence of the global education crisis; on the other hand, it is a consequence of the environmental and economic crisis taking place in Russia. The notorious conservatism had a significant impact on the development of the crisis: the teacher is more difficult to change than the student, as in the overwhelming majority of cases the higher-education teaching personnel are the established personalities who have chosen higher education as an area for their talent and work [13]. At the same time, adherence to traditions predetermined the positive aspects of the Russian higher school. It is obvious that:

- it traditionally focused on professional activities and has a close relationship with practice, developed system of career guidance, vocational training and retraining, as well as a system of employment;
- it is capable of training personnel in all areas of science, technology and production, popular in the modern world, and, if necessary, responds quickly to changed demands in the context of the "range of specialties" provided;
- in terms of graduate training volume, the quality of vocational training and staffing is one of the leading places in the world;
- it is fundamental, in particular in the natural sciences, and also provides a high level of technological training.

Why reform is an urgent need? Could it be only a tribute to the "pan-European" fashion? Let us recognize that the changes taking place in society are increasingly objectifying the shortcomings of Russian higher education, which were previously considered as indisputable advantages:

- an obvious "bias" towards the professionalization of higher education, to the detriment of the development of social competence;
- the status of a higher education diploma holder due to the system of free preparation does not mean being a part of social elite, it is perceived as a given;
- "dumbing-down" associated with averaging: the individual's desire to express themselves in a narrow industry stumbles upon an insurmountable set of obstacles to self-realization;

- excessive attention of higher inspection bodies, leading to averaging of the teaching staff.

This and some other factors do show the need for specific reforms.

In the authors' opinion, the breakdown of technical and humanitarian training, the ever-increasing "technicalization" and dehumanization of training, caused by the extensive nature of mastering knowledge, skills and abilities (and with increasing amounts of information, this is unproductive and disastrous) lead to a depletion of the content of education, a decrease in the creativity of graduates, and result in the reduce of the science and production potential.

The total alienation of man from the natural world should also be regarded as the most important sign of the crisis in the education system: he does not see himself as a part of this world, he opposes nature and culture, he is deprived of a sense of authenticity, meaningfulness, he loses his connection with the world, which strengthens him in his own being. The existing educational system in Russia, still one of the best in the world, still lags far behind the rapidly changing conditions of the existence of a technocratic society of the 21st century [14].

The main reason for this conclusion is that the education system, which has always been an important civilizational factor, should not develop spontaneously today, under the conditions of Russia's ecological and economic slowdown. The correct strategy for the orientation of this system, taking into account the long-term development goals of the country, directed to the NOOSPHERIC ENVIRONMENTAL IMPERATIVE, is extremely relevant right now, since it can become a decisive factor in Russia's transition to a model of sustainable and safe development. After all, the education system largely forms that public consciousness which will determine the future of the civilized development of Russia.

"The National Doctrine of Education in the Russian Federation" sets a number of priorities for creating a basis for sustainable socio-economic and spiritual development of Russia, ensuring a high quality of life for the people and national security. The Doctrine gives an idea of the general outlines, goals and objectives of the development of Russian education in the coming years, being a concept of transition from the Soviet model of education to a market-democratic one. It should provide conditions that ensure the development of a new model of education for the 21st century. The new system of education, in the authors' deep conviction, should be oriented toward the model of the development of civilization and sustainable development of the world community. The transition to a new civilizational model of education will allow not only to ensure human rights and freedoms, to increase the priorities of the moral and fair mind, spiritual values and ecohumanism, but also to achieve the overall security of development, and the harmony of man, society and nature.

According to A.D. Ursul, sustainable development is understood as development ensuring a balanced solution of demographic problems, socio-economic tasks, questions of maintaining positive environment and natural resource potential in order to meet the needs of present and future generations of people [4].

The transition to a model of sustainable development of education is caused by fundamental contradictions of both the world community and the education itself. As noted by the experts of the International Commission on Education for the 21st Century, headed by J. Delors, education lags behind the demands of the time, contradicting not only the present, but also the future. According to experts, almost all the educational systems do not have a clear focus on the future. The value system that operates in the educational sphere is not adapted to the future, which can satisfy both the humankind and the individual. It is possible to say that the current education system does not suit neither society nor nature; it aggravates the systemic crisis of civilization and contributes to the further degradation of the biosphere.

One predicts that there must be a cardinal turn to the future, determined by the goals of sustainable development in the entire educational system of the world community. As A. Toffler pointed out, the consciousness of people should be focused on the future, a "futurization" of consciousness is necessary. It is necessary to ecologize and futurize the consciousness; moreover - to form the ecological consciousness as the core and the dominant component of the future noospheric thinking. Change in public consciousness is possible only as a result of using the most mass social process, called education. Education should deal with the same specific task that Russia faces. Precisely, the education system, implementing market-democratic transformations, should now turn to the principles of a sustainable development model functioning. In comparison with other spheres of social activity, education should move faster. According to A.D. Ursul, the share of knowledge about the past is equal to 95% in modern education.

The new component of education will no longer be translational and communicative; it will be generative and superior. In this regard, the need arises for a transition to a whole new round of the educational process - a noospheric education, correlated with the aims of education, coordinated with other structural components of the system of higher education, and most importantly - with the interests, aspirations and opportunities of students and teachers. Noospheric education puts intellectual-spiritual and rational-information factors and resources at the forefront.

Noospheric education, based on the views of outstanding thinkers of the present, such as V.I. Vernadsky (theory of the biosphere and noosphere), K.N. Wentzel (space pedagogics), N.N. Moiseev (ideas of coevolution and ecological imperative), attracts more and more attention of the pedagogical science, which develops pedagogical systems that are adequate to the modern development of civilization. The concept of the noosphere, developed by Vernadsky, has a great methodological and heuristic significance in understanding the interaction of society and nature and plays a big role in the postmodern thematization of sustainable development. In it: 1) society and nature are considered in indissoluble connection, within the framework of an integrated system; 2) alongside with unity, the fundamental qualitative change in society and the rest of nature is fixed, society is considered not as one of the equal elements, but as the higher, organizing element of the system.

Based on the ideas of collective creation, it supposes a harmonious interaction of all spheres of the personality of the students, filling the process of education and upbringing with moral and ethical norms, universal human values, and philosophical sense of being. Thus, it is quite obvious that education needs to be refocused on new universal principles, values and objectives that correspond to the new civilization paradigm [1,2,3].

Noospheric education is aimed at the formation of the intellectual elite of society, with a developed self-regard, the consciousness of its free, creative essence, the sense of responsibility for the possible choice within the boundaries of this freedom and understanding of the moral duty to oneself, other people, the Motherland, and humankind.

Teaching students new worldview and social attitudes is based on humanistic ideas, universal values and ideals, priority of a healthy way of life, axiological aspect, scientific knowledge, philosophical world view, research skills and abilities, ways of cognizing the surrounding world that meet the requirements of nature-conserving imperatives (the system of norms and rules of interaction "man-society-nature").

In the authors' opinion, the required quality of an educational process can be achieved if the process of education, upbringing and development is directed towards the formation of a person who is conscious of one's responsibility for the active evolution (co-evolution) of the biosphere and society. At the same time, the pedagogical system is built on the basis of creative activity that allows us to master, first of all, the ways of interaction between Nature and Reason which forms the substantive and methodical aspects of noospheric education:

1. **Substantive**: professional, ecological, scientific-research, valeological, social, civil-patriotic, moral-legal, moral-axiological, humanistic, ethical, cultural-aesthetic, cultural-historical, philosophical, information-bibliographic.
2. **Methodical**: developing, integrative, active-play, self-creative, sociometric, self-management.

3. Conclusion

The above-mentioned landmarks of noospheric education, means of their implementation can be used in the educational and extra-curricular educational process as an example of the formation of a student of the modern pattern, as the basic paradigm of modernization of education. The socio-ecological, ecohumanistic approaches and the concept of sustainable development of modern society should be fundamental for it, and the model of noospheric education that realizes the principle "the modern generation is not the last link in human civilization" should become its methodological basis.

References

[1] Wentzel K N 1993 *Notes on space education.* (Free education. M.)
[2] Vernadsky V I 1989 *General concept of the biosphere. The Beginning and Eternity of Life. comment.* (Moscow: Sov. Russia)
[3] Moiseyev N N 1998 Once again on the problem of co-evolution. *Problems of Philosophy* **8**
[4] Ursul A D 2003 The concept of sustainable development. Sustainable development of noospheric orientation. *Scientific and methodical journal "Biology at school"* **1**
[5] Shantarin V D, Zemenkova M Y, Zemenkov Y D et al. 2016 Development of Thermophysical Hydrocarbon Wastes Pyrolysis Model (in the Case of Wood) *IOP Conference Series: Materials Science and Engineering* **154** DOI: 10.1088/1757-899x/154/1/012012
[6] Shantarin V D, Zemenkova M Y, Zemenkov Y D et al. 2016 Development of Nature Protection Technologies of Hydrocarbon Wastes Disposal on the Basis of High-Temperature Pyrolysis. *IOP Conference Series: Materials Science and Engineering* **154** DOI: 10.1088/1757-899x/154/1/012016
[7] Zemenkova M, Zemenkov Y, Gladenko A, Podorozhnikov S 2016 Estimation of Emissions During Monitoring of Pipelines in the Dynamic Mode of Operation. 5th International Scientific Conference Integration, *Partnership and Innovation in Construction Science and Education,* 86 DOI: 10.1051/matecconf/20168604053
[8] Zemenkova M, Shalay V, Zemenkov Y, Kurushina E 2016 Improving the Efficiency of Administrative Decision-Making when Monitoring Reliability and Safety of Oil and Gas Equipment *MATEC Web of Conferences* **73** DOI: 10.1051/matecconf/20167307001
[9] Vlasenko V S, Slesarenko V V, Gulkov A N and Lapshin V D 2015 *Preparation of arctic oil and gas condensate deposit formation fluid for storage and transport in the form of hydrocarbon hydrate-containing dispersed system* Proceedings of the International Offshore and Polar Engineering Conference 1863
[10] Vlasenko V S, Slesarenko V V, Gulkov A N 2016 *Using of triple-flow vortex tubes in associated petroleum gas preparation units.* Proceedings of the International Offshore and Polar Engineering Conference, **2016**, 46-51
[11] Gulkov A N, Lapshin V D, Slesarenko V V, Morozov A A, Solomennik S F 2015 *Adiabatic conversion method for the development of marine hydrate deposits.* Proceedings of the International Offshore and Polar Engineering Conference, 2015-January, **2015**, 73-76
[12] Brand A E, Chekardovskiy S M, Akulov K A 2017 The Analysis of the Development Dynamics and Structural Balance of Solar Energy in the World. *IOP Conference Series: Earth and Environmental Science* **50** DOI: 10.1088/1755-1315/50/1/012034
[13] Konev V, Polovnikov E, Krut O, Merdanov S, Zakirzakov G 2017 Investigation and Development of the Thermal Preparation System of the Trailbuilder Machinery Hydraulic Actuator *IOP Conference Series: Materials Science and Engineering* **221**, DOI: 10.1088/1755-1315/221/1/012001
[14] Polyakov D V, Merdanov Sh M 2005 Equipment for snow utilization using electric power *Stroitel'nye i Dorozhnye Mashiny* **12** 24-26

IOP Conf. Series: Materials Science and Engineering **357** (2018) 012002 doi:10.1088/1757-899X/357/1/012002

Gas hydrate suspensions formation and transportation research

A N Gulkov[1], S Gulkova[1] Yu D Zemenkov[2] and V D Lapshin[1]

[1]Department of Oil and Gas Engineering and Petrochemistry of the Engineering School of the Far Eastern Federal University, Vladivostok, Ajax, Russia
[2] Industrial University of Tyumen, 70 Volodarskogo St., Tyumen, 625000, Russia

E-mail: alexdvgtu@mail.ru, muzemenkova@mail.ru

Abstract. An experimental unit for studying the formation of gas hydrate suspensions and their transport properties is considered. The scheme of installation and the basic processes, which can be studied, are described. The results of studies of gas hydrates and a gas hydrate suspension' formation in an adiabatic process in a stream of seawater are given. The adiabatic method of obtaining gas hydrates and forming gas hydrate suspensions is offered to use. Directions for further research are outlined.

1.Introduction

One of the main research areas of the Oil and Gas Engineering and Petrochemistry Department of the Engineering School of Far Eastern Federal University in the last three years is the study of the gas hydrates, gas hydrate suspensions formation' processes and the possibility of their transportation.

The use of gas hydrates for the purpose of utilization of associated petroleum gases is a promising area. Earlier, the theoretical possibility of using this technology in oil fields for utilization of associated petroleum gases under certain gas factors was shown (V.D. Lapshin, A.N. Gulkov et al., 2015). Furthermore, a method of delivering natural gas to consumers in the form of a gas hydrate suspension was offered and patented (A.N.Gulkov. Patent RU MKI7S 07 C 5/02, Patent RU MKI7S 07 C 5/02, Patent RU MKI7 F17D1/02) [5, 6].

One of the starting points for the studies of transportation of gas hydrate suspensions is work on the Cold Flow technology of Norwegian colleagues from the Norwegian University of Science and Technology (NTNU) (E.Bokin et al., 2010). Also, in the early elaboration of the theoretical part of the guesses, materials on the kinetics of hydrate formation (D.Sloan, C.Koh et al., 2011) and studies of the factors affecting the clathrate structures fill rate were considered (R.Sun, Z. Duan, 2005).

It is proved that in order to successfully implement the method offered, it is necessary to investigate the main factors affecting the formation of gas hydrates, gas hydrate suspension and its stability in pipeline transport. To study these processes, an experimental unit was created consisting of a refrigerating machine, an evaporator reactor, a circulating loop, a gas measuring and feeding system, and an automation system supporting specified parameters [2, 3, 4, 7].

2. Materials and methods

In the Laboratory of Gas Hydrates, the experimental unit UOTG 1416.05-01 was created and put into operation (Figure 1). The unit allows studying the processes of formation, dissociation and transportation of gas hydrate suspension and changes in the rheological characteristics of the fluids.

The unit consists of a refrigerating machine, a reactor-generator of a refrigerated machine of a flooded type, a circulating hydraulic circuit, a system for measuring and feeding the gas of the hydrating agent, an automation system supporting the set parameters, and a system for recording the measured parameters.

Figure 1. Experimental unit UOTG 1416.05-01 for the study of formation and transportation of gas hydrate suspensions

A schematic diagram of the hydraulic circulation circuit of the unit used is shown in Fig. 2.

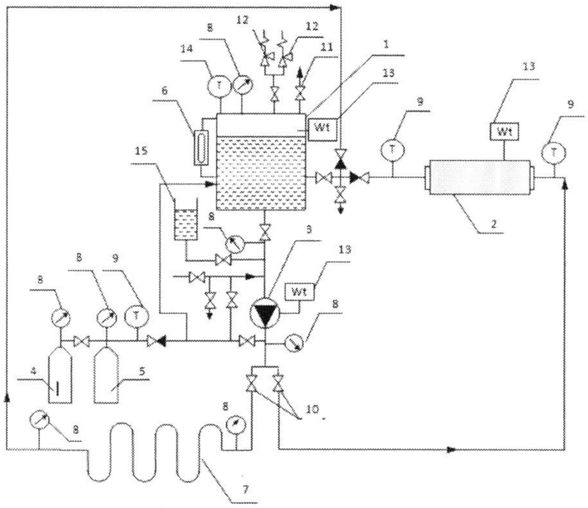

Figure 2. A schematic diagram of the laboratory scientific research unit UOTG 1416.05-01. 1 Homogenizer, 2 Reactor-generator (evaporator of the refrigerating machine of the flooded type), 3 Circulation pump, 4 Cylinder with hydrate-forming agent, 5 Measuring cylinder with hydrating agent, 6 - Inspection column, 7 Coil, 8 Pressure gauges, 9 Thermometers, 10 Isolation valves, 11 Atmospheric valve, 12 Safety valve, 13 Watt meter, 14 Thermometer, 15 Feeding tank with saltwater.

The hydraulic circuit is equipped with circulation pump 3, which allows moving both gas-saturated solutions and gas hydrate suspensions. In reactor-evaporator 2, the gas hydrate phase is formed due to the removal of the heat of hydrate formation by the refrigerating machine. Then the gas hydrate suspension is finally formed in the vessel of homogenizer 1. After that, isolation valve 10 is switched to the pumping mode through coil 7 simulating the field pipeline to investigate the transport properties of the resulting suspensions.

When setting up the installation systems, cyclic tests of the hydraulic circuit and the chiller were carried out. According to the test results, dependencies of the system chilling time on the flow rate in the hydraulic circuit were obtained (Fig. 3).

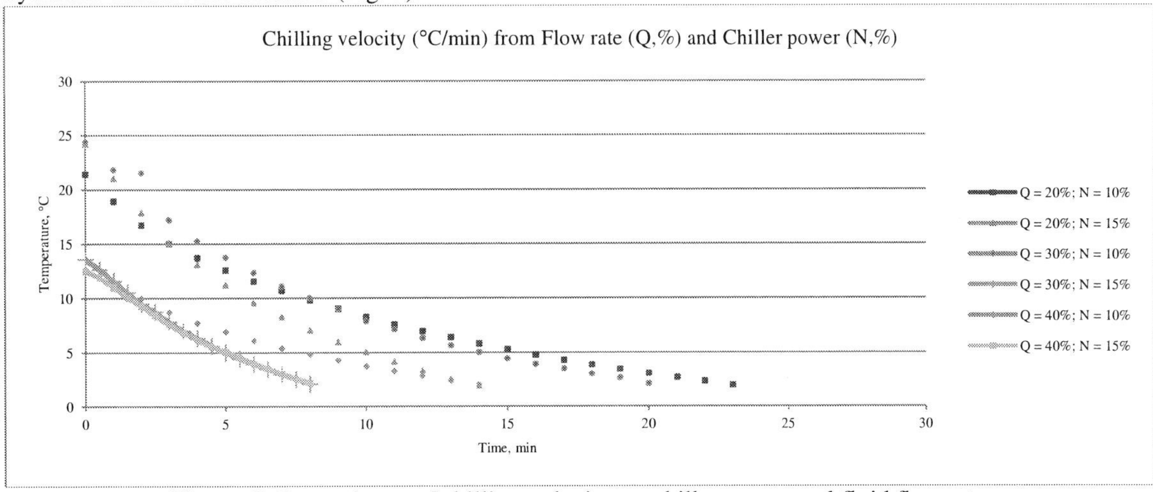

Figure 3. Dependence of chilling velocity on chiller power and fluid flow rate

The unit under consideration makes it possible to carry out investigations in the pressure range from 0.1 to 5.0 MPa and temperatures from -10 to 25 °C. The initial plan provides the implementation of two fundamentally different methods for obtaining a gas hydrate suspension.

The first method: obtaining a suspension at the apparent interface between the liquid and gas phases, by gradually chilling the pumped medium and sharply raising the pressure in the pipeline to transfer the entire system to the coordinates of the imaging point of system condition $H_2O - CO_2$.

The second method: obtaining a suspension from the saturated solution of the hydrating agent by cooling it at a pressure sufficient to transfer the whole system to the coordinates of the imaging point of system condition $H_2O - CO_2$.

When obtaining a gas hydrate suspension, it is necessary to take into account the possibility of forming large aggregates of crystalline structures on the heat exchange surfaces and bends of pipelines. To prevent the hydrate crystals from adhesion, a homogenizer is provided in the preparation of the suspension by the first method, mechanically destroying the hydrate aggregates. In the second method, adhesion of hydrate crystals is prevented by grinding them in a reactor having a flat scraper that removes growing crystals from a cylindrical heat exchange surface. The experiments carried out showed the possibility of obtaining gas hydrate suspensions under certain system parameters [9, 10, 11].

3. Results and Discussion

Using the unit, studies of the gas-hydrate phase CO_2 adiabatic formation' processes in high-mineralized water and in normally mineralized water (tap water) due to changes in pressure and temperature in the system were held. As water of high mineralization, the seawater of the Peter the Great Bay and the Sea of Japan was used.

In the first experiment, the mineralized water circulating in the circuit was saturated with carbon dioxide and cooled by a refrigeration machine in a reactor-generator. Then, the pressure in the circulation system increased due to the supply of carbon dioxide, resulting in the formation of a gas hydrate in the form of a finely dispersed suspension that could be observed in the inspection windows (Fig. 4).

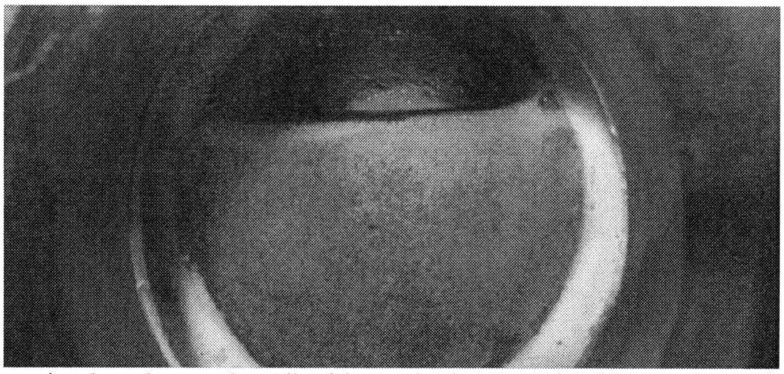

Figure 4. A suspension based on carbon dioxide gas hydrate obtained in a reactor of a laboratory unit

Let us consider the processes occurring on the P-T diagrams. At initial Point 2 (P = 2 MPa, t = + 12 □, P-T diagram, Figure 5), the water was saturated with carbon dioxide to a state of complete saturation (3.3 g CO_2/ 100 g г H_2O) (E. Perkins, 2003) and then cooled to a temperature of + 4 °C (Point 2 '). At Point 2, temperature reduction ceased despite the fact that the refrigeration unit continued to operate in the same mode with hydrate particles appearing in the sight glass installed in the plant's high-pressure circuit.

It should be noted that the termination of the decrease in the temperature of the $H_2O - CO_2$ system on the hydration line (Point 2 ') under the operating refrigeration system is explained by the fact that the formation of all gas hydrates, and CO_2 hydrate in particular, is accompanied by a powerful thermal effect. The formation of hydrate CO_2 at a pressure of 1 MPa is accompanied by the release of 430-450 kJ / kg of thermal energy (E.D. Sloan, C. Koh, 2008). Of special note is the fact that hydrate formation at Point 2 '(Fig. 5) occurs inside the heat exchanger, from which the heat of hydration was removed by the refrigeration machine due to the boiling of coolant R22 [8].

As a result of gas hydrate phase formation, hydraulic resistance of circulation loop system $H_2O - CO_2$ in the annular gap of the reactor dramatically increased. It led to an increase in the electric power consumption on the circulation pump drive and the rotor of the reactor, which was recorded by means of wattmeter having a channel for outputting information to the computer. In general, such change in the rheological characteristics of a saturated hydrating solution converted to a gas hydrate suspension is expected and described in previous studies (V.D. Lapshin, A.N. Gulkov et al, 2015).

The second experiment, conducted as part of the adiabatic formation of the gas hydrate phase, was carried out at a pressure of 1 MPa (Process 1-6, Fig. 5). The water saturated with carbon dioxide was pumped by the circulation pump through the reactor-evaporator, as was mentioned above, under a pressure of 1 MPa.

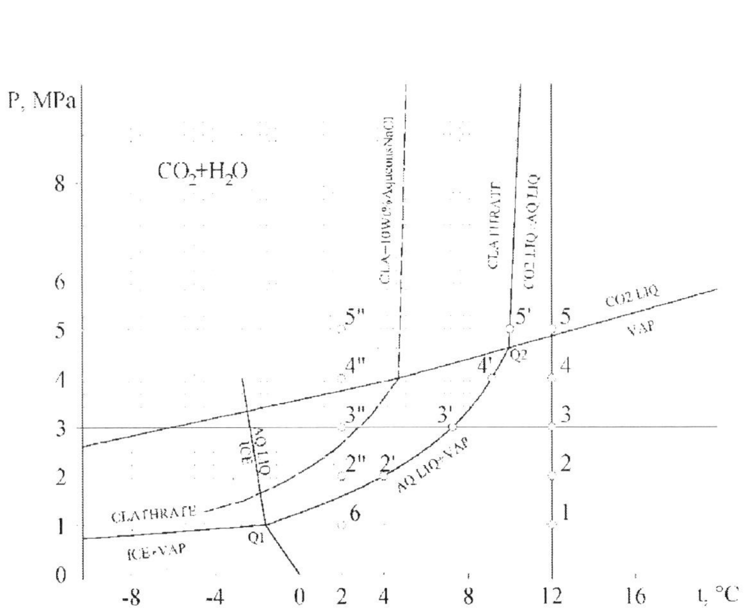

Figure 5. Processes in the laboratory unit shown in the Stackelberg-Rosenbom diagram

Thus, through the operation of the refrigeration machine, the temperature of the homogeneous system $H_2O - CO_2$ was reduced from 12 °C to 2 °C, with significantly less energy consumption than in the previously described case, when a suspension with a continuously increasing fraction of the dispersed phase was formed in the circulation loop.

After reaching a temperature of 2 °C, the pressure of circulating water, saturated with carbon dioxide, in the hydraulic circuit increased sharply to 2 MPa (Point 2", Fig. 5), which led to the instant formation of the gas hydrate phase in the entire circuit simultaneously (Fig. 6).

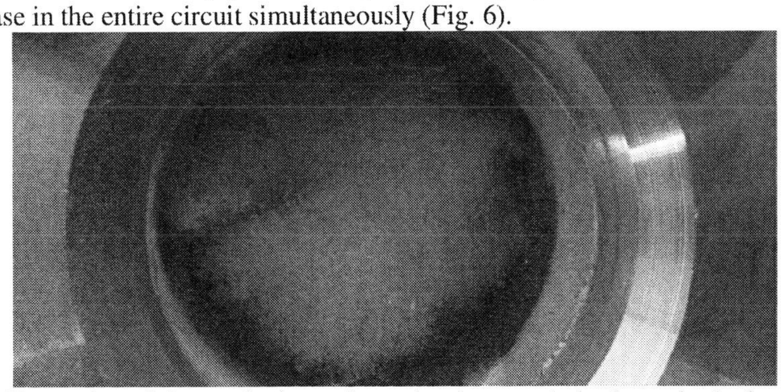

Figure 6. Formation of the gas hydrate phase simultaneously in the entire hydraulic circuit

Further, the experiments were carried out on isobars 3, 4 and 5 MPa, which led to an increase in the content of the gas hydrate phase in the circulating hydraulic circuit with the increase of pressure in it. In this case, each time the imaging point of system state $H_2O - CO_2$ returned to the equilibrium line (at Points 3 ', 4', 5 '). It is characteristic that an increase in the content of the gas hydrate phase was observed with increasing pressure only up to a value corresponding to the coordinates of quadrupole point Q_2. It should be noted that above quadrupole point Q_2, carbon dioxide passes from the vapor state to the liquid state, which changes the character of the thermobaric conditions of hydrate formation on its basis (E.D. Sloan, C. Koh, 2008) [1].

4.Conclusions

The conducted experiments testify to the operability of the unit offered.

The possibility of realizing two fundamentally different methods of gas hydrate suspensions' formation has been confirmed.

Studies confirm the validity of the assumption that there is no need for mechanical destruction of large hydrate crystals to prevent the formation of hydrate plugs in pipelines at certain phase and environment ratios in the dispersed system.

The validity of the assumptions about the transport properties of gas hydrate suspensions has been experimentally demonstrated.

Further studies will be focused on studying the influence of various factors on the stable gas hydrate suspensions' formation and their transport properties.

References

[1] Sloan D, Koh C, Amadeu K Sum, Adam L Ballard, Creek J, Eaton M, Lachance J, McMullen, Palermo T, Shoup G, Talley L 2011 *Natural Gas Hydrates In Flow Assurance* (Elsevier) ISBN 978-1-85617-945-4

[2] Dandy Sloan E, Koh C 2008 *Clathrate hydrates of natural gases. – 3rd ed.* (CRC Press)

[3] Bokin E, Febrantini F, Khabibullin E 2010 Carlos Eduardo Sanches Perez. *Flow Assurance And Sour Gas In Natural Gas Production* (Throndheim)

[4] Ernie Perkins 2003 *Fundamental Geochemical Processes Between CO2, Water and Minerals.* (Edmonton, Alberta)

[5] Gulkov A N, Lapshin V D 2007 *Device for preparation of natural gas for transportation. Patent MKI7S 07 C 5/02*

[6] Gulkov A N, Lapshin V D 2002 *Device for transportation of natural gas. Patent MKI7 F17D1/02*

[7] Sun R, Duan Z 2005 *Prediction of CH4 and CO2 hydrate phase equilibrium and cage occupancy from ab initio intermolecular potentials* (Elsevier)

[8] Lapshin V D, Gulkov A N, Morozov A A, Nikitina A V 2015 *Utilization Of Associated Petroleum Gas And Transport It As A Hydrate Dispersed System,* Proceedings of the Scientific Conference Modern Technologies and the Development of Polytechnic Education (WIT Press)

[9] Podorozhnikov S, Shabarov A, Shirshova A, Berdyshev A 2016 Method for Diagnosing Formation of Gas Hydrates in Gas Pipelines *MATEC Web of Conferences* **73** DOI: 10.1051/matecconf/20167302025

[10] Kurushina V, Kurushina E, Shabarov A 2016 Technology Audit: Assessment of Innovative Portfolio *MATEC Web of Conferences* **73** DOI: 10.1051/matecconf/20167307028

[11] Nelyubov D, Shabarov A, Podorozhnikov S 2016 Complex-action reagent for oil-infield and oil-trunk pipelines *Solid State Phenomena* **871** 185 DOI: 10.4028/www.scientific.net/MSF.871.185

IOP Conf. Series: Materials Science and Engineering **357** (2018) 012003 doi:10.1088/1757-899X/357/1/012003

Reduction of movement resistance force of pipeline in horizontal curved well at stage of designing underground passage

V S Toropov and S Yu Toropov

Industrial University of Tyumen, 38 Volodarskogo St., Tyumen, 625000, Russia

E-mail: vladimir.s.toropov@gmail.com

Abstract. A method has been developed to reduce the resistance to movement of a pipeline in a horizontal curved well in the construction of underground passages using trenchless technologies. The method can be applied at the design stage. The idea of the proposed method consists in approximating the trajectory of the designed trenchless passage to the equilibrium profile. It has been proved that in order to reduce the resistance to movement of the pipeline arising from contact with the borehole wall, the profile of its initial and final sections must correspond, depending on the initial conditions, to the parabola or hyperbolic cosine equation. Analytical dependences are obtained which allow supplementing the methods of calculation of traction effort in trenchless construction for the case when the profile of the well is given by an arbitrary function.

1. Introduction

The use of trenchless technologies for the construction of pipeline passages requires careful attention to the design of the well profile in which one or another form of communication is to be laid. Parameters of the profile of trenchless pipeline passages are closely related to the relief of the terrain, the ground conditions in the longitudinal section, the location of the boundary of the zone of possible channel changes, and of course the technology of production.

The key parameters of the passage itself are usually the length of a trenchless pass and the greatest depth of the pipeline section, especially if one constructs a passage under a water obstacle.

The listed parameters are obtained by calculation at the design stage, and as a rule, they are specified at the place of construction of the underground or underwater passage during technological design.

Erroneous methods of calculation and design have to be compensated in practice by the capabilities of drilling rigs and auxiliary equipment [1]. Statistics of emergency situations in trenchless construction indicates that the design and construction methods for passages are far from perfect [2]. In Russia, each industry sector develops standard documentation on this issue separately.

Obviously, in this situation, there is an industry demand for improving trenchless technologies, including methods of calculation and design. This potential is not fully realized, and improvement of modern design methods can well provide significant savings in resources and production capacity [3].

2. Materials and methods

Unlike steel pipelines, the buoyancy of which depends on the diameter and wall thickness, plastic pipes always have positive buoyancy. This fact does not allow optimizing the profile of the well by the criterion of weight or distributed load [4]. At the same time, the main force component of the pipeline movement resistance in a curved well is the normal force acting on the borehole wall from buoyancy forces [5].

When ballasting with water or drilling mud, the small values of the rigidity of plastic pipes and the relatively large areas of the pipeline burial, both during drilling wells and when entering the day surface, in these areas allow us to consider a ballasted plastic pipeline as a flexible thread of the third form of sagging. In this case, the point of access to the day surface and the point of conjugation of curvilinear and horizontal sections of the borehole are considered suspension supports. In the absence of a horizontal section, the pipeline can be considered as a thread of the first form of sagging.

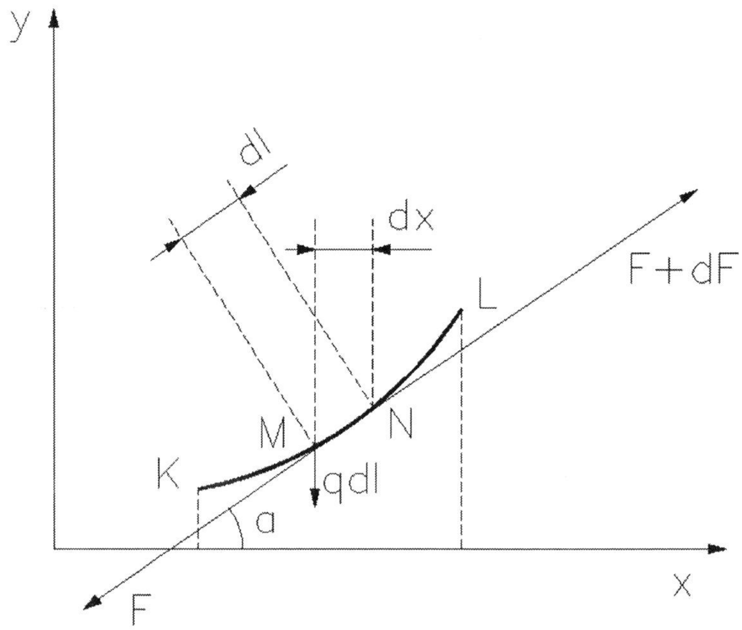

Figure 1. Load distribution along the length of the pipeline: *KL* – the considered section of the curved well; *MN* – the infinitesimal section of the well; *dl* – the infinitesimal increment of the trajectory along the length of the profile; *dx* – the infinitesimal increment of the trajectory along the horizontal projection of the profile; qdl – the magnitude of unit load applied to an infinitesimal section of the profile length; F – the pipeline pulling force; dF – the pulling force of an infinitesimal section of the pipeline; *a* – the slope angle of an infinitesimal section of the profile to the horizontal at the given point of the profile.

Such assumption allows the well to be profiled from the condition of sagging of the pipeline pulled along the axial well, without touching the walls. Obviously, in real conditions, because of the uncertainty of the "suspension" conditions, complete alignment of the well and pipeline cannot be achieved. Therefore, the task of profiling involves reducing the component of frictional forces on the side walls by approximating the shape of the borehole profile to the equilibrium profile [6, 7].

When calculating the well trajectory for the equilibrium profile depending on the adopted scheme, two variants of load distribution are possible [8]:

1. Load distribution along the length of the pipeline.
2. Load distribution along the pipeline projection.

In the first case, according to the calculation scheme shown in Figure 1, let us obtain the equation of the well trajectory in the form:

$$y = \frac{H_h}{2q}\left(e^{\frac{q}{H_h}x} + e^{-\frac{q}{H_h}x}\right) \quad (1)$$

where $\mathbf{H_h}$ - horizontal component of the pipeline tension force;
q - unit load distributed along the length of the pipeline;
x – current coordinate of the passage profile;
y – elevation mark of the current coordinate of the passage profile.

Let us take into account that the ratio of the horizontal component of the pipeline tension force to the unit load will be equal to the slope angle of the infinitesimal section of the profile to the horizontal at the given point of the profile:

$$a = \frac{H_h}{q}. \quad (2)$$

Then, (1) has the form:

$$y = \frac{a}{2}\left(e^{\frac{x}{a}} + e^{-\frac{x}{a}}\right) = a \cdot ch\frac{x}{a}. \quad (3)$$

Thus, let us obtain the well-known equation of catenary. That is, from the point of view of load distribution along the length of the pipeline, the well profile corresponding to the hyperbolic cosine equation will be optimal to minimize the resistance to pipe movement [9].

In the other case, if one distributes the load along the horizontal projection of the borehole profile, it is possible to use the calculation scheme shown in Figure 2 and compose the following equation:

$$\frac{d^2y}{dx^2} = \frac{q}{H_h}. \quad (4)$$

Solving this equation, one gets:

$$y = \frac{1}{H_h}\int dx\left[\int q \cdot dx\right] + C_1x + C_2 \quad (5)$$

where C_1 and C_2 – constant values.

Let us introduce the notation:

$$\varphi(x) = \frac{1}{H_h}\int dx\left(\int q \cdot dx\right). \quad (6)$$

And let us reduce equation (5) to the form:

$$y = \varphi(x) + C_1x + C_2. \quad (7)$$

In equation (6), let us denote the integral by $\varphi(x)$ and keep in mind that the profile of the well must pass through points ***K*** and ***L***. Then one obtains the limiting conditions:

$$\begin{cases} \varphi(x_0) + C_1x_0 + C_2 = y_0 \\ \varphi(x_1) + C_1x_1 + C_2 = y_1 \end{cases}. \quad (8)$$

From these conditions, one can determine both constants.

In a particular case, with a constant load along the projection of the well, the profile will have the form of a parabola.

IOP Conf. Series: Materials Science and Engineering **357** (2018) 012003 doi:10.1088/1757-899X/357/1/012003

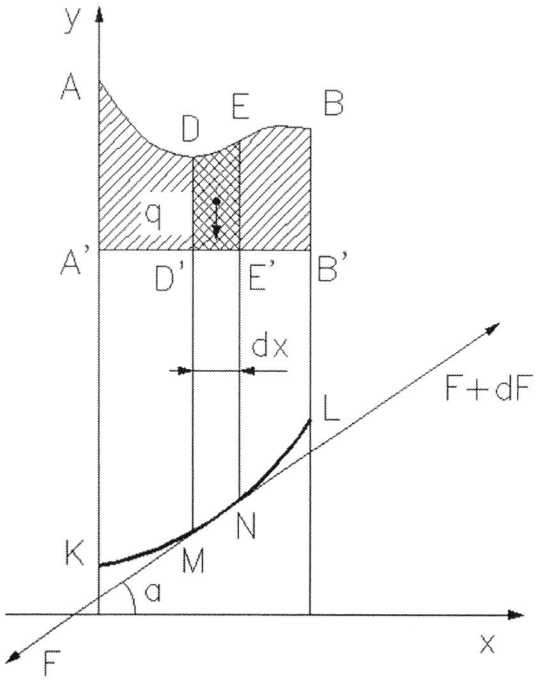

Figure 2. Distribution of load along the horizontal projection of the pipeline: *ADEBB'E'D'A'* – distributed load diagram for the pipeline.

Thus, in order to reduce the resistance force from contact with the borehole wall, the profile of its initial and final sections must correspond, depending on the initial conditions, to the parabola or hyperbolic cosine equation.

For practical application of the obtained dependences, it is necessary to supplement the method of calculation of the resistance force for the case when the well profile is given by an arbitrary function:

$$y = F_{bh}(x). \quad (9)$$

Figure 3 shows the profile of such well.

Drawing a tangent to the profile at point *B*, let us obtain the angle of inclination *a* of the profile at a given point to axis *Ox*, i.e. to the day surface. This angle will be:

$$\alpha = \left| arctg\left(F'_{bh}(x)\right)\right|. \quad (10)$$

For further calculations, it is necessary to set the accuracy value. Since the profile of the well is given by an arbitrary function, the radius of curvature can vary continuously throughout the entire length of the passage. So it usually happens in practice. Just as the process of drilling and expanding the pilot well, the process of pulling the working pipeline has a cyclic character [10], since the string consists of separate drill pipes connected to each other. They have the property of bending within certain limits, but the most significant change in the radius of curvature of the profile occurs just at the junction of two adjacent drill pipes.

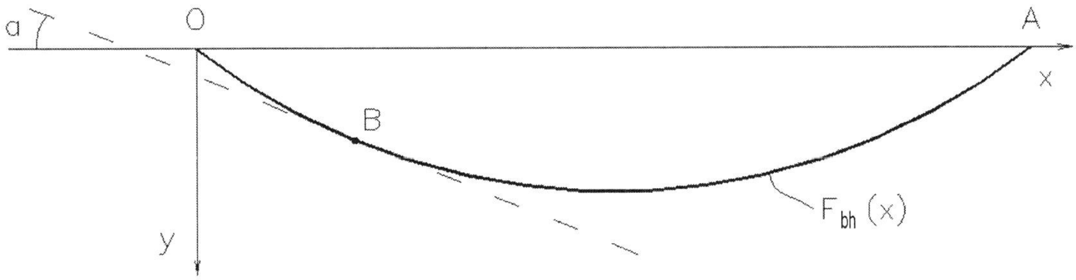

Figure 3. Trenchless passage profile given by an arbitrary function: O – spudding point; A – exit point to the day surface; y – depth of profile at the calculated point; B – current profile point for calculating the radius of curvature; $F_{bh}(x)$ – arbitrary function with a passage profile; a – angle of curvature of the profile at the current point.

This fact allows us to set the calculation accuracy in the case of an arbitrary curvature function in two ways:

1. For extended passages, let us take this value divisible by the linear dimension of one drill pipe.
2. For short-range passages, let us take the accuracy value divisible by a certain fraction of the linear dimension of the drill pipe used.

Thus, by setting calculation accuracy t_c in one way or another, let us calculate angle a through every t_c of the values of function (9). The difference between the current and previous values of a gives us angle a_w:

$$\alpha_w = \alpha_i - \alpha_{(i-1)} \tag{11}$$

where a_i - current value of the angle of inclination of the well profile at a given point to the day surface;

$a_{(i-1)}$ - previous value of the angle of inclination of the well profile at distance t_c from the current value.

Whence, the current radius of curvature of the profile will be:

$$r = \frac{t_c}{2\sin\frac{\alpha_w}{2}}. \tag{12}$$

This dependence of the variable radius of the curvature of the profile on the current coordinate with a given accuracy can be used to calculate any components of the resistance force to movement of the pipeline in a horizontal well. Movement can be carried out by pulling or pushing [11], or using a combination of basic and auxiliary machines and mechanisms.

3. Results and Discussion

As a result of the study, a designer using any specific technique for designing trenchless pipeline passages has the opportunity to use the obtained analytical dependences to minimize the calculated pulling force of the working pipeline into the well [12].

In horizontal directional drilling, the overall result is formed taking into account a combination of various factors, both minor and major [13]. And the first thing to do when designing a passage is to approach the profile building responsibly.

With due consideration of all details and a careful approach to designing and planning of operations, it is possible to obtain such design effort of pulling a working pipeline at the design stage that will allow using a less powerful and therefore cheaper installation from the existing standard size range [14]. Thus, there is a possibility of a situation in which there will be a particular case of the implementation of the law of transition of quantity to quality. This, naturally, will mean a significant reduction in the cost of the pipeline passage construction and an increase in the overall economic efficiency of construction.

4. Conclusion

The methods discussed in the paper can be used in the design of underwater and underground trenchless pipeline passages through any kinds of obstacles.

Any model of the pipeline pulling process used to calculate horizontally directed wells is provided in addition to taking into account properties of soils and the rheological model of the drilling mud, also taking into account the curvature of the profile and its effect on the amount of force for pulling the pipeline into the well. And the effect is negative [15].

The obtained dependences allow minimizing the negative effect of the very fact of the well curvature in the construction of the passage. Moreover, the use of this approach is not limited to wells having only curved areas.

Various special cases of designing a passage profile can be, for example, the following. The profile can have a central horizontal section and two curved sections of the entrance and exit. In another case, the central section of the profile can be curved, and the entrance and exit areas - rectilinear.

The breakdown of the profile into sections, in general, is typical for the majority of calculation and design methods, but this does not prevent the use of the dependences obtained in the work to reduce the resistance to movement of the pipeline from contact with the walls of the well. The results of the work can be applied to individual sections of the profile.

As shown in the paper, the profile of such sections, depending on the loading conditions, must correspond to the parabola or hyperbolic cosine equation.

The methods proposed in the work can also calculate the pulling force of the pipeline in the event that the profile of the well is specified by an arbitrary function, which is important for the practical implementation of the results of minimizing the pulling force.

References

[1] Toropov V S, Toropov E S, Podorozhnikov S Yu, Seroshtanov I V 2015 The use of additional equipment in horizontal directional drilling technology. *Mining information and analytical bulletin.* **36** 36-44

[2] Seroshtanov I V, Toropov V S, Toropov S Yu 2015 *Accidents and complications in trenchless construction of pipelines.* Oil and gas of Western Siberia Materials of the International scientific and technical conference dedicated to the 90th anniversary of the birth of Anatoly Nikolaevich Kosukhin. pp. 98-102

[3] Zemenkova M Yu, Shipovalov A N, Dudin S M, Zemenkov Yu D 2007 System analysis of oil and gas facilities monitoring and control processes. *News of Higher Schools. Oil and gas.* **5** 116-119

[4] Toropov V S, Tamer O S , Toropov S Yu and Nikiforov N A 2016 Dynamics of pipeline pulling process by horizontal directional drilling. *IOP Conf. Series: Materials Science and Engineering* **154**

[5] Zemenkov Yu D 2008 *Monitoring of hydrodynamic and technical characteristics of pipeline systems: instructional aid.* (Tyumen: publishing house "Vector Book")

[6] Voronin K S, Ogudova E V 2016 The Effect of Dynamic Processes in the System "Pipe-Soil" on the Pipeline Deviation from Design Position. Transport and Storage of Hydrocarbons. *IOP Conf. Series: Materials Science and Engineering* **154** 012019 doi:10.1088/1757-899X/154/1/012019

[7] Voronin K S 2016 Forecasting and Evaluation of Gas Pipelines Geometric Forms Breach Hazard. Transport and Storage of Hydrocarbons. *IOP Conf. Series: Materials Science and Engineering* **154** 012020 doi:10.1088/1757-899X/154/1/012020

[8] Toropov V S, Ponomareva T G, Toropov S Yu, Toropov E S 2016 Quantitative assessment of changes in the pipeline - wellbore system in the event of an emergency situation in horizontal

directional drilling. *News of higher educational institutions. Oil and gas* **5** 71-77

[9] Dudin S, Bahmat G, Zemenkov Y, Voronin K, Shipovalov A 2017 Strategy for monitoring and ensuring safe operation of Russian gas transportation system. *MATEC Web of Conferences* **106** DOI: 10.1051/matecconf/201710606004

[10] Shalay V, Zemenkova M, Zemenkov Y, et al. 2016 Modeling Parameters of Reliability of Technological Processes of Hydrocarbon Pipeline Transportation. *MATEC Web of Conferences* **73** 01029 DOI: 10.1051/matecconf/20167301029

[11] Zemenkov Y D, Shalay V V, Zemenkova M Y 2015 Expert systems of multivariable predictive control of oil and gas facilities reliability *Oil and Gas Engineering Procedia Engineering* **113** 312-315 DOI: 10.1016/j.proeng.2015.07.271

[12] Zemenkova M, Shalay V, Zemenkov Y, Kurushina E 2016 Improving the Efficiency of Administrative Decision-Making when Monitoring Reliability and Safety of Oil and Gas Equipment *MATEC Web of Conferences* **73** DOI: 10.1051/matecconf/20167307001

[13] A A Ibragimov, C Yu Podorozhnikov, A B Shabarov, M V Medvedev, M Yu Zemenkova. Calculation model and algorithm for determining residual resource of the pipeline under the periodic variations of stress and corrosion. *Mining informational and analytical bulletin* **S4** 199-206

[14] Zemenkov Yu D 2008 *Technique and technologic processes at transport of energy resources: two volume instructional aid.* (Tyumen: publishing house "Vector Book")

[15] Smirnov V V, Zemenkov Yu D 2013 Using the finite element method for calculation of intense deformed condition of above ground pipelines constructed on permafrost soil. *Pipeline transport: theory and practice.* **4** 18-23

IOP Conf. Series: Materials Science and Engineering **357** (2018) 012004 doi:10.1088/1757-899X/357/1/012004

Determination of relative phase permeabilities in stochastic model of pore channel distribution by diameter

M Y Zemenkova[1], A Shabarov[1,2], A Shatalov[2] and L Puldas[1]

[1]Industrial University of Tyumen, 38 Volodarskogo St., Tyumen, 625001, Russia
[2]Tyumen State University, 6 Volodarskogo St., Tyumen, 625003, Russia

E-mail: muzemenkova@mail.ru, kaf_mms@utmn.ru

Abstract. The problem of the pore space description and the calculation of relative phase permeabilities (RPP) for two-phase filtration is considered. A technique for constructing a pore-network structure for constant and variable channel diameters is proposed. A description of the design model of RPP based on the capillary pressure curves is presented taking into account the variability of diameters along the length of pore channels. By the example of the calculation analysis for the core samples of the Urnenskoye and Verkhnechonskoye deposits, the possibilities of calculating RPP are shown when using the stochastic distribution of pores by diameters and medium-flow diameters.

1. Introduction

Modern methods of forecasting oil production are based on solving the equations of multiphase filtration [1, 2 et al]. A significant uncertainty in the calculation of oil flow rates is the calculation of the relative phase permeabilities (RPP) for oil and water, depending on the structure of the pore space, saturation and properties of the fluids.

Russian and foreign literature [3-14] provides results of ongoing studies of methods for describing the structure of pore space, as well as methods for calculating pressure losses in pore-network structures, with the help of which the filtration characteristics of the formation can be found and generalized (absolute permeability, RPP and others). Stationary and non-stationary experimental methods are widely used for determining RPP on a core scale. Numerous versions of empirical and approximate analytical methods have been developed. At the initial stage of development there are methods based on solving the inverse problems of underground hydrodynamics with the use of field data.

2. Experimental research

The most promising methods for determining RPP include calculation and experimental methods [3-11, 13, 14 et al]. These methods include two stages: construction of a geometric model of pore space and simulation of a multiphase flow in a certain pore space. Models of a pore space are constructed on the basis of curves of capillary pressure, analysis of thin slices, tomography of core samples, etc. [3-6, 9-14 et al]. Methods for simulating multiphase flows include: semi-empirical channel methods (non-overlapping channels) [9, 10, 11]; quasi-one-dimensional modeling in channels with a network structure [3, 4, 5, 6, 13]; 2D and 3D modeling in spatial structures [7, 8, 14 et al].

IOP Conf. Series: Materials Science and Engineering **357** (2018) 012004 doi:10.1088/1757-899X/357/1/012004

The authors' approach is based on the representation of pore space as unconnected clusters, each of which contains one "large" capillary and $\beta = 4 - 12$ "small" capillary channels connected by bridges to a "large" channel. Diameters of "large" and "small" channels are assumed to be different for different sections in accordance with the capillary pressure curve known from the experiment. The sequence of locations of sections with different diameters along the flow is specified multiple times randomly. In calculations, the value of local resistances is corrected by the value of local resistances at the boundary values of water and oil saturation [4]. It turns out that in "large" capillaries in the "train flow" mode, oil and water move, and in small capillaries and crosspieces only water moves (Fig. 1, 2).

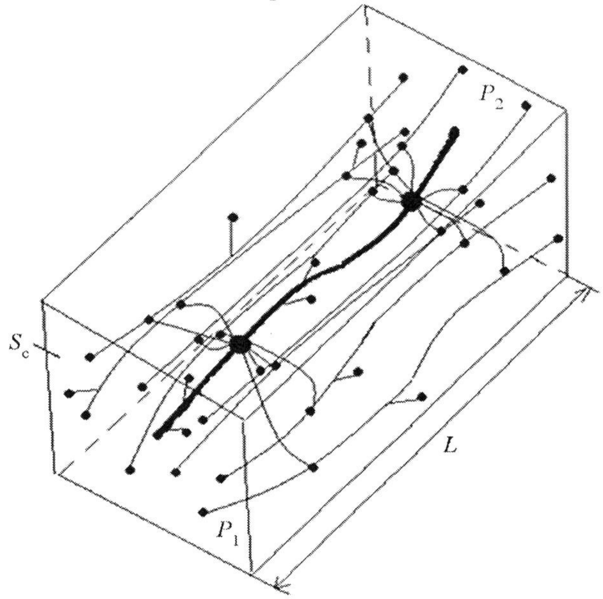

Figure 1. Capillary cluster. [4, 5, 6]
Sc – cluster section, P1, P2 – pressures at the entrance and exit from the cluster, L – cluster length equal to the core sample length.

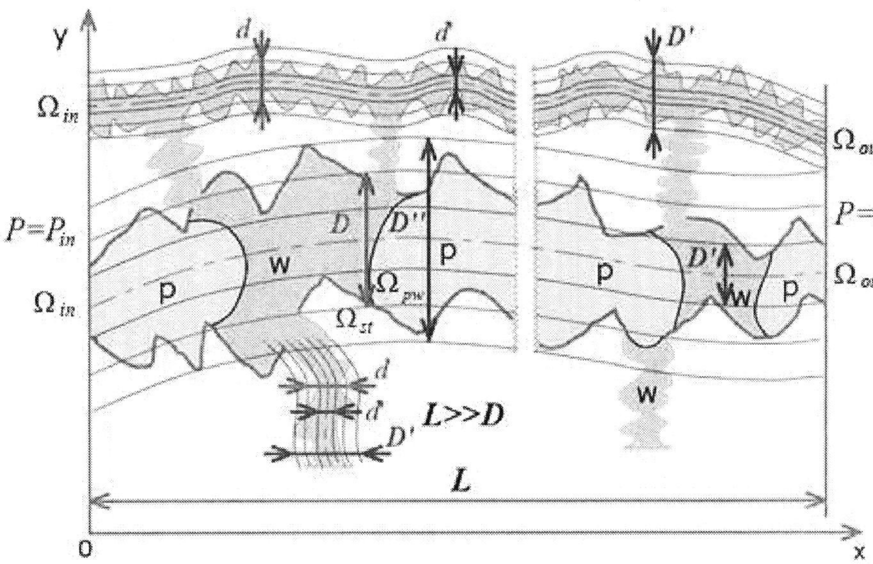

Figure 2. Pattern of fluid flow within the cluster [5, 6].

L — cluster length, *d* — medium-rate diameters of peripheral (small, crosspieces) capillaries, *D* — diameters of central (large) capillaries; *d'*, *D'*,*D''* — diameters in the histogram of pore distribution corresponding to the beginning of the range, the boundary value, the end of the range (see Fig. 3); P_{in}, P_{out} — pressures at the entrance and exit from the cluster; p, w — oil and water; Ω_{st}, Ω_{pw},$P=P_{in}$,$P=P_{out}$— cluster wall, oil-water interface, input and output cross sections of the cluster.

Thus, within a single capillary, the diameter distribution corresponds to the distribution along the diameters extracted from the capillary pressure curve (Figure 3).

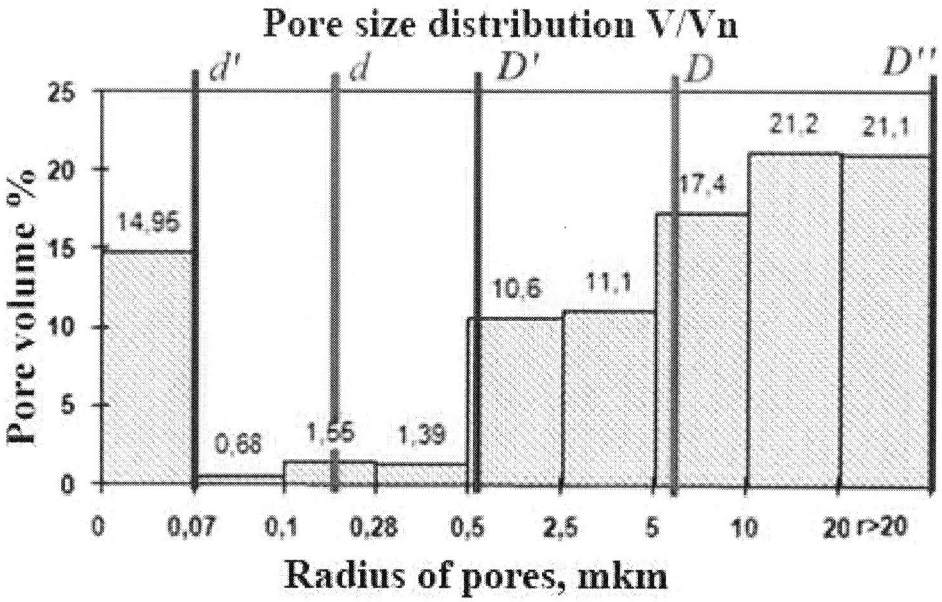

Figure 3. Histogram of pore distribution by diameters for sandstone deposits of Western Siberia [5, 6]. (Pore volume, pore distribution by dimensions, pore radius, mkm)
d', *D'*,*D''* — diameters corresponding to the beginning of the range, the boundary value, the end of the range for "small" and "large" channels, *d*, *D* - medium-rate diameters of peripheral (small, crosspieces) and central (large) capillaries.

An alternative deterministic approach [4, 5, 6] is the introduction of "small" and "large" channels for calculating the average diameters:

$$\begin{cases} d = \left(\frac{1}{N_M}\sum_{jmin}^{NINTM} N_j d_j^4\right)^{0.25}, \\ D = \left(\frac{1}{N_B}\sum_{NINTM+1}^{NINT} N_j D_j^4\right)^{0.25}, \\ \sum_{jmin}^{NINTM} N_j \le N_M \le \sum_{jmin}^{NINTM+1} N_j. \end{cases}, \quad (1)$$

where N_M, N_B – the number of "small" and "large" channels in the cluster, $\frac{N_M}{N_B} = \beta$

Hydraulic losses in the *i* section of a large capillary of the cluster are described on the basis of a system of equations derived from the generalized Bernoulli equation [4, 5, 6]:

$$\begin{cases} |\Delta P|_B^i = Q_{1WB}^i C_{BW}^i + Q_{1P} C_{BP}^i + \overline{P}_{MF}(S^k) \cdot |\Delta P|_B^i, \\ C_{B\psi}^i = \frac{K_\psi 128 \overline{L}_B^i \mu_\psi}{\pi D^{i^4}}, \psi = W, P, \end{cases} \quad (2)$$

IOP Conf. Series: Materials Science and Engineering **357** (2018) 012004 doi:10.1088/1757-899X/357/1/012004

Here Q^i_{1WB}, Q_{1P} – flow of water, oil, m^3/s, $\tilde{L}^i_B$ – length of the *i* capillary section, m, $\overline{P}_{MF}(S^k) = \Delta P^i_{MF}(S^k) / |\Delta P|^i_B(S^k)$, $\overline{P}_{MF}(S^k) \in [0;1]$ – dimensionless ratio of losses at interfacial interactions to total losses, W – water, P – oil, K_P, K_W, - scale coefficients showing how many times the pressure drop across the core sample at the water-saturation boundary points in the RPP measurement experiment is greater than the corresponding design pressure drops due to friction on the cluster.

Pressure drops in the *i* section of a small capillary, crosspiece:

$$\begin{cases} Q^i_{1\chi} = \dfrac{\pi d^4 |\Delta P|^i_\chi}{K_W 128 \tilde{L}^i_\chi \mu_W} = \dfrac{|\Delta P|^i_\chi}{C^i_\chi} \\ C^i_\chi = \dfrac{K_W 128 \tilde{L}^i_\chi \mu_W}{\pi d_i^4} \end{cases} \tag{3}$$

where index χ=*M* – corresponds to small capillaries, χ=*PER* – crosspieces.

The system of equations (2-3) contains $3n_c$ linear equations, where n_c – the number of closed hydraulic circuits in the diagram in Fig. 4 (without taking into account the pump circuit).

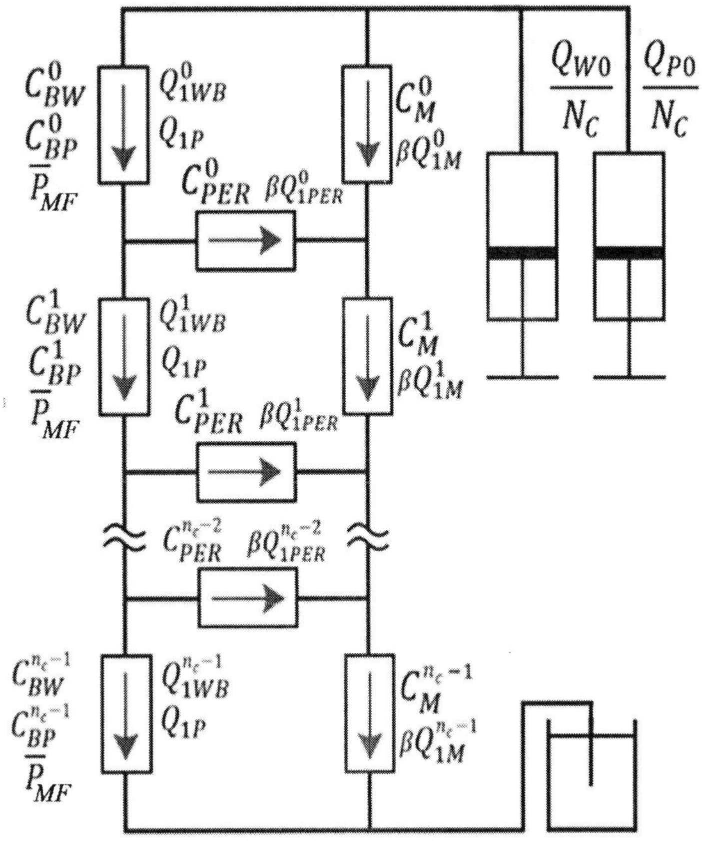

Figure 4. A hydraulic diagram of the cluster.

Fig. 4 denotes: C^i_{BW}, C^i_M, C^i_{PER} — coefficients that combine friction losses and "local" for water in the i section of a large capillary, a small capillary, a crosspiece; C^i_{BP} — coefficient that combines friction losses and "local" for oil in the i section of a large capillary; Q^i_{1WB} , Q^i_{1M}, Q^i_{1PER}; Q_{1P} — flow of water and oil; $\overline{P}_{M\Phi}$ — interphase losses referred to the total; Q, N_C — total liquid flow through the core sample and the number of clusters in the core sample.

Setting the total flows of fluids $Q_{P0}(S^j)$, $Q_{W0}(S^j)$ as a result of the solution, let us obtain fluid flows in all the components of the cluster, as well as the pressure drop across the cluster. RPPs are determined by the formulae:

$$\begin{cases} K_{rP}(S^j) = \frac{Q_{P0}(S^j)\mu_P L}{\Delta P_B(S^j) F_K F_{PM}}, \\ K_{rW}(S^j) = \frac{Q_{W0}(S^j)\mu_W L}{\Delta P_B(S^j) F_K F_{PM}}, \end{cases} \tag{3}$$

where F_{PM} – absolute normalization permeability of the core sample for oil at the point of bound water, m^2; F_K – sample section area, m^2, μ_W, μ_P – viscosity of water, oil, Pa*s.

Bell-shaped function $\overline{P}_{MF}(S^k)$ has two parameters for which the empirical dependences [5] have been developed on the basis of experimental data on three deposits in Western Siberia using the correlation analysis apparatus by dimensionless criteria $KSS = K_0/(mD^2)$, $Ca = \mu_P v_P/\sigma$, where K_0 – absolute permeability, m^2, m – porosity, σ – interfacial tension coefficient, v_P – rate of oil filtration at the point of bound water.

For six core samples from Verkhnechonskoye (1.1-1.6) and five core samples from Urnenskoye (2.1-2.5) deposits, a calculated analysis of the dependence of the normalized difference in RPP at the interior points of the water saturation range (between bound water S_{wc} and residual oil S_{or}) was carried out using two algorithms. Fig. 5 shows the dependence of the relative difference in RPT of oil obtained on the basis of flow calculation with variable diameters of capillary sections $\widetilde{K}_{rP}$ and with average (1) diameters K_{rP}. $\Delta\overline{K}_{rP} = \frac{\widetilde{K}_{rP} - K_{rP}}{K_{rP}}$.

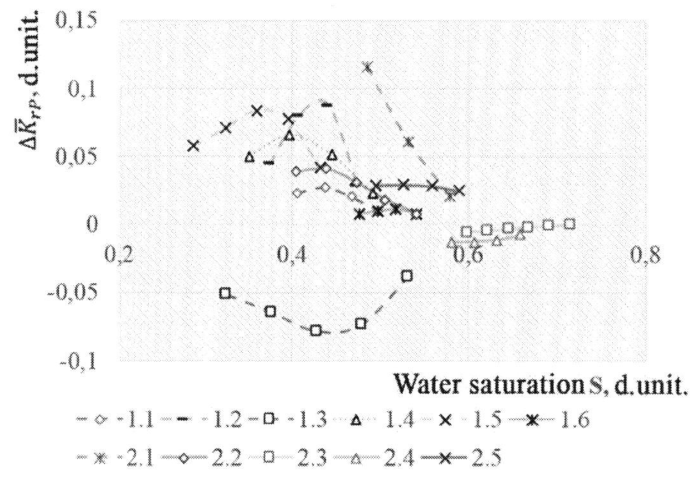

Figure 5. Error in determining the OPP for oil under the assumption of the constancy of diameters with the correction of local losses by the experimental data. (unit fractions, Water saturation)

From the calculation results, presented in Fig. 5, it follows that the error associated with the replacement of diameter distribution obtained from the capillary pressure curves for average diameters depends on the saturation and on the conditions considered $\Delta\overline{K}_{rP} \leq 0{,}1$.

3.Conclusions

The results of the research are as follows.

A model of the network structure of pore channels with variable length diameters corresponding to the pore distribution found from the capillary pressure curves is proposed. An algorithm for the stochastic description of variable channel diameters has been developed.

With respect to the generated network structure of pore space, based on the known model of pressure losses in capillary channels including pressure losses due to friction, local losses and losses due to interfacial interaction, relative phase permeabilities are determined when the water saturation is changed.

It has been established that replacement of variable diameters of "large" and "small" channels by medium-rate diameters with local losses corrected by the experimental data leads to design errors within ±10% for oil RPP.

References

[1] Basniev K S, Kochina I N, Maksimov V M 1993 *Underground hydromechanics. Textbook for high schools.* (Moscow: Nedra)

[2] Nigmatulin R I 1987 *Dynamics of multiphase media. P2.* (Moscow: Nauka)

[3] Blunt M J, Jackson M D, Piri M, Valvatne P H 2002 Detailed physics, predictive capabilities and macroscopic consequences for pore-network models of multiphase flow. *Advances in Water Resources.* **25 8-12** 1069–1089

[4] Altunin A E, Sokolov S V, Stepanov S V, Cheremisin N A, Shabarov A B 2013 Calculation method for obtaining RPP based on the solution of the generalized Bernoulli equations for the pore channel system. *Oilfield engineering* **8** 40-46

[5] Shabarov A B, Shatalov A V 2016 Loss of pressure during flow of water-oil mixture in pore channels. // Bulletin of the Tyumen State University. *Physico-mathematical modeling. Oil, Gas, Energy.* **2** 50-72

[6] Shabarov A B, Shatalov A V 2016 *Geometrical model of porous space for the calculation of oil and water filtration. Collection of articles of the IX school-seminar of young scientists Thermophysics, heat engineering, hydrodynamics. Innovative technologies led by A.B. Shabarov.* (Tyumen: Publishing house of Tyumen State University)

[7] Igoshin D.E., Maksimov A.Yu. Numerical and analytical estimates of the permeability of a porous medium formed by channels having rotational symmetry. // Bulletin of the Tyumen State University. Physico-mathematical modeling. Oil, Gas, Energy. – 2015. – Vol 1. - №3. - pp. 112-121.

[8] Gubaidullin A A, Igoshin D E, Khromova N A 2016 Generalization of Kozeny approach to determining the permeability of model porous media from solid spherical segments. *Bulletin of the Tyumen State University. Physico-mathematical modeling. Oil, Gas, Energy.* **2-2** 105-120

[9] Fatikhov S Z 2012 On the calculation of relative phase permeabilities. *Oil and gas: electronic scientific journal* **1**

[10] Dudin S, Bahmat G, Zemenkov Y, Voronin K, Shipovalov A 2017 Strategy for monitoring and ensuring safe operation of Russian gas transportation systems *MATEC Web of Conferences* **106** DOI: 10.1051/matecconf/201710606004

[11] Stepanov S V, Shabarov A B, Bembel G S 2016 Computational technology for determining the function of interfacial interaction on the basis of flow simulation in a capillary cluster. *Bulletin of the Tyumen State University. Physico-mathematical modeling. Oil, Gas, Energy.* **2 - 1** 63-71

[12] Joekar-Niasar V, Hassanizadeh S M 2012 Analysis of fundamentals of two-phase flow in porous media using dynamic pore-network models: a review. *J. Crit. Rev Environ. Sci. Technol.* **42 18** 1895-1976

[13] Al-Gharbi M S 2004 *Dynamic Pore-Scale Modelling of Two-Phase Flow. A thesis submitted in fulfillment of the requirements for the degree of Doctor of Philosophy of the University of London and the Diploma of Imperial College.* (London)

[14] Raeini A Q, Bijeljic B, Blunt M J 2014 Numerical Modelling of Sub-pore Scale Events in Two-Phase Flow Through Porous Media. *Transport in Porous Media* **101 2** 191-213

Study of method for reducing fuel consumption and amount of specific emissions of harmful substances with exhaust gases of passenger cars when using "climate control" system

L N Burakova, I A Anisimov, A D Burakova and O D Burakova

Industrial University of Tyumen, 38 Volodarskogo Str., Tyumen, 625000 Russia

E-mail: tkcc@list.ru, burakova.1992@mail.ru, burakova.96@bk.ru

Abstract. The article deals with the issue of improving fuel efficiency and the ecological nature of passenger cars, servicing the administrative and management personnel of the oil and gas complex. It is established that fuel consumption and the amount of specific emissions of harmful substances with exhaust gases of cars when using the "climate control" system depend on the effective ambient temperature, the color of the opaque car body elements, the power of the car engine and the interior volume. However, the simplest controlled factor is the color of the opaque car body elements, which is characterized by the coefficient of light reflection. In the course of experimental studies, the authors established the dependences of a change in fuel consumption and a share of reducing emissions of harmful substances with exhaust gases of passenger cars with the "climate control" system on the coefficient of light reflection. A method has been developed to reduce fuel consumption and the amount of specific emissions of harmful substances with the exhaust gases of passenger cars using the "climate control" system, which involves painting the vehicle roof white and allows reducing fuel consumption by 5.5-10.3%, and the amount of specific emissions of harmful substances by 0.8-2.3%.

1. Introduction

Currently, in the field of transport services for administrative and management personnel of the oil and gas industry, there is an increase in the number of vehicles. Limited reserves of natural energy sources cause a rise in prices for all types of fuel. This causes the need to improve fuel efficiency of vehicles in use [1, 2, 3, 4]. Reducing fuel consumption leads to increased environmental friendliness of vehicles. According to the Transport Strategy of the Russian Federation until 2030, the ecological nature of the transport system plays an important role in the socio-economic development of the country, so reducing the harmful impact of transport on the environment is one of its main tasks. GOST R 52033-2003 "Cars with gasoline engines. Emissions of pollutants with exhaust gases. Norms and methods of control in assessing the technical condition (Amend. No. 1 of 01.07.2012)" indicates values of the content of carbon monoxide and hydrocarbons in exhaust gases.

Thus, cars servicing the administrative and management personnel of the oil and gas complex must achieve minimum fuel consumption and emissions of harmful substances with exhaust gases. [6-12].

Currently, there are some specific features of the passenger car operation in urban conditions: low density of the street and road network, rapid growth in the number of vehicles, separation of pedestrian and traffic flows at intersections. This reduces the throughput and increases the average idle time of vehicles. As a result of the analysis of previous studies, it has been established that for the last 30 years the idle time of cars with a running engine has increased by 18% and in modern conditions - 30 ... 35% [13-16].

When identifying features of the passenger car operation in modern cities, one should also take into account the effect of high temperatures. In the Russian Federation, the number of car-days in the operation of cars at temperatures above +20 ° C is more than 51%. Deviation of operating conditions from standard affects the motor vehicles by changing the thermal state of the elements of their design. In particular, in the warm period of the year, there is an increased temperature of the ambient air (above + 20 ° C) and an increase in the amount of solar radiation entering the Earth's surface. This causes an increase in the amount of heat entering the interior of the car, hence increasing the air temperature inside the vehicle. According to the norms of POT RM-008-99 "Inter-industry rules for labor protection" in the operation of industrial vehicles, the air temperature at the driver's workplace during the warm season should be 20-25°C or no more than 3°C above the average outside air temperature at 13.00 of the hottest month, but not more than 28°C. To create the required microclimate in the car, "climate control" systems are applied, which are the auxiliary equipment of the car. The inclusion of this in the work leads to an increase in the proportion of the car engine power spent on the work of the auxiliary equipment, which causes an increase in the consumption of fuel by the vehicle. The greatest impact is observed during the idle time of the car with the running engine; it is caused by the absence of streams of air cooling the interior during movement, and the minimum value of the effective capacity of the car engine.

The increase in fuel consumption of cars with the use of "climate control" during the idle time with the engine running increases the amount of exhaust gases. [19-22]. Their composition is 95 ... 98% products of complete combustion, unused oxygen and nitrogen of the air, and 2 ... 5% - several hundred components that determine the degree of a harmful impact of the vehicle on the environment. However, in the products of complete combustion, there are chemical compounds that have a negative effect on the ecological state of the atmosphere, for example, carbon dioxide, which contributes to the development of the greenhouse effect. Reducing the concentration of this compound in the exhaust gas can be achieved by reducing the fuel consumption of cars.

Thus, the purpose of this study is to improve the ecological nature and fuel efficiency of cars when using the "climate control" system during idle time with the engine running.

2. Methodology

The ecological nature of the transport system can be characterized as the ability of cars to carry out transport work or transportation of passengers with the minimum possible emissions of harmful substances in strictly defined conditions. Vedenyapin G.V., Govorushchenko N.Ya., Gutarevich Yu.F., Evtin P.V., Gelagin O.I., Zvono V.A., Magaril E.R. Reznik L.G. studied the problems of reducing the toxicity of exhaust gases and increasing fuel economy. Fuel consumption and the content of harmful substances in the exhaust gases of cars depend on many factors, in particular, on incorporating the auxiliary equipment and operating conditions.

An increase in the ambient air temperature leads to an increase in the number of heat gain to the car interior, therefore, the microclimate inside the vehicle changes. Assessment of the impact of operating conditions on the microclimate in the vehicles was carried out by Palutin Yu.I., Kormin A.M., Semikin S.N., Shelyakin A.S. and others. The given works proposed a set of natural and climatic factors of the conditions of vehicle operation to evaluate the effective temperature of the ambient air with the help of a complex index, the numerical value of which corresponds to the true temperature of the still and saturated air. However, the effective temperature of the ambient air is the determining factor only when heat is supplied to the car interior by heat transfer.

When assessing the heat balance of the vehicle interior, it should be taken into account that heat gain from external sources is realized not only by heat transfer, but also by thermal radiation. The amount of heat entering the passenger car interior through the opaque car body elements with the help of thermal radiation depends on the reflectivity of the paintwork. The color characterizes the reflectivity of the opaque body elements. For example, an absolutely white body has the coefficient of light reflection equal to 1, and absolutely black (soot) -0. However, it is impossible to achieve absolutely white or black colors when painting, therefore, as a rule, cars with a white color of the opaque body elements have the coefficient of light reflection equal to 0.95, and with a black color - 0.05. Thus, the amount of heat entering the interior of the car depends on the effective ambient temperature and the coefficient of light reflection of the opaque body elements and influences the microclimate inside the vehicle. An increase in the amount of heat input causes an increase in the temperature of the air inside the car; in order to reduce and maintain it, within the limits corresponding to comfort conditions, a "climate control" system is used. The main characteristics of its operation affecting the fuel consumption of a car are the cooling capacity and power consumption which depend not only on the amount of heat entering the interior of the car, but also on the interior volume. An increase in the interior volume, as well as the amount of incoming heat, leads to an increase in the power consumption of the "climate control" system. It is established that the increase in the fuel consumption of a car when the "climate control" system is switched on is determined by an increase in the proportion of the power take-off of the automobile engine for the operation of this auxiliary equipment, which is characterized by the ratio of the cooled volume of the interior to the car engine power. This value is characterized by the specific climatic power parameter, which determines the part of the power of the automobile engine spent for cooling the unit volume of the car interior.

The fuel consumption of a passenger car when using the "climate control" system that affects the amount of specific emissions of harmful substances with exhaust gases depends on the effective ambient temperature, the coefficient of light reflection of the opaque car body elements and the specific climatic power parameter (the ratio of the passenger car interior volume to its engine power). However, the effective ambient temperature is an uncontrolled factor, and the coefficient of light reflection of the opaque car body elements is a controlled input indicator that can be quite simply changed by giving the car the color characterized by the maximum coefficient of light reflection. Thus, it is necessary to develop a method for reducing the increase in fuel consumption of a passenger car using the "climate control" system.

2.1. Equipment and devices used in studies

The experiment involved cars of three makes - Ford Focus, Toyota Corolla, and Mitsubishi Lacer. The light reflection coefficient was measured with a thermal imager. Illumination of the site was assessed using a luxmeter. The effective ambient temperature was taken into account by measuring the three factors: ambient air temperature and humidity - by the Assmann psychrometer and wind velocity - by the Fuss anemometer. The air temperature in the car was measured with mercury and digital thermometers. Fuel consumption was measured using a BT-ECU CAN adapter connected to the diagnostic socket of the car with an adapter and a computer program Check-Engine. To monitor the volume content of harmful substances in the exhaust gases of engines, a gas analyser GIAM-29M-1 was used. The resulting volume concentrations were converted into specific CO and CH emissions.

2.2. Methodology of research

Passenger cars participating in the experiment consisted of a series of pairs of cars (black, silver, grey and white), one of which had an original roof color and the other was given a white color with the help of a vinyl film. Initially, the area of the opaque car body elements, in particular the roof, was determined.

Then, the vehicles were installed on a uniformly illuminated platform without shading. At the time of the maximum warm-up of the car with the "climate control" system switched off, the air temperature in the car was measured.

Then, the "climate control" system was switched on. When the constant thermal conditions were achieved, the air temperature in the car was measured and checked for compliance with its set value in the "climate control" system. The coefficient of light reflection for opaque body elements of different colors was determined with an illumination of 60,000 lux (illumination at noon on a cloudless summer day) using a thermal imager.

Fuel consumption data of a passenger car equipped with a "climate control" system were measured during the idle time at a steady coolant temperature in the operating range of 80-100°C from the vehicle ECU using the BT-ECU CAN adapter and the Check-Engine computer program. Together with the measurement of fuel consumption, carbon dioxide and hydrocarbons in the exhaust gases were measured with a gas analyzer.

3. Results and discussion

Analysing the state of the matter, it has been revealed that the factors that have the greatest effect on the fuel consumption of a passenger car using the "climate control" system are the effective ambient temperature, the power of the car engine, the interior volume and the color of the opaque car body elements (the light reflection coefficient). Of the presented factors only the power of the car engine, the interior volume and the color of the opaque car body elements are controllable, but the change in the first two factors is difficult. Thus, one of the ways to reduce fuel consumption of a car with the use of the "climate control" system is to impart a color that is characterized by the largest coefficient of light reflection (white color). A change in the increase in fuel consumption of a Ford Focus passenger car using the "climate control" system during the idle time with the engine running from the coefficient of light reflection of the opaque body elements is shown in Fig. 1.

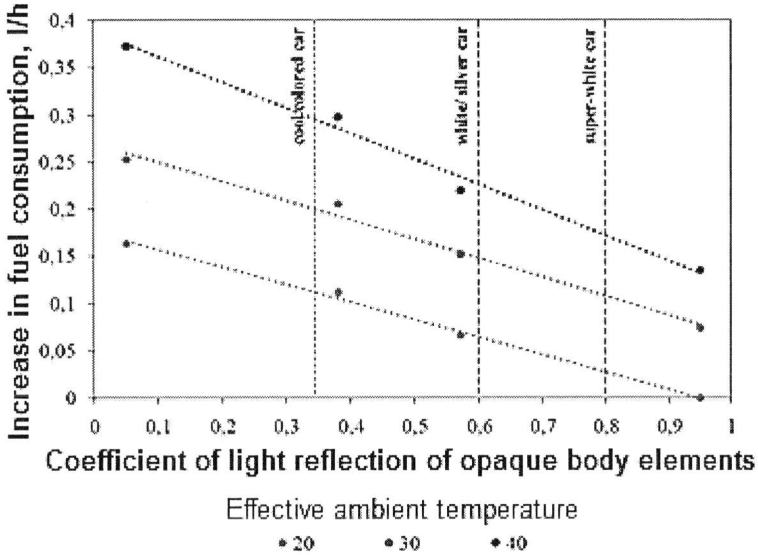

Figure 1. A change in the increase in fuel consumption of a Ford Focus passenger car using the "climate control" system during the idle time with the engine running from the coefficient of light reflection of the opaque body elements

An increase in the fuel consumption of a car with the use of the "climate control" system during the idle time with the engine running becomes greater with an increase in the effective ambient temperature and a change in the color of the opaque body elements from light to dark characterized by a smaller coefficient of light reflection. This causes an increase in specific emissions of harmful substances with exhaust gases into the atmosphere, which is shown in Figures 2 and 3.

IOP Conf. Series: Materials Science and Engineering **357** (2018) 012005 doi:10.1088/1757-899X/357/1/012005

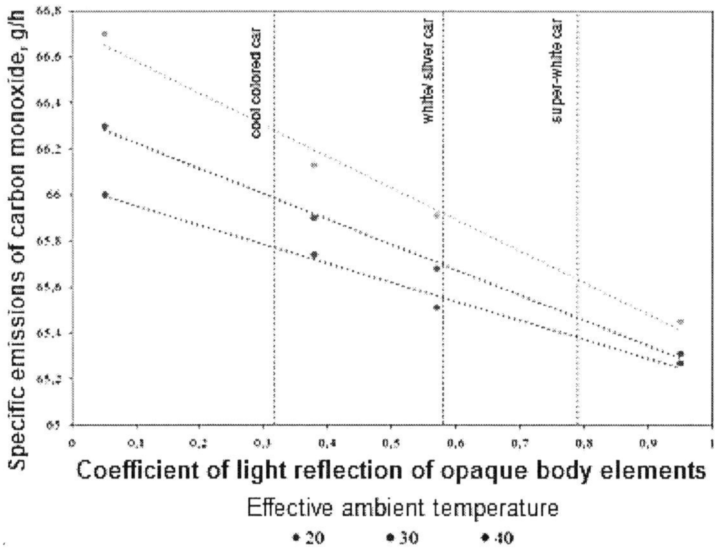

Figure 2. Reduction in specific emissions of carbon monoxide with exhaust gases of a passenger car when using the "climate control" system during the idle time with the engine running from the coefficient of light reflection of the opaque body elements

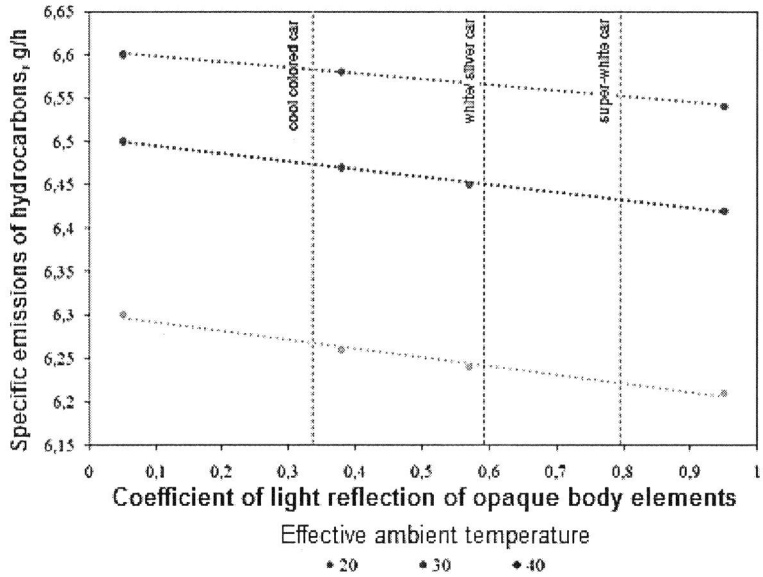

Figure 3. Reduction in specific emissions of hydrocarbons with exhaust gases of a passenger car when using the "climate control" system during the idle time with the engine running from the coefficient of light reflection of the opaque elements

An increase in the coefficient of light reflection leads to a decrease in specific emissions of carbon monoxide and hydrocarbons with exhaust gases. Thus, in order to reduce fuel consumption and specific emissions of harmful substances with exhaust gases, opaque body elements must be painted in colors characterized by a high coefficient of light reflection. However, the amount of heat entering the interior of the car is also affected by the internal arrangement of the opaque body elements. The diagram of the internal arrangement of the opaque body elements is shown in Fig. 4.

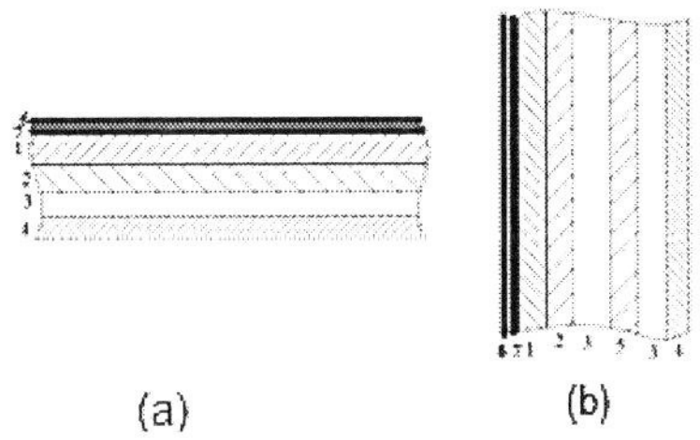

1 steel; 2 - noise, vibration and thermal insulation; 3 - air;
4 - plastic; 5 - foam polystyrene; 6 - paint; 7 - primer

Figure 4. A diagram of the internal arrangement of the (a) roof and (b) doors

The sequence of layers inside the roof structure is: steel, noise and vibration insulating film, air, plastic. In the doors, there are additional layers of expanded polystyrene and air. Each of them is characterised by the coefficient of thermal conductivity which is the ability of a substance to conduct heat. Depending on this indicator, some of the heat will be absorbed by air interlayers and noise, vibration and thermal insulation. The absence of additional layers inside the roof causes an increased amount of heat entering the interior through this opaque body element. When studying the peculiarities of heat gain inside the car, it is established that the area of this structural element of a passenger car is between 45 and 52%. The results of measurements of the area of vehicle structural components are presented in Table 1.

Table 1. Areas of structural elements of cars

	Ford Focus	Mitsubishi Lancer	Toyota Corolla
Roof area, m^2	1.69	1.66	1.56
Door area, m^2	1.50	1.57	1.70
Area of opaque body elements, m^2	3.19	3.23	3.26

The greatest amount of heat enters the interior through the roof, due to its internal features, which consists in the absence of two additional layers (foam polystyrene and air) in comparison with the structural design of doors, a large area, and a perpendicular location relative to solar radiation at the warmest time of day (from 10:00 to 16:00). Thus, a reduction in fuel consumption and specific emissions of harmful substances with exhaust gases of a passenger car using the "climate control" system can be achieved by painting the roof of a car white.

To prove the proposed assumption, calculation of the amount of heat entering the passenger car interior through a roof painted in various colors was initially performed. The results of the calculation are presented in Table 2.

Table 2. The amount of heat coming into the car interior, at an ambient temperature of +30°C and a temperature inside the car +24°C (due to heat transfer and heat radiation)

Car	Roof area, m^2	Amount of heat entering the passenger car interior, kW/day			
		white (ρ=0.95)	silver (ρ=0.57)	silver-gray (ρ=0.38)	black (ρ=0.05)
Ford Focus	1.69	0.51	3.41	6.04	7.39
Toyota Corolla	1.56	0.48	3.16	5.59	6.83
Mitsubishi Lancer	1.66	0.51	3.36	5.94	7.26

As a result of the analysis of the received data, it is established that the roof of the car equipped with the "climate control" system should be white to reduce fuel consumption and the amount of specific emissions of harmful substances with the exhaust gases. In the course of experimental studies, the results that attest to the feasibility of introducing the developed method are obtained. Experimental data on the effect of the white color of the roof on the change in fuel consumption and the amount of specific emissions of harmful substances with the exhaust gases are presented in Figures 5 and 6.

Application of the developed method allowed reducing the fuel consumption of a passenger car using the "climate control" system by 5.5-10.3%. Making the roof of a silver car white led to the achievement of fuel consumption corresponding to that of a white car.

Reducing the fuel consumption of a car with the introduction of the developed method allowed reducing the amount of specific carbon monoxide emissions by 0.37-1.13% and hydrocarbons - by 0.47-1.08% with the exhaust gases.

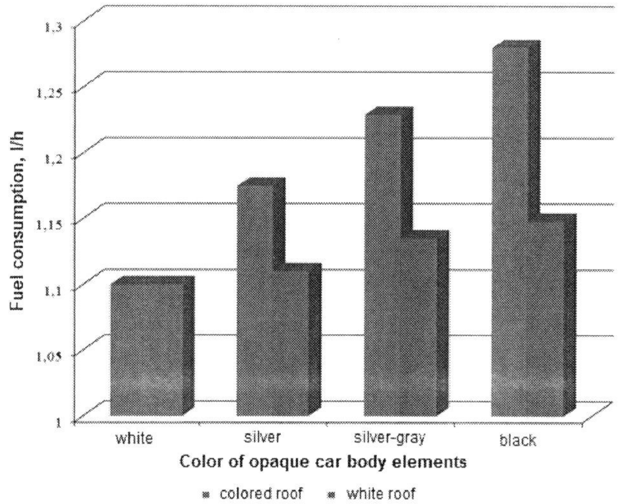

Figure 5. Change in the fuel consumption of a car equipped with the "climate control" system with the introduction of the developed method

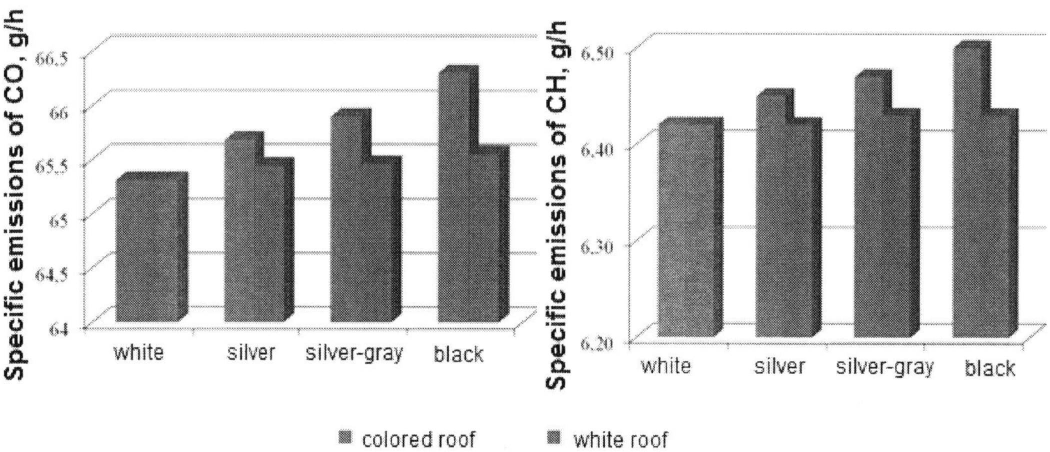

Figure 6. Decrease in specific emissions of carbon monoxide and hydrocarbons with exhaust gases of cars equipped with the "climate control" system

Thus, painting the roof of a vehicle white will reduce the fuel consumption of a car by 5.5-10.3% and the total amount of specific emissions of harmful substances with the exhaust gases - by 0.8-2.3%.

4. Conclusions

In the course of experimental studies, it has been established that when the color of the opaque body elements changes from light to dark (reducing the coefficient of light reflection), fuel consumption and concentration of carbon monoxide and hydrocarbons in the exhaust gases of passenger cars increase when using the "climate control" system.

The area of the opaque body elements and their internal arrangement are analyzed. This made it possible to reveal that the greatest amount of heat enters the interior through the roof, due to its internal features, which consists in the absence of two additional layers (foam polystyrene and air) in comparison with the structural design of doors, a large area, and a perpendicular location relative to solar radiation at the warmest time of day (from 10:00 to 16:00). Based on the results of the preliminary stage of the experiment, a method has been developed for reducing the fuel consumption of a passenger car using the "climate control" system. It is to give a white color to the roof of the vehicle.

According to the obtained results of the experiment, it is established that the application of the developed method allows reducing the fuel consumption of a car by 5.5-10.3% and the total amount of specific emissions of harmful substances with the exhaust gases - by 0.8-2.3%.

References

[1] Anisimov I, Magaril E, Magaril R, Chikishev E, Chainikov D, Gavaev A, ErtmanS, Ertman Yu and Ivanov A 2016 Improving vehicle adaptability to the operating conditions of "smart" cities in the northern regions *E3S Web of Conf. Rev.* **6** 02003

[2] Panuccio P, Amodeo L, D'agostino P, Lamari D and Scattarreggia T 2015 Urban Regeneration and smart sity according to EU strategies: an urban distribution center in city ogistics *WIT Transactions on The Built Environment* **146** 313-324

[3] Monfaredzadeh T and Berardi U 2014 How can cities lead the way towards a sustainable, competitve and smart future? *WIT Transactions on Ecology and the Environment* **191** 1063-1074

[4] Russo F, Rindone C and Panuccio P 2014 The process of smart city definition at an EU level *WIT Transactions on Ecology and the Environment* **191** 979 989

[5] Government of the Russian Federation 2014 *Transport Strategy of the Russian Federation until 2030* (Moscow: Government of the Russian Federation)

[6] Chikishev E, Chainikov D and Anisimov I 2017 Increasing the use of natural gas on motor transport by an efficient location of the NGV RS *IOP Conf. Ser.: Earth Environ. Sci. Rev.* **50** 012004

[7] Chikishev E, Chikisheva A, Anisimov I and Chainikov D 2017 Natural gas use on minibuses, engaged in the carriage of passengers and baggage on the regular routes, as a measure for decrease I harmful environment effects *IOP Conf. Ser.: Earth Environ. Sci. Rev.* **50** 012008

[8] Chainikov D, Chikishev E, Anisimov I and Gavaev A 2016 Influence temperature on the CO_2 emitted with exhaust gases of gasoline vehicles *IOP Conf. Ser.: Mater. Sci. Eng. Rev.* **142** 012109

[9] Chikishev E, Ivanov A, Anisimov I and Chainikov D 2016 Prospects of and problems in using natural gas for motor transport in russia *IOP Conf. Ser.: Mater. Sci. Eng. Rev.* **142** 012110

[10] Ertman J, Ertman S, Anisimov I, Chainikov D and Chikishev Y 2015 Complex characteristic estimation of vehicle adaptability for low temperature operation conditions *Research Journal of Pharmaceutical, Biological and Chemical Sciences Electronic Materials* **6** 1761-70

[11] Anisimov I, Ivanov A, Chekishev E, Chainikov D and Reznik L 2014 Assesment of gas cylinder vehicles adaptability for operation at low ambient temperature conditions *WIT Transactions on Ecology and the Enviroment* **190** 685-695

[12] Ertman S, Ertman J and Zaharov D 2016 Raising of operating a motor vehicle effects on environment in winter *Innovative Technologies in Engineering Rev.* **142** 12119
[13] Cloke J and Layfield R 1996 The environmental impacts of traffic management schemes *WIT Transactions on The Built Environment* **26** 11
[14] Ntzeremes P, Kirytopoulos K and Tatsiopoulos I 2016 Management of infrastructure's safety under the influence of normative provisions *WIT Transactions on The Built Environment* **164** 51-59
[15] Langeland A 2015 Sustainable transport in cities: learning from best and worst practice? *WIT Transactions on The Built Environment* **146** 93-103
[16] Langer D and Mattrish G Future trends and driving forces in transport and logistics *WIT Transactions on the Built Environment* **75** 10
[17] Galderisi A 2014 Adapting cities for a changing climate: an integrated approach for sustainable urban development *WIT Transactions on Ecology and the Environment* **191** 549-560
[18] Alisov B and Poltaraus B 1974 *Climatology* (Moscow: Publishing House of Moscow University)
[19] Jacyna M and Merkisz J 2015 Estimations in real urban traffic conditions as a determinant of shaping sustainable urban development *WIT Transactions on The Built Environment* **146** 219-231
[20] Gong R and Waring P 1996 Investigation of emission behaviour in an urban driving cycle with a remote sensing system *WIT Transactions on The Built Environment* **26** 10
[21] Tong H and Hung W 1999 Vehicular modal emission and fuel consumption factors in Hong Kong *WIT Transactions on Ecology and the Environment* **37** 8
[22] Kazopoulo M, El-Fadel M and Kaysi I 2004 Emissions Standards Development for an I/M program *WIT Transactions on Ecology and the Environment* **74** 10
[23] Bogoslovski V 1985 *Air conditioning and refrigeration* (Moscow: Stroiizdat)
[24] Ertman S, Ertman J and Zakharov D 2016 Raising of operating a motor vehicle effects on environment in winter *Innovative Technologies in Engineering Rev.* **142** 12119

IOP Conf. Series: Materials Science and Engineering **357** (2018) 012006 doi:10.1088/1757-899X/357/1/012006

Practical calibration of design data to technical capabilities of horizontal directional drilling rig

S Yu Toropov and V S Toropov

Industrial University of Tyumen, 38 Volodarskogo St., Tyumen, 625000, Russia

E-mail: vladimir.s.toropov@gmail.com

Abstract. In order to design more accurately trenchless pipeline passages, a technique has been developed for calculating the passage profile, based on specific parameters of the horizontal directional drilling rig, including the range of possible drilling angles and a list of compatible drill pipe sets. The algorithm for calculating the parameters of the trenchless passage profile is shown in the paper. This algorithm is based on taking into account the features of HDD technology, namely, three different stages of production. The authors take into account that the passage profile is formed at the first stage of passage construction, that is, when drilling a pilot well. The algorithm involves calculating the profile by taking into account parameters of the drill pipes used and angles of their deviation relative to each other during the pilot drilling. This approach allows us to unambiguously calibrate the designed profile for the HDD rig capabilities and the auxiliary and navigation equipment used in the construction process.

1. Introduction

One of the criteria for choosing a drilling rig for constructing a pipeline passage by the method of horizontal directional drilling is the effort necessary to pull a pipeline of a given diameter, taking into account the properties of pipe material [1]. Therefore, in order to minimize energy costs for the construction of the passage, it is necessary to investigate the value of the effect of all factors affecting the pipeline pulling process and the ability to control the pulling process [2].

For practical application of the results of the study of the dependence of the amount of force needed to pull the pipeline on process parameters, it is necessary to calibrate the design data for the technical capabilities of the drilling rig.

In real construction conditions, the characteristics of the HDD rig, by means of which the construction of the pipeline passage is performed, can sufficiently restrict the design possibilities and cause certain difficulties in the implementation of the technical project in full-scale conditions [3, 4]. These parameters generally include not only the power of the rig, but also the possible range of drilling angles, as well as the properties of the drill pipes that can be used with this HDD rig.

The solution of the problem of calibrating theoretical design data to the characteristics of the rig itself can greatly increase the accuracy of work and maximally approximate the working draft to the data of the executive documentation for the facility [5].

IOP Conf. Series: Materials Science and Engineering **357** (2018) 012006 doi:10.1088/1757-899X/357/1/012006

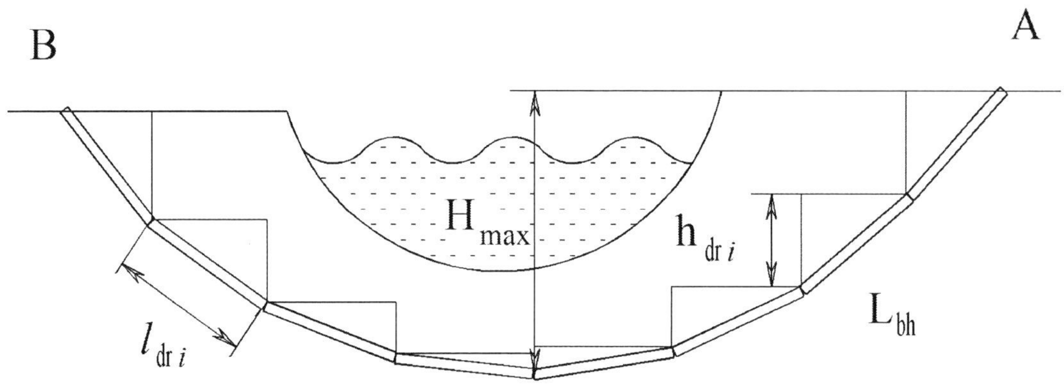

Figure 1. Well profile diagram: *A* – spudding point; *B* – point of entry of the pulled pipeline into the well; L_{bh} - borehole length; l_{dri} - length of the i-th drill pipe; H_{max} - depth of the lower point of the passage profile; h_{dri} - burial depth per one pipe.

From the point of view of practical application, calculation of the well profile must be carried out based on the geometric parameters of the drill pipes used, i.e. the length and angle of deviation of two adjacent drill pipes relative to each other [6]. Since the length of the drill pipe is known and constant, and the deflection angles can be measured or specified at the design stage, the calculation made on the basis of these parameters seems most appropriate [7].

2. Materials and methods

Figure 1 shows a well profile schematically divided into sections equal in length to the drill pipes used in HDD rigs.

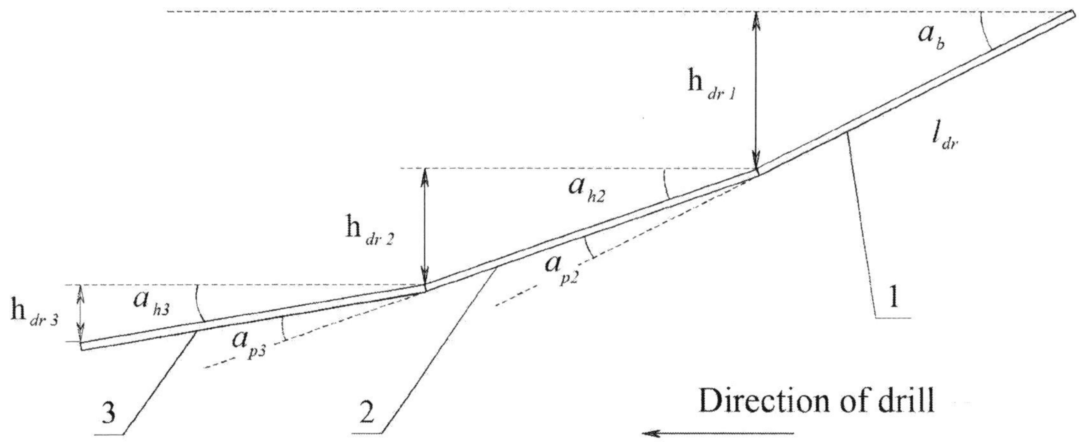

Figure 2. A fragment of the well profile: *1, 2, 3* – drill pipes; a_b – angle of spudding; a_{h2}, a_{h3} - angles of deviation of drill pipes *2* and *3* from the horizontal; a_{p2}, a_{p3} - angles of deviation of drill pipes *2* and *3* relative to the previous one; l_{dr} - drill pipe length, h_{dr1}, h_{dr2}, h_{dr3} - vertical burial depth of the drill pipe.

The maximum burial depth to the well axis is denoted by H_{max}. The axis of the well is taken as the trajectory of the pipe movement.

A section equal in length to one drill pipe is denoted by l_{dr}. The depth per one current drill pipe is denoted by $h_{dr\,i}$. The total well length is L_{bh}.

In accordance with the design scheme, drilling is performed from right to left, and pulling, respectively, is done in the opposite direction.

Figure 2 shows a fragment of the well profile performed during the construction of an underground passage. Due to deviation of current drill pipe *2* relative to previous drill pipe *1* by angle a_{p2}, drill pipe *2*, after completely entering the ground, will be buried at h_{dr2} relative to drill pipe *1*, and so on.

Since the process of passage construction begins with pilot drilling, let us calculate the geometric parameters of the profile in a sequence from the point of spudding [8].

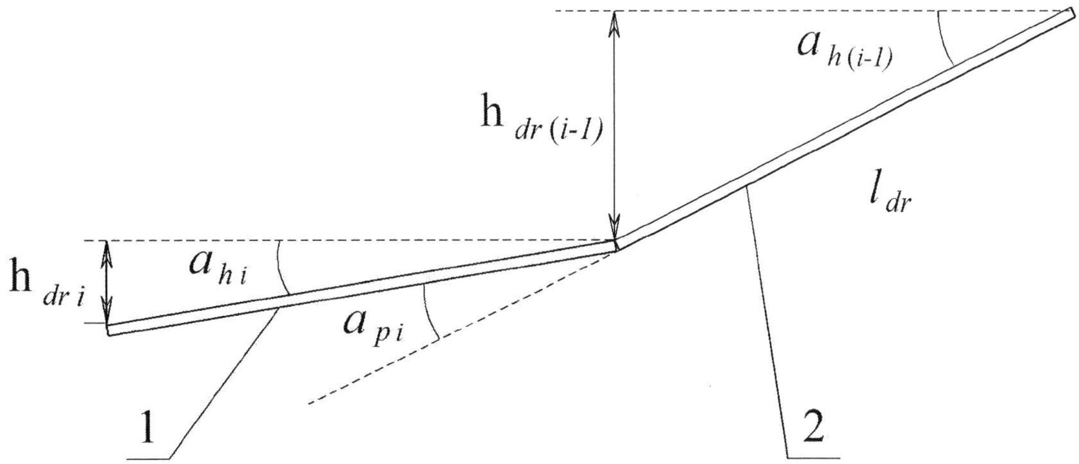

Figure 3. A fragment of the descending section of the borehole profile: *1* – current drill pipe; *2* – previous drill pipe.

There can be three cases in the sequential calculation of the burial depth of drill pipes [9]:

1. Both the current and previous drill pipes are in the descending section of the profile (see Figure 3).
2. Profile passing through the lowest point (see Figure 4).
3. Both current and previous drill pipes are in the ascending section of the profile after passing the lowest point (see Figure 5).

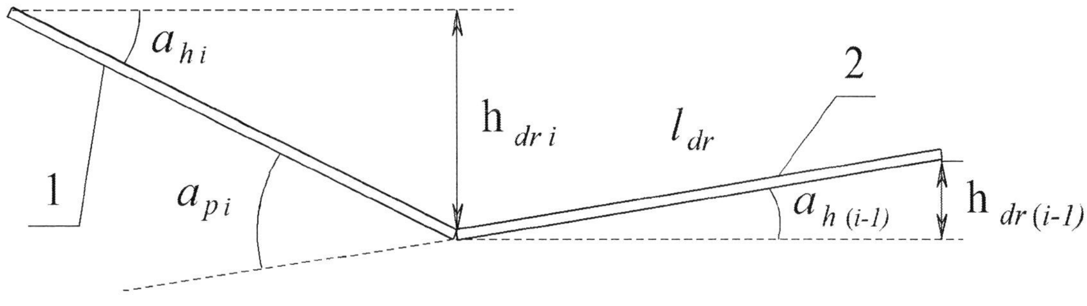

Figure 4. Fragment of the well profile at the lowest point.

Thus, for the first drill pipe:

$$\alpha_{h(1)} = \alpha_b \tag{1}$$

where $a_{h(1)}$ - angle of deviation of the first drill pipe relative to the horizon;
a_b - angle of deviation of the drill mount from the horizon (angle of spudding).
And burial depth $h_{dr\,i}$ for the first drill pipe will be equal to:

$$h_{dr1} = l_{dr} \cdot \sin\alpha_b \tag{2}$$

where l_{dr} – one drill pipe length.

For subsequent drill pipes, the burial depth for current drill pipe $h_{dr\,i}$ relative to the previous one is found:

In the first case:

$$h_{dri} = l_{dr} \cdot \sin \alpha_{hi} \tag{3}$$

where a_{hi} - angle of deviation of the current drill pipe relative to the horizon.

In the second and third cases:

$$h_{dri} = -l_{dr} \cdot \sin \alpha_{hi} \tag{4}$$

The minus sign in the formula indicates that the well profile bends in the second and third cases toward the earth's surface, i.e. depth reduction.

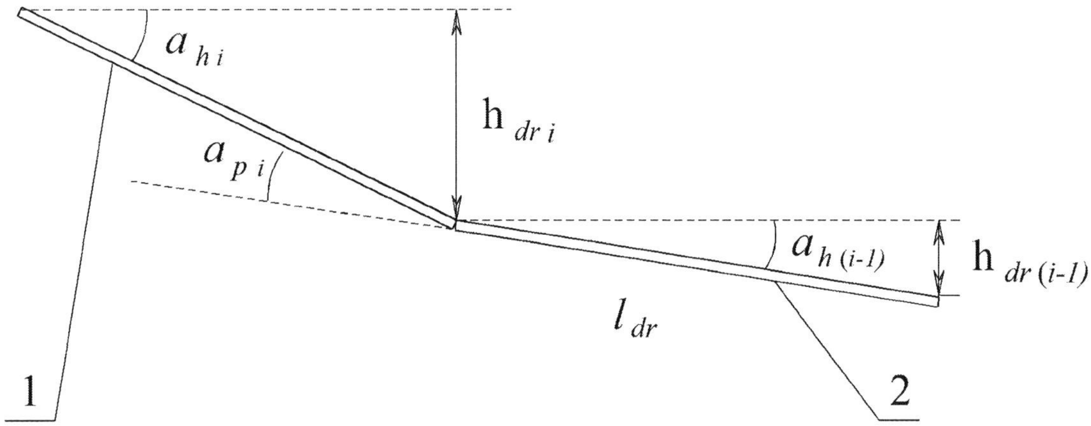

Figure 5. A fragment of the ascending section of the borehole profile.

Angle a_{hi} in the three cases above can be found in different ways.

In the first case:

$$\alpha_{hi} = \alpha_{h(i-1)} - \alpha_{pi}. \tag{5}$$

In the second case:

$$\alpha_{hi} = \alpha_{pi} - \alpha_{h(i-1)}. \tag{6}$$

In the third case:

$$\alpha_{hi} = \alpha_{h(i-1)} + \alpha_{pi}. \tag{7}$$

Then the difference of elevation marks between the drilling point and the current profile point will be equal to:

$$\Delta H = \sum_{i=1}^{n} h_{dri} \tag{8}$$

where n - number of drill pipes in the string.

During summation, the value of ΔH will increase until the value of H_{max} is reached, and then it will gradually decrease and at the point of exit to the surface it will take a value that is equal to the difference in the elevation marks of drilling point A and the exit point to surface B (see Fig.1).

To calculate pulling parameters, it is necessary to use the obtained values of ΔH in the reverse order with reference to the corresponding drill pipes [10].

In case the well profile is given by a constant-radius curve, which may be convenient, values of the borehole bottom burial depth relative to the earth's surface depending on the current well profile can be calculated as follows.

The maximum burial depth of the well profile is determined as:

$$H_{max} = R_c \cdot \left(1 - \sin\left(\frac{\pi}{2} - \alpha_b\right)\right) \tag{9}$$

where R_c - radius of curvature of the well profile.

In case of curvature with a constant radius, the well profile will be described by the equation (see Figure 6):

$$x^2 + (\Delta H - H_{max} + R_c)^2 = R_c^2 \tag{10}$$

where x - horizontal coordinate.

For the convenience of calculations, let us assume that the burial depth of the profile is indicated by an increase in ΔH, and the exit to the day surface is indicated by a decrease in ΔH.

The center of the imaginary circle describing the well profile is placed at point $O(0;H_{max} - R_c)$.

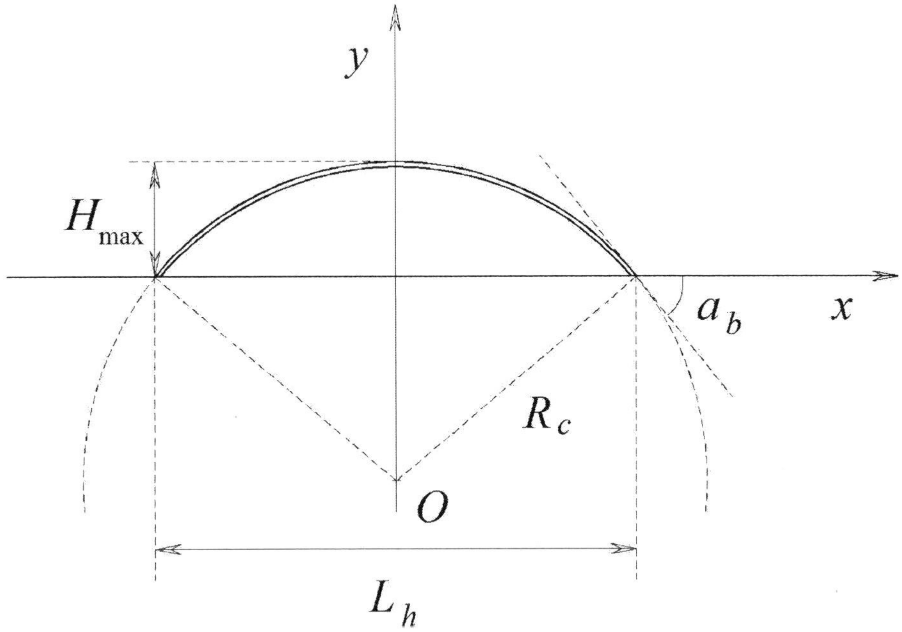

Figure 6. A scheme for calculating the curve profile of a well with a constant radius: O – center of the imaginary circle of the radius R_c; L_h – horizontal well profile length.

Then the burial depth to the current point of the profile will be equal to:

$$\Delta H = \sqrt{R_c^2 - x^2} + H_{max} - R_c. \tag{11}$$

This will be the burial depth of the current drill pipe at this stage of the pilot well drilling [11].

3. Results and Discussion

If it is required to design a well profile of required horizontal length L_h and maximum burial depth H_{max}, then by selecting the angle of spudding from a range of possible values for a particular rig, it is possible to calculate the radius of curvature of the borehole profile at these parameters:

$$R_c = \frac{L_h}{2\cos\left(\frac{\pi}{2} - \alpha_b\right)} \tag{12}$$

where L_h - horizontal well profile length.

In (10), the current horizontal coordinate is denoted by x. To find the current coordinate along the curved well profile, let us determine the angle of the corresponding section of the profile:

$$\alpha_x = \arcsin\left(\frac{L_h}{2R_c}\right) - \arcsin\left(\frac{\frac{L_h}{2} - x}{R_c}\right). \tag{13}$$

Current coordinate L along the well profile will be equal to:

$$L = R_c \alpha_x . \tag{14}$$

Thus, the current coordinate of the well profile and the length of the drilling site along the day surface were linked.

4. Conclusion

Studying the process of pipeline pulling into the well cannot be done without taking into account the features of the technology [12]. The technology of directional drilling involves a cyclic movement of the pipeline in the well due to the need to consistently untwist drill pipes that come to the surface when the drill string is pulled out by the rig [13].

Therefore, the entire technological process of pulling is expediently divided into cycles. One cycle is the movement of the pipeline and the drill string to the length of one drill pipe [14, 15].

The paper shows a practical method for increasing the accuracy of designing a profile of a curved well for a particular drilling rig with a specific set of equipment and drill pipes.

References

[1] Kiryanov I E, Toropov V S and Toropov S Yu 2015 Determination of dependences of pipeline movement resistance forces in curve borehole from physical parameters and rheological model of using mud *Oil equipment* **8**

[2] Willoughby D 2005 *Horizontal Directional Drilling Utility and Pipeline Applications* (New York; The McGraw-Hill Companies) 24

[3] Seroshtanov I V, Toropov V S, Toropov S Yu 2015 Accidents and complications in trenchless construction of pipelines. In the collection: *Oil and gas of Western Siberia Materials of the International scientific and technical conference dedicated to the 90th anniversary of the birth of Anatoly Nikolaevich Kosukhin*, pp. 98-102

[4] Zemenkov Yu D 2004 *Transporting and storage of oil and gas in examples and sums* (St. Petersburg; Nedra)

[5] Zemenkova M Yu, Shipovalov A N, Dudin S M and Zemenkov Yu D 2007 System analysis in oil and gas objects control processes *News of higher school. oil and gas* **5** 116-9

[6] Toropov V S, Ponomareva T G, Toropov S Yu, Toropov E S 2016 Quantitative assessment of changes in the pipeline - well system in the event of an emergency situation in horizontal directional drilling. *News of higher educational institutions. Oil and gas* 5 71-77

[7] Zemenkov Yu D 2008 *Technique and technologic processes at transport of energy resources* (Tyumen; Vector Book) **1** 380

[8] Toropov V S, Temirbaev R M, Toropov E S and Toropov S Yu 2016 Interaction between the pipeline and additional equipment for trenchless technologies. International Scientific-Practical Conference of Students, Graduate Students and Young Scientists *"Transport and Storage of Hydrocarbons"* **154**

[9] Voronin K S 2016 Forecasting and Evaluation of Gas Pipelines Geometric Forms Breach Hazard. Transport and Storage of Hydrocarbons. *IOP Conf. Series: Materials Science and Engineering* **154** 012020 doi:10.1088/1757-899X/154/1/012020

[10] Voronin K S, Ogudova E V 2015 The Effect of Dynamic Processes in the System "Pipe-Soil" on the Pipeline Deviation from Design Position / Transport and Storage of Hydrocarbons. *IOP Conf. Series: Materials Science and Engineering* **154** 012019 doi:10.1088/1757-899X/154/1/012019

[11] Dudin S, Bahmat G, Zemenkov Y, Voronin K, Shipovalov A 2017 Strategy for monitoring and

ensuring safe operation of Russian gas transportation systems *MATEC Web of Conferences* **106** DOI: 10.1051/matecconf/201710606004

[12] Zemenkova M, Shalay V, Zemenkov Y, Kurushina E 2016 Improving the Efficiency of Administrative Decision-Making when Monitoring Reliability and Safety of Oil and Gas Equipment *MATEC Web of Conferences* **73** DOI: 10.1051/matecconf/20167307001

[13] Smirnov V V, Zemenkov Yu D Using the finite element method for calculation of intense deformed condition of above ground pipelines constructed on permafrost soil. *Pipeline transport: theory and practice.* **4** 18-23

[14] Zemenkov Y D, Shalay V V, Zemenkova M Y 2015 Expert systems of multivariable predictive control of oil and gas facilities *reliability Oil and Gas Engineering Procedia Engineering* **113** 312-315 DOI: 10.1016/j.proeng.2015.07.271

[15] Shalay V, Zemenkova M, Zemenkov Y, et al. 2016 Modeling Parameters of Reliability of Technological Processes of Hydrocarbon Pipeline Transportation. *MATEC Web of Conferences* **73** 01029 DOI: 10.1051/matecconf/20167301029

IOP Conf. Series: Materials Science and Engineering **357** (2018) 012007 doi:10.1088/1757-899X/357/1/012007

Physical and mathematical modeling of process of frozen ground thawing under hot tank

M Y Zemenkova[1], U Shastunova[2], A Shabarov[1,2], A Kislitsyn [2], A Shuvaev [2]

[1]Industrial University of Tyumen, 38 Volodarskogo St., Tyumen, 625001, Russia
[2]Tyumen State University, 6 Volodarskogo St., Tyumen, 625003, Russia

E-mail: muzemenkova@mail.ru, kaf_mms@utmn.ru

Abstract. A description of a new non-stationary thermophysical model in the "hot tank-frozen ground" system is given, taking into account mass transfer of pore moisture. The results of calculated and experimental data are presented, and the position of the thawing front is shown to be in good agreement with the convective heat transfer due to moisture migration in the thawed ground.

1. Introduction

Processes, united by the general influence of the cryogenic factor, are very diverse and complex in nature [1]. Freezing, frost heaving and thawing of moist soil are complex thermodynamic processes in freezing bases. In the process of thawing as the temperature field changes, moisture migration to the freezing front takes place [2].

In this regard, the study of the processes of heat and mass transfer in the ground and their interaction with engineering structures acquire particular urgency. It is important to solve the joint task "environment - construction - ground".

The experimental setup and the results of the experiments on thawing of frozen ground under a hot oil tank are presented in [4, 5].

It was experimentally discovered that when a frozen moist ground is heated from above, a vertical filtration flow of thawed water may occur, and, owing to this, a significant (approximately twofold) increase in the mean velocity of the thawing front movement.

This can be explained by the fact that since the volume of water is less than the volume of ice, pores appear in the thawed ground through which melt water moves down to the center of the melt front, convective heat transfer occurs, which increases the thawing rate [4].

The dependences of the temperature change of the gas space and the hot heat-transfer medium over time have also been obtained [5]. At the beginning of filling an empty tank, the temperature of oil fell as it was heating the walls of a cold metal tank. After reaching the thermal balance of the metal wall and oil, the temperature of the heat-transfer medium reached the value to which it was heated in the oil tank - 57 °C. In the stand-by mode, the temperature of the heat-transfer medium decreased exponentially (according to Shukhov's formula).

A change in the temperature of the gas space was slightly different: in the injection mode, the temperature increased sharply, but when the oil reached the maximum temperature, the temperature of the gas space continued to grow for some time. The increase in the temperature of the gas space is evidently due to evaporation of light fractions of oil, as well as a complicated nature of convective gas flows in the gas space inside the tank.

2. Analytical solution

The authors in [4] found an approximate analytical quasi-stationary solution of the two-dimensional Stefan problem in cylindrical coordinates *r, x* in the classical formulation, both taking into account the transport of pore moisture and without taking it into account.

The system of equations for the ground thawing problem (the two-dimensional Stefan problem in cylindrical coordinates *r, x* in the classical formulation without convection) has the form [4]:

$$\frac{\partial T_{1,2}}{\partial t} = a_{1,2}\left(\frac{\partial^2 T_{1,2}}{\partial r^2} + \frac{1}{r}\frac{\partial T_{1,2}}{\partial r} + \frac{\partial^2 T_{1,2}}{\partial x^2}\right), \quad (1)$$

where subscripts 1 and 2 refer to the thawed and frozen zones, respectively. These zones are separated by a moving surface (the melt front) $F(r,x,t) = 0$. The initial condition is the absence of a thawed zone and the equality of the temperature in the entire region to initial value T_0 (in this case $T_0 =$ -9°C). The boundary conditions on the fixed boundaries have the form:

$$T_{1,2}\left(x=0, r>R\right) = T_0,\ -\lambda_1 \left.\frac{\partial T_1}{\partial x}\right|_{x=0,r<R} = q = \frac{W}{\pi R^2},\ \left.\frac{\partial T_{1,2}}{\partial r}\right|_{r=0} = 0,\ T_2(r\to\infty, x\to\infty)\to T_0,$$

where R - radius of the tank, W - heater wattage (in this model R = 12cm, W = 500W). At the melt front, the temperatures are equal $T_1 = T_2 = 0$°C, and the energy balance condition is given, often called the Stefan condition:

$$\left(\lambda_1\left|gradT_1\right| - \lambda_2\left|gradT_2\right|\right)\Big|_{F=0} = L\frac{\partial F/\partial t}{\left|gradF\right|}, \quad (2)$$

where $\lambda_{1,2}$ - thermal conductivity coefficients in the thawed and frozen zones, L - volumetric heat of melting (J/m^3).

There is no exact analytical solution to this problem, but approximate quasi-stationary solutions of a number of related problems are known, published in [11, 12]. Following the procedure outlined in these works, let us seek the solution of equation (1) in form $T(F)$, where $F = const$ - isothermal surface equation; in particular, $F = 0$ - zero isotherm equation. Let us calculate function F in the form:

$$F = x - f(r) - \int_0^t g(t)dt, \quad (3)$$

where $f(r)$ and $g(t)$ - unknown functions in advance. Differentiating $T(F)$ by r, x, t, one will find:

$$\frac{\partial T}{\partial r} = -T'\cdot f',\ \frac{\partial^2 T}{\partial r^2} = f'^2\cdot T'' - T'\cdot f'',\ \frac{\partial^2 T}{\partial x^2} = T'',\ \frac{\partial T}{\partial t} = -g\cdot T', \quad (4)$$

where $T' = \frac{dT}{dF}$, $T'' = \frac{d^2T}{dF^2}$, $f' = \frac{df}{dr}$, $f'' = \frac{d^2f}{dr^2}$. Substituting (4) into (1) and neglecting the difference between a_1 and a_2 (that is, let us assume $a_1 \approx a_2 = a$), let us obtain an ordinary differential equation with respect to function $f(r)$:

$$P\left(1+f'^2\right) - f'' - \frac{1}{r}f' + \frac{g}{a} = 0, \quad (5)$$

where

$$P = T''/T'. \quad (6)$$

For (5), an analytical solution can be obtained if one assumes that g and P are constants. However, after the temperature field is found, when it is substituted into the Stefan's condition (2), let us assume that g and P are functions, which allows us to find the velocity of the front motion. Physically, this means that the melt front moves so slowly that one can consider the temperature field to differ little from the stationary one (the quasi-stationary approximation). Introducing notations $P = B^2$, $B^2 + g/a = A^2$, $y = ABr$, $f' = -w'(y)/(B^2 w(y))$, let us transform (5) to the form:

$$w'' + \frac{1}{y}w' + w = 0, \tag{7}$$

the solution of which, as we know, is the Bessel function of the first kind of zero order $J_0(y)$. Returning to previous notations, let us obtain a solution for function $f(r)$:

$$f(r) = -\frac{1}{B^2}\left\{\ln\frac{273 + T_1(r)}{273} + \ln\left[J_0(ABr)\right]\right\}, \tag{8}$$

where $T_1(r)$ – ground surface temperature. Obviously, the solution obtained makes sense if:

$$ABr < 2.4 \tag{9}$$

where 2.4 - the first zero of the Bessel function J_0. Hence, assuming $ABR = 2.4$, there is:

$$g = a\left(\frac{2.4}{BR}\right)^2 - aB^2. \tag{10}$$

On the other hand, integrating (6), let us find:

$$T' = PT_1 e^{PF}, \quad T = T_1 e^{PF} + T_0. \tag{11}$$

Substituting in (2), one obtains the second relation between g and $P = B^2$:

$$g = \frac{B^2 T_1 |\lambda_1 - \lambda_2|}{L}. \tag{12}$$

Excluding B^2 from (10) and (12), let us obtain the relation between g and T_1:

$$g = \frac{2.4}{R}\sqrt{\frac{a\left(T_1|\lambda_1 - \lambda_2|\right)^2}{L\left(T_1|\lambda_1 - \lambda_2| + aL\right)}}. \tag{13}$$

The experimental dependence of the temperature of the ground surface at point $r = 0$ on time is shown in Figure 1 (curve 1 approximates the experimental points). Fig. 1 shows calculated curve 2 obtained as a result of numerical integration of function $g(T_1(t))$. As can be seen from the figure, the coincidence with the experimental points (denoted by the symbols +) is observed only for the beginning of warm-up, when the transfer of heat by melt water is negligible.

Let us write the equation of thermal conductivity taking into account the convective heat transfer. Here let us assume that only the convective transfer along the x axis is essential, and the transfer along the r axis is negligible:

$$\frac{\partial T}{\partial t} + v(r,t)\frac{\partial T}{\partial x} = a\left(\frac{\partial^2 T}{\partial r^2} + \frac{1}{r}\frac{\partial T}{\partial r} + \frac{\partial^2 T}{\partial x^2}\right), \tag{14}$$

where $v(r,t)$ - average projection of the melt water rate on vertical axis x.

As it was mentioned above, it is necessary to seek a solution in form $T(F)$ and function F in the form:

$$F = x - f(r) - h(r)\int_0^t g(t)dt = x - f(r) - h(r)\cdot z(t), \quad v(r,t) = g(t)\cdot h(r), \tag{15}$$

where function $f(r)$ is assumed to be known from the previous solution for a stationary medium, and function $h(r)$ - unknown. Differentiating $T(F)$ by r, x, t, and substituting in (14), one obtains a differential equation with respect to functions $f(r)$ and $h(r)$:

$$P\left[1 + (f' + h'z)^2\right] - (f'' + h''z) - \frac{1}{r}(f' + h'z) = 0, \tag{16}$$

where P, as it was mentioned above, is determined by (6). Replacing $y = P\cdot r$:

$$f' + h'z = -u'/(Pu), \tag{17}$$

let us obtain an equation for function u:

$$\frac{d^2u}{dy^2} + \frac{1}{y}\frac{du}{dy} + u = 0, \tag{18}$$

coinciding in the form with equation (7). The solution of this equation is the Bessel function $u = J_0(y) = J_0(Pr)$, and $u' = -P \cdot J_1(P \cdot r)$, where J_1 - the Bessel function of the first kind of the first order. From (17) let us find:

$$h'(r) = -\frac{1}{z}\left(\frac{u'}{Pu} + f'\right) = \frac{1}{z}\left(\frac{J_1(P \cdot r)}{J_0(P \cdot r)} - f'\right). \tag{19}$$

Using the experimental values of function $z(t) = \int_0^t g(t)dt$, one can numerically integrate (19). The result of the calculations is shown in Fig. 1, 2 as function $F(r,x,t) = 0$ at different times t, as well as in Fig. 1 as the dependence of the coordinate of the center of the front on time (curve 3). It can be seen that, in contrast to the solution in a stationary medium, it is possible to obtain a satisfactory agreement with the experimental data.

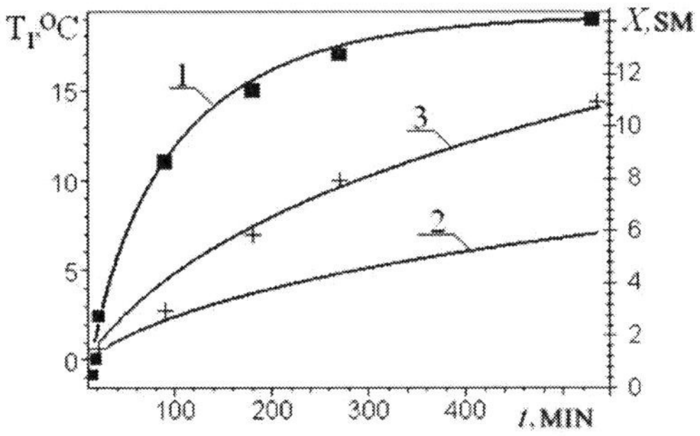

Figure 1. The temperature of the ground surface; 2 - the coordinate of the thawing front center without taking into account convective heat transfer; 3 - the same taking into account convective heat transfer; + - experimental points.

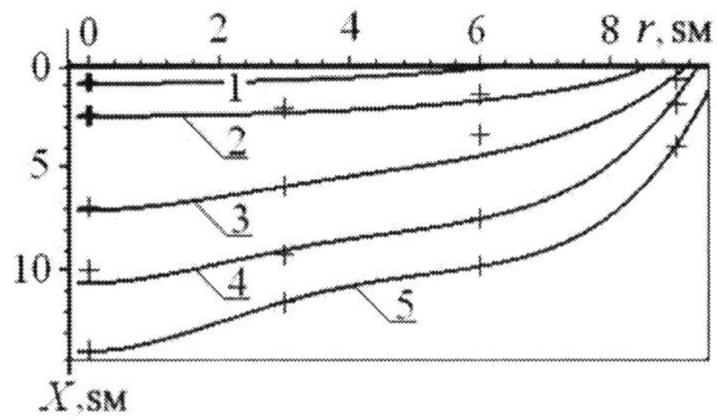

Figure 2. Position and shape of the thawing front taking into account convective heat transfer:
1 - t = 10 min; 2 - t = 90 min; 3 - t = 180 min; 4 - t = 270 min; 5 - t = 540 min.

3. Numerical modeling

Computer modeling of this problem was considered by several authors [6, 7, 8] who confirm the conclusion obtained experimentally by the authors of the article that the movement of the liquid in the ground creates an additional thermal flux that must be taken into account when solving the thermal conductivity equation. In low-water soils, this mechanism can be neglected, which does not apply to soils in the north of Russia, through which most of the existing and under construction pipelines and facilities of the fuel and energy complex pass [7].

Physical and mathematical modeling is based on a numerical solution of a system of equations consisting of thermal balance equations for the gas and oil phases, the heat-transfer medium flow equation, the thermal conductivity equation in a two-dimensional setting, taking into account phase transformations and migration of pore moisture.

The non-stationary thermophysical model can be represented in the following form [5]:

$$\begin{cases} Q_g = -Q_{g1} + Q_{go1} + Q_{go2} - Q_l \\ Q_l = Q_{l1} + Q_{lo} + Q_{ls} + Q_{\mathrm{lg}} \\ \dfrac{\partial T_s}{\partial t} = a_s \cdot \left(\dfrac{1}{r} \cdot \dfrac{\partial}{\partial r} \cdot \left(r \cdot \dfrac{\partial T_s}{\partial r} \right) + \dfrac{\partial^2 T_s}{\partial z^2} \right) + \vartheta_r \cdot \dfrac{\partial T_s}{\partial r} + \vartheta_z \left(\dfrac{\partial T_s}{\partial z} \right), \\ \vartheta_r = -K \cdot \dfrac{\partial w}{\partial r}, \vartheta_z = -K \cdot \dfrac{\partial w}{\partial z} \\ G = \pi R^2 \cdot \dfrac{dh}{dt} \end{cases} \quad (20)$$

$$Q_g = C_g \cdot \rho_g \cdot \pi R^2 \cdot (H-h) \cdot \frac{d(T_g - T_0)}{dt}; \qquad Q_l = C_l \cdot \rho_l \cdot \pi R^2 \cdot (h) \cdot \frac{d(T_l - T_0)}{dt};$$

$$Q_{g1} = C_g \cdot \rho_g \cdot \pi R^2 \cdot (T_g - T_0) \cdot \frac{d(H-h)}{dt}; \qquad Q_{l1} = C_l \cdot \rho_l \cdot \pi R^2 \cdot (T_{lo} - T_l) \cdot \frac{d(h)}{dt}$$

$$Q_{go1} = k_{go1} \cdot (T_g - T_0) \cdot \pi R^2; \qquad Q_{lo} = k_{lo} \cdot (T_l - T_0) \cdot 2\pi R h$$

$$Q_{go2} = k_{go2} \cdot (T_g - T_0) \cdot 2\pi R \cdot (H-h); \qquad Q_{ls} = k_{ls} \cdot (T_l - T_s) \cdot \pi R^2$$

$$Q_l = \alpha_{gl} \cdot (T_g - T_l) \cdot \pi R^2 \qquad Q_{\mathrm{lg}} = \alpha_{\mathrm{lg}} \cdot (T_l - T_g) \cdot \pi R^2$$

where T_0 - ambient temperature, K; T_{lo} - temperature of the heat-transfer medium at the inlet to the tank at a constant flow rate G, K; k_{go1} - coefficient of heat transfer through the roof of the tank, $W/(m^2 \cdot K)$; k_{go2} - coefficient of heat transfer through the side wall of the tank in the gas space region, $W/(m^2 \cdot K)$; α_{gl} and α_{lg} - coefficient of heat exchange from the gas-air mixture to the "hot" heat-transfer medium and back, $W/(m^2 \cdot K)$; k_{lo} - coefficient of heat transfer through the side wall of the tank in the liquid space region, $W/(m^2 \cdot K)$; k_{ls} - coefficient of heat transfer through the tank bottom from the "hot" heat-transfer medium to the foundation of the tank, $W/(m^2 \cdot K)$; $C_{g,l,s}$ and $\rho_{g,l,s}$ - heat capacity and density of the gas-air mixture, heat-transfer medium and ground, respectively, $J/(kg \cdot K)$, m; R, H - radius and height of the tank, m; h - height of the heat-transfer medium filling, m; G - flow rate of the heat-transfer medium, m^3/s, a_s - coefficient of thermal diffusivity of ground, m/s^2, ϑ_r and ϑ_z - filtration rate of pore moisture, K – isothermal coefficient of moisture conductivity, m/s^2, w - moisture content by weight (subscripts g, l, s denote, respectively, gas, liquid and ground components of the system).

The technique for finding coefficients of heat transfer is described in [9, 10, 11]. System (1) is solved using an implicit scheme [12] with the following initial and boundary conditions:

Initial conditions:

at $t = 0$ $h = 0, T_g = T_0, T_l = T_{lo}$,

$J = k_{ls} \cdot (T_l - T_s) \cdot \pi R^2 = 0$

at $z = 0$ $T_s = T_0$

Boundary conditions:

$z > 0, T_s = T_s(z)$

$z = L, T_s = T(L) = const$

$$J = k_{ls} \cdot \pi R^2 \cdot (T_l - T_s) = \lambda \left. \frac{\partial T_s}{\partial z} \right|_{z=0}$$

4. Discussion of the numerical solution

The physical and mathematical modeling of the proposed non-stationary model in the "hot tank-frozen ground" system taking into account the effect of the migration of pore moisture gives a fairly accurate agreement with the experimental data, unlike the variant without taking it into account [4].

5. Conclusion

A significant influence of pore moisture migration on the shape and velocity of the thawing front during thermal interaction in the "hot tank-frozen ground" system has been established experimentally.

A non-stationary thermophysical model is developed that takes into account the migration of pore moisture, the numerical solution gives a fairly satisfactory agreement with the experimental data.

The solution of the non-stationary thermophysical system can be adapted to predict the halo of thawing when storing oil and oil products on frozen soils.

References:

[1] Gorelik Ya B 2002 *Physics and modeling of cryogenic processes in the lithosphere* (Novosibirsk: Publishing house of the SB RAS)

[2] Kudryavtsev S A 2012 Numerical modeling of the migration of moisture depending on the freezing speed of soils. *Engineering and construction journal* **4** 33-38

[3] Mikhailov P Yu 2012 *Dynamics of heat-and-mass transfer processes and heat-force interaction of freezing grounds with an underground pipeline: dissertation* (Tyumen)

[4] Kislitsyn A A 2015 Experimental and theoretical study of the process of frozen ground thawing under a tank for storage of petroleum products *Bulletin of the Tyumen State University. Physico-mathematical modeling. Oil, gas, power engineering.* **1 – 4** 15-25

[5] Shastunova U Yu 2016 Investigation of the temperature regime of a tank with oil products in winter time. *Bulletin of the Tyumen State University. Physico-mathematical modeling. Oil, gas, power engineering* **2** 24-30

[6] Vaganova N A 2008 *Modeling of heat transfer processes in moist soil with convection considered. Vaganov. Problems of theor. and appl. Mathematics: coll. of 39th All-Russia. Youth. Conf.* (IMM UrB RAS)

[7] Bashurov V V, Vaganova N A, Filimonov M Yu 2011 Numerical modeling of heat exchange processes in the ground considering fluid filtration. *Computational technologies.* **16-4** 3-18

[8] Gishkelyuk I A, Grinchik N N, Kundas S P 2008 Mathematical modeling of convective diffusion of soluble compounds in soil under nonisothermal moisture transfer. *Engineering-Physics Journal.* **81-5** 924–935

[9] Shastunova U Yu, Yanbikova Yu F 2015 Calculation of the cooling system of frozen soil under a tank with fuel oil / U.Yu. Shastunova. *Achievements and problems of modern science: a collection of articles of the international scientific and practical conference.* pp. 48-54

[10] Bakhmat G V, Kislitsyn A A, Shastunova U Yu 2011 Methodology for calculating the thermal conditions of a tank of RVS type in winter during operation. Proceedings of the International Scientific and Practical Conference on Permafrost Engineering, dedicated to the 20th anniversary of the foundation of the LLC Fundamentstroyarcos. (Tyumen: City-Press) pp. 301–306

[11] Gishkelyuk I.A., Stanislovskaya Yu.V., Evlanov D.V. Forecasting the thawing of permafrost soils around a long underground pipeline. Moscow, Science and Technologies of Pipeline Transportation of Oil and Oil Products, 2015, №1 (17), pp. 20-25
[12] Paskonov V.M., Polezhaev V.I., Chudov L.A. Numerical modeling of heat and mass transfer processes. M., Nauka, 1984. – 288 p.

IOP Conf. Series: Materials Science and Engineering **357** (2018) 012008 doi:10.1088/1757-899X/357/1/012008

Study of influence of various factors on electrochemical signal of lead in water solutions

Yu N Zhikharev[1], L I Andrianova[2] and E V Ogudova[2]

[1]Tyumen high military-engineering college, 1 L. Tolstogo St., Tyumen, 625000, Russia
[2] Industrial University of Tyumen, 38 Volodarskogo St., Tyumen, 625000, Russia

E-mail: specular@mail.ru

Abstract. The conditions for obtaining a reproducible signal of lead in water solutions of indifferent electrolytes on various substrates (working electrodes) for analytical purposes were studied. Attention was also paid to studying the regularities of the initial stage of formation of lead sediments by the method of inversion voltammetry. The possibility of using different working electrodes to obtain stable current-potential curves is shown depending on the conditions of electrolysis, pH of the medium, the electrolysis potential and impurities.

1. Introduction

The study of the processes taking place at the electrode and solution interface has received a lot of attention in recent decades [1-7]. This concerns both the issues of quantitative determination of elements in natural environments with the aim of increasing the sensitivity, and the development of methods for studying the initial stages of the formation of new phases.

The electrochemical useful signal is a curve with an acute maximum (Fig. 1).

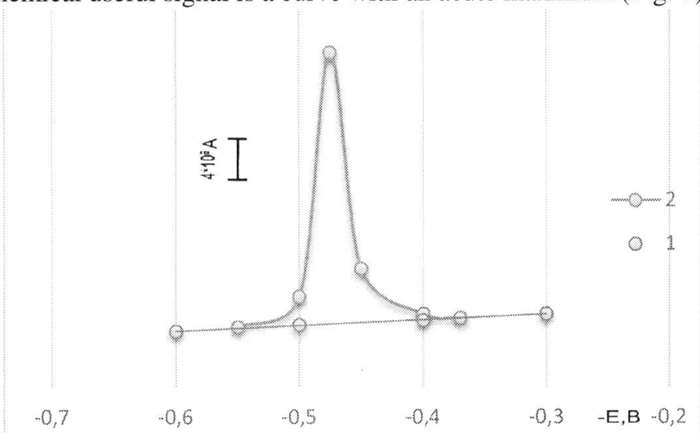

Figure 1. A general view of the polarization curve of the electrical dissolution of the lead plate. curve 1 – background line; curve 2 – peak of lead plate dissolution Background – 0.2M KCl, pH = 3,5, $C_{Pb}^{2+} = 4.7 \cdot 10^{-5}$ mole/l; $\tau_e = 2$ min; $E_э = -0.9$V, w = 40 mV/s for a glassy-carbon electrode.

2. Materials and methods

Dependences for the peak-point currents of the process of electrical dissolution of plates concentrated on the glassy-carbon, copper and silver electrodes, on the concentration of the depolarizer, on the time of the electro-accumulation, on the electrolysis potential, and on the pH of the solution were obtained. The influence of various factors on the structure of the lead plate has been studied as well as the number of nucleating centers and their radius has been calculated. The effect of the adsorption of chloride and iodide ions on the process of the anodic dissolution of lead sediments concentrated on a silver electrode was studied.

The dependence of the peak-point current and the amount of electricity of anodic dissolution of the lead plate, electro-deposited on a glassy-carbon electrode from a 0.2M KNO_3 solution with pH = 2.75 on the concentration, was obtained. It was found that in the concentration range from $7.1*10^{-6}$ to $1.4*10^{-4}$ mole/l, the I-C and Q-C dependences had linear character, which could be used for analytical purposes. Deviation from linearity was observed at concentrations $>1.4*10^{-4}$, when volumetric layers of the plate form on the electrode surface.

The dependence of the current and the amount of electricity of the anodic dissolution of the lead plate on the accumulation time (I-t, Q-t) has a similar nature, which is related to the amount of the plate on the electrode.

On the silver electrode in 0.2M KNO_3, during anodic dissolution of the plate, there are two peaks (Figure 2) with different potentials. At low concentrations or accumulation times, an adsorption peak ($E_п$=-0.28V) is registered, with increasing in concentration or accumulation time, a more electronegative, analytical peak appears ($E_п$=-0.44V), which is used in inversion voltammetry for quantitative and qualitative analysis.

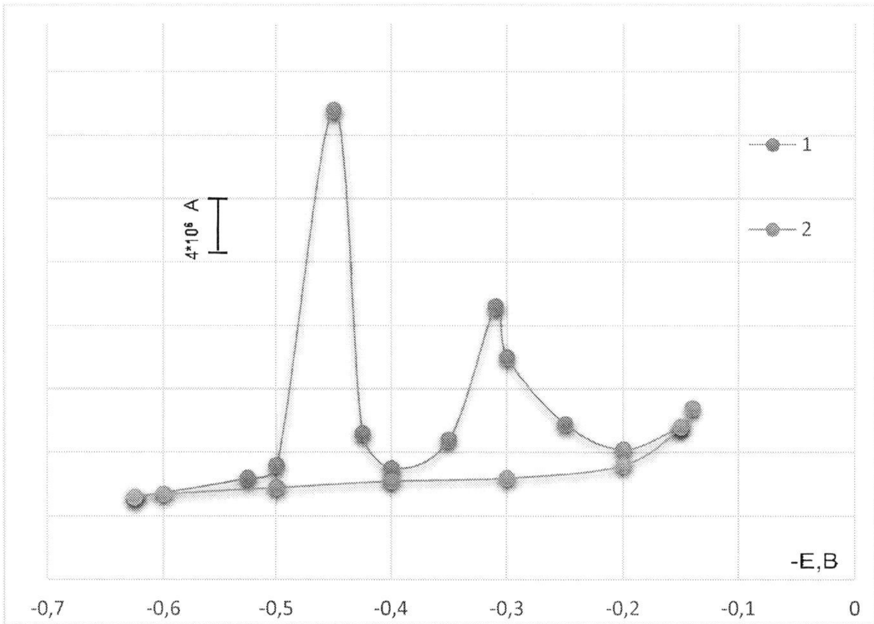

Figure 2. A general view of the polarization curve of the electrical dissolution of the plate; curve 1 – background line; curve 2 – peak-point of lead plate.
The first peak-point is an analytical signal; the second peak-point is an adsorption signal.
Background – 0.2M KCl, pH = 3.5, $C_{Pb}{}^{2+} = 4.7 \cdot 10^{-5}$ mole/l; $\tau_э$ = 2 min;
$E_э$ = -0.9 V, w = 40 mV/s for a copper electrode.

Further, when studying the dependencies on various factors, let us deal with an analytical signal.

Dependences of anode peak-points on glassy-carbon, silver and copper anodes on the potential of electro-accumulation have been studied (Fig. 3, 4). These dependences come through the maximum, which is due to the hydrogen evolution, which reduces the current output of lead and blocks the surface for lead plate.

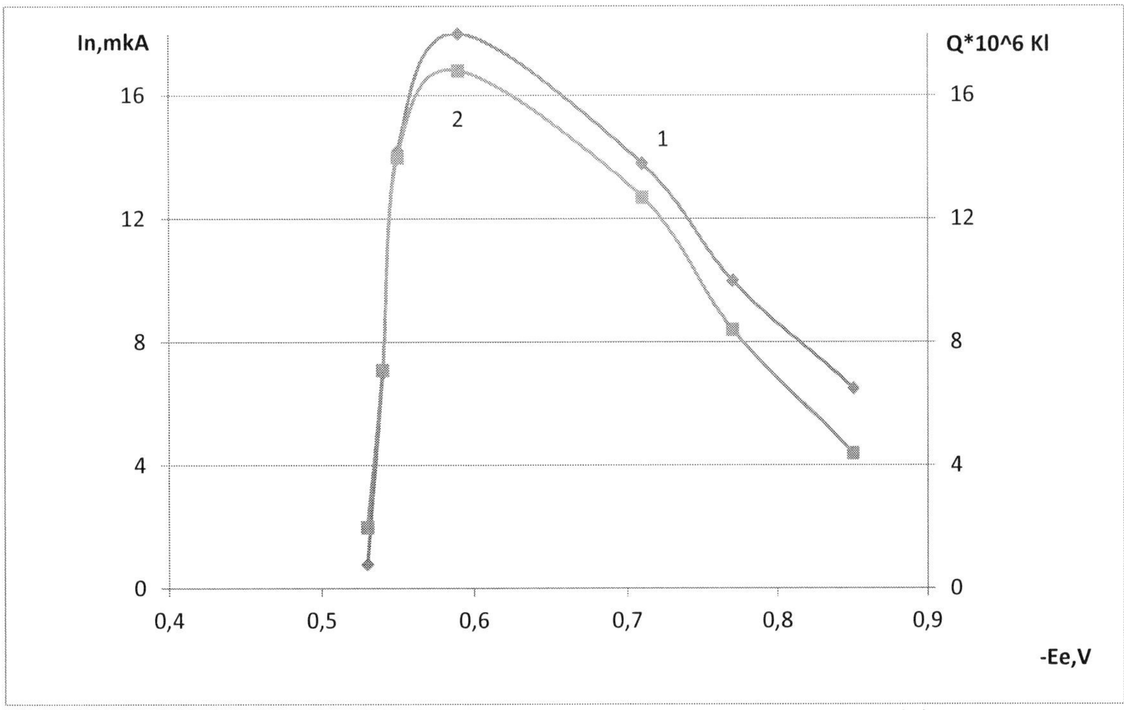

Figure 3. Dependence of the current peak-point of lead plate dissolution (Curve 2) and the amount of the electricity plate (Curve 1) on the electrolysis potential.
Background – 0.2M KNO_3, pH = 3.5, $C_{Pb}{}^{2+}$ = 4.7 mole/l; τ_e = 2 min; w = 40 mV/s for a copper electrode.

The optimum value of the potential of electro-accumulation at 0.2M KNO_3 on the copper electrode is in the range of -0.6V, and 0.2M KCl in the solution in the range from -0.9 to -1.2V. Considering the size and width of the anodic lead peak-point, the nature of the background affects the reversibility of the process. Because of the formation of complexes that affect the discharge energy, the shape of the peak (the width of the half-peak) changes.

Dependence of the current peak-point and the amount of electricity on the pH of the indifferent electrolyte was studied (Fig. 3). As the acidity of the solution increases, the dissolution current peak-point decreases due to the effect of the hydrogen evolution process. Optimal pH values of indifferent electrolyte for studying the physicochemical properties of the plate, as well as for analytical purposes on various electrodes, are:

The number of lead sediment nucleating centers and their sizes from the electrolysis potential are shown in Fig. 4.

- glassy-carbon electrode: pH= 3.75;
- silver electrode: pH= 3.5-3.7;
- copper electrode: pH= 3.0-3.5.

IOP Conf. Series: Materials Science and Engineering **357** (2018) 012008 doi:10.1088/1757-899X/357/1/012008

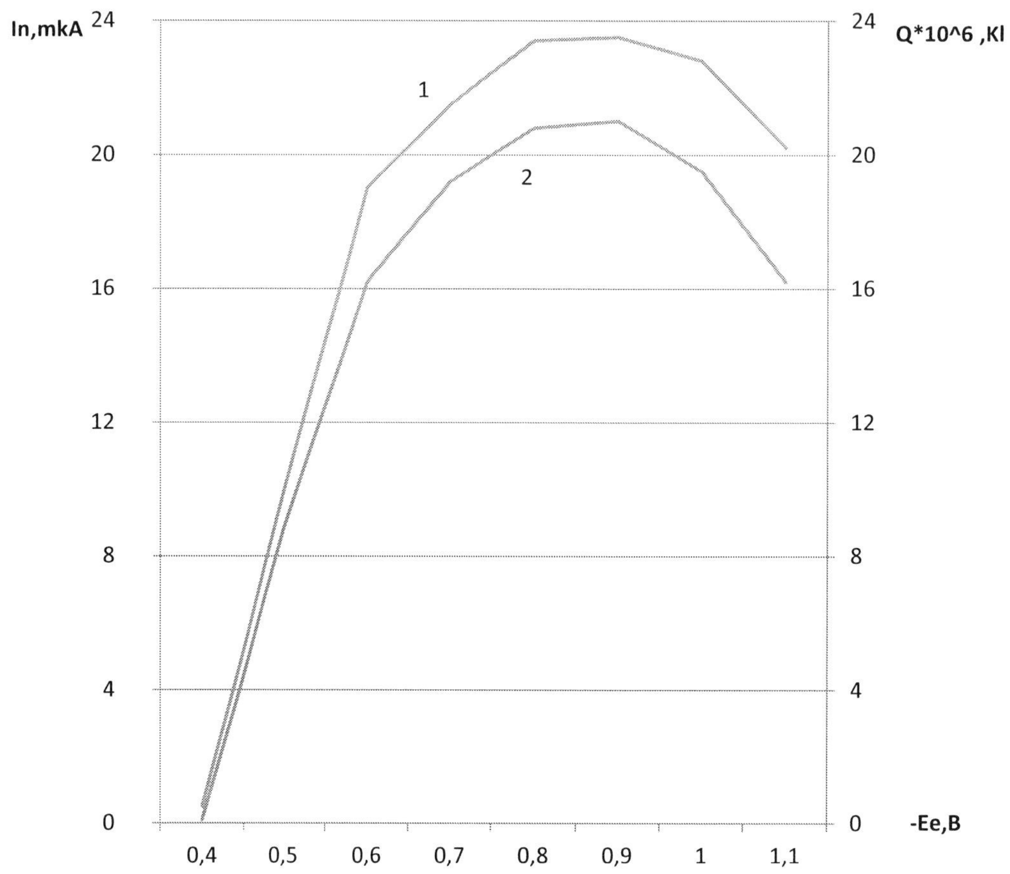

Figure 4. The number of lead sediment nucleating centers and their sizes from the electrolysis potential.

In order to study the structure of the lead plate, the experimental data were processed on the basis of the concepts of nucleation of plates developed in the papers [8, 9] with reference to the conditions of inversion voltammetry.

In the process of electrical dissolution of embryos, current-potential curves passing through the maximum are recorded. The magnitude of the current at any time is described by the equation:

$$\bar{I} = \widehat{M} N^{\frac{1}{3}} q^{\frac{2}{3}} z(\delta t), \quad (1)$$

where

$$\widehat{\mathrm{M}} = \mathrm{nF}\bar{\gamma}\mathrm{C}^{\circ}(\mathrm{D}\delta)^{1/2} \quad (2)$$

$$\bar{\gamma} = 2\pi\left(\frac{3\widehat{\mathrm{M}}}{2\pi}\right)^{2/3} \quad (3)$$

$$\widehat{\mathrm{M}} = \frac{\mathrm{M}'}{\mathrm{n}\cdot\mathrm{F}\cdot\rho} \quad (4)$$

$$\delta = \cdot\, \mathrm{w} = \frac{\mathrm{hF}}{\mathrm{RT}}\mathrm{w}\,. \quad (5)$$

M' – molar mass, g/mole;
ρ – plate density, g/cm^3;
F – Faraday number, C/g-e;
D – diffusion coefficient, cm^2/s;
C° - concentration of ions in the depth of the solution, mol/cm^3;
n – number of electrons participating in the electrode process;

IOP Conf. Series: Materials Science and Engineering **357** (2018) 012008 doi:10.1088/1757-899X/357/1/012008

N – number of nucleating centers on the electrode.

By equation 1, one can calculate the number of nucleating centers at any point of the polarization curve. The solution of equation 1 with respect to value J can be expressed by the function:

$$X_m = \delta \cdot t_m, \tag{6}$$

where t_m – half-peak width, s.

There were calculated function curves X_m of $\lg \bar{g}$ in papers [27, 28], where:

$$\bar{g} = nF\bar{\gamma}C^{\circ}N^{1/3}D^{1/2}\delta^{-1/2}q_0^{-1/3} \,. \tag{7}$$

From the above-mentioned expressions, the number of nucleating centers on the electrode can be found:

$$N^{1/3} = \frac{\delta^{1/2} \cdot q_0^{1/3} \cdot \bar{g}}{n \cdot F \cdot \bar{\gamma} \cdot C^{\circ} \cdot D^{1/2}}. \tag{8}$$

The average size of nucleating centers r_m can be found from the expression:

$$q_0 = \frac{4}{3}\pi r_m^3 \cdot N; \tag{9}$$

$$r_m = \frac{3q_0}{4\pi N} \,. \tag{10}$$

Knowing the half-peak width, one can find function X_m by formula 6.

Under optimal conditions for the dependencies described earlier the number and size of the nucleating centers formed on the electrode were analyzed.

The dependence of number *N* (curve 1) and radius r of the nucleating centers (curve 2) for the glassy-carbon electrode in 0.2M KNO_3 on the concentration was obtained (Fig. 5).

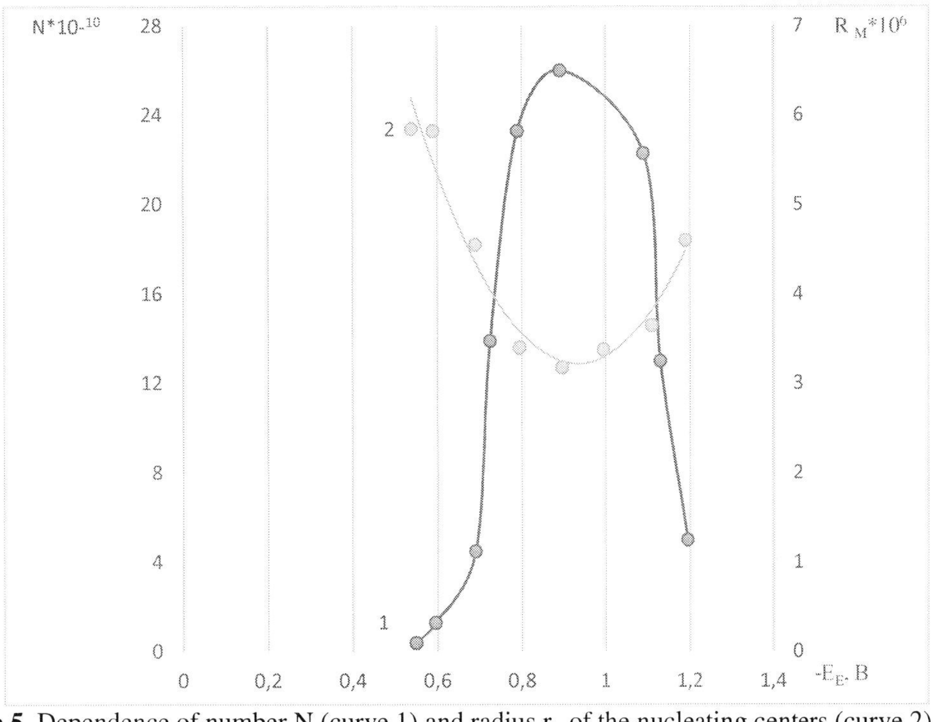

Figure 5. Dependence of number N (curve 1) and radius r_m of the nucleating centers (curve 2) on the electrolysis potential. Background – 0.2M KNO_3, pH = 3.5; $C_{Pb}^{2+} = 4.7 \cdot 10^{-5}$ mole/l; τ_e = 2 min; w = 40 mV/s. for a glassy-carbon electrode.

With an increase in the concentration, the lead plate growth occurs due to overlapping of nucleating centers into larger formations, while their radius increases and the number of embryos decreases.

Dependence of the number of nucleating centers and their radius on the time of electrical accumulation on a glassy-carbon electrode on the background 0.2M KNO_3, containing $4.74 \cdot 10^{-5}$

$Pb(NO_3)_2$ shows that lead plate with an increase in time of electrolysis occurs due to the merging of crystallization centers into larger formations, which radius increases.

Dependence of the number of embryos as well as their radius deposited from 0.2M KNO_3 and 0.2M KCl on the potential of electrolysis was studied. The dependence of the number of nuclei passes through a maximum, and the dependence of the radius of the embryos on the potential of the electric accumulation passes through a minimum. The points of the maximum and minimum of the curves were observed at practically the same value of the potential of electro-accumulation. In the potential range, more positive than the maximum of the electrolysis potential, the plate growth occurs due to an increase in the number of nucleating centers on the surface of the electrode, and not at the expense of an increase in their sizes.

In the range of potentials, which is more negative than the maximum of the potential for accumulation, the growth of the sediment is mainly due to the fusion of nuclei into larger ones due to the fact that part of the electrode surface is blocked by evolved hydrogen.

The dependence of the number of nucleating centers and their dimensions on the nature of the electrolyte is shown. Thus, for example, the number of nucleating centers of lead deposited from 0.2M KCl is larger than that from 0,2M KNO_3. In both cases, the electrolysis was carried out under optimal conditions (E_e, pH, w).

The electrode material influences the nature of the plate. For example, on the glassy-carbon electrode, the number of nucleating centers is 2 times and the size is 1.5 times larger than those on the copper electrode, which is due to the difference in the substrate structure, the hydrogen overvoltage on these substrates, and also the structure of the double electric layer on the electrode surface as well as with the magnitude of the potential of zero charge.

The dependence of the number of nucleating centers N and the radius of nucleating centers r_m on glassy-carbon, copper and silver electrodes from pH of the solution on the background 0.2M KNO_3 is obtained. The growth of the lead sludge in the range of pH 6.8 – 4.0 is due to an increase in the number of nucleating centers. In the range of pH 3.5 – 2.13, the growth of the plate is due to the fusion of nucleation and the increase in their size, due to the fact that a part of the surface of the electrode is blocked by evolved hydrogen bubbles.

On a silver electrode, it was necessary to take into account the influence of impurity ions Cl^- and I^-, since they are tension-active, and silver forms hardly soluble compounds with them. This especially applies to iodide ions. The experiment showed that at a concentration of an iodide ion, the useful signal, commensurate with the concentration of lead, decreased noticeably, which was due to the achievement of the solubility product. At similar concentrations of chloride ions, a decrease in the peak-point of lead was not observed. Only the iodide ion exerts a noticeable influence on the current peak-point of lead for a copper electrode under the same conditions. On the glassy-carbon electrode, the influence of impurity ions was much less.

3. Conclusion

1. The possibility of obtaining well-reproduced signals of anodic dissolution of lead plates for various electrodes is shown.
2. The effect of the nature of the background electrolyte on the magnitude of the useful signal is studied.
3. Dependences of the value of the useful signal on the electrolysis potential, on the electrolysis time, on the pH of the solution and on the nature of the substrate were obtained.
4. Dependences of the plate structure on the conditions of electrolysis are reveled.
5. Optimum conditions for obtaining a stable and well reproducible useful signal are found.

References

[1] Sleptchenko G B, Pikula N P, Dubova N M, Shchukin T N 2009 Electrochemical quality control of water. *Bulletin of Tomsk Polytechnic University.* **3** 314

[2] Mai G, Sia L, Van V 2013 Electrochemical determination of lead in the blood in the presence of copper ions. *Electrochemistry* **5-49** 502-508

[3] Sheremet A A, Ermakov S S, Chezhina N N 2008 *Ethanol-free electrochemical method for content determination of lead, copper and cadmium in water solutions. Proc. of 7th All-Russian Conference on Electrochemical Methods of Analysis "EMA-2008"* (Ufa)

[4] Zakharova E A, Slepchenko G B 1995 *Technique of quantitative chemical analysis of water for the content of zinc, cadmium, lead and copper by inversion voltammetry. MVI 08-47/008* (Tomsk, TPU)

[5] Temery S V, Loginova O B 2014 *Extraction and current-potential method for detection of zinc, cadmium, lead and copper. Patent Ru 2476853*

[6] Extraction and current-potential method for the detection of zinc, cadmium, lead and copper in natural water. *Patent of the Russian Federation 2383014*

[7] Stromberg A G, Kaplin A A, Karbainov Yu A 2000 Inversion the voltammetry in the papers of Tomsk Scientific School. *Izvestiya vuzov. Series "Chemistry and Chemical Technology"* **43-3** 8

[8] Pnev V V 1983 *The processes of electric crystallization in inversion electrochemical methods. Abstract of the doct. Diss.* (Sverdlovsk)

[9] Pnev V V, Lyngrin D D 1977 *Study of the processes of electrical dissolution of metal embryos from the graphite surface by the potential and dynamic method. Coll. of art. "Development of physical and chemical methods and their application in research and analysis"* (Tyumen)

Innovations in education and approaches to assessment of regional educational systems

V I Bauer, V D Iljinyh, A V Bazanov and E S Kozin

Industrial University of Tyumen, 38 Volodarskogo Str., Tyumen, 625000, Russia

E-mail: artyom777@mail.ru, bauervi@tyuiu.ru

Abstract: The analysis of the latest innovations in the Russian education and trends in the development of educational regional systems is provided. Approaches to assessing the effectiveness of the operation of the company-university system are discussed by an example of motor engineering training for the oil and gas industry.

1. Introduction

The innovative development of the global and domestic economy has led to an unprecedented rate of knowledge renewal. According to some experts [5], 5% of theoretical and 20% of professional knowledge are updated annually. This means that today it is impossible to train a specialist for professional activities only once.

In response to these challenges, the structure of educational levels, the requirements for the teaching staff, the methodological support, the material and technical base of educational institutions have fundamentally changed in the system of Russian education [9], which is stated in the new Law on Education [10], a series of educational standards "3+" [8].

2. Methodology

Implemented in 2011, the transition of Russian education to a multi-level training system, on the one hand, was to add flexibility in training highly qualified personnel. On the other hand, training of engineering personnel is impossible without the industry participation in the development of educational programs for undergraduate and graduate programs. To do this, the production sector must understand the characteristics of the new training system and realize the possibilities of its use for solving production tasks for training and retraining personnel.

Education has yet to adapt to a two-level system. Standards and educational programs of higher education fall short because bachelor's programs are extremely reduced training programs for engineers of the 2000s. The analysis of the main educational programs for engineering training in several universities (Table 1) shows that there has been a significant decrease in the absolute and relative number of hours allocated to disciplines of humanities, mathematical and natural sciences, as well as disciplines that form the basis of engineering training. (Theoretical Mechanics, Theory of Machines and Mechanisms, Resistance of Materials, Technology of Structural Materials, Machine Parts). At the same time, the load in the disciplines of the professional cycle increased substantially in absolute and relative terms with the same number of practice hours.

Table 1. Ratio of training load by groups of disciplines

Educational program	Year of approval	Distribution of training load (hour)							Total amount of training load (hour)
		Humanities, Socio-economic	Mathematical and natural sciences	General engineering	Special	Specialization (profilization)	Practice	Others	
Engineer	2002	1796	1715	1834	1990	570	1080	1112	10097
Bachelor	2016	1732	828	1404	2484	1008	1080	1112	8640
Master	2016	504	936	-	1224	-	672	1488	4320

The Law on Education [10] and educational standards of the generation 3+ [8] give ample opportunities for improvisation in the development and implementation of basic educational programs.

New opportunities consist in a reasonable combination of competence-based and socio-cultural approaches, dual educational technologies, project training methods, modular principle of organizing the learning process. All this allows us to maximally individualize the educational process, make it as efficient as possible, including from the point of view of generation and transfer of new knowledge.

Starting from 2011, there has been a serious trend toward the transfer of authority to manage education, including higher education, to the regions. If this policy from the side of federal organizations proves to be long-term, it will entail a number of consequences that will need to be taken into account in the management of regional budgets, economy of the region as a whole and educational system in particular. Regions will need to predict the need for human resources at all levels of training, depending on the prospects for the development of the region [6]. It will be necessary to optimize the development of the infrastructure of educational institutions of all levels in the areas of training [11]. It will be necessary to learn how to evaluate the effectiveness of the regional system to form competences in terms of the effectiveness of the company-educational institution system.

In this system, let us put company first, since it has the task of developing requirements for training and retraining personnel, the goodwill of using high-tech equipment of the company to obtain working professions, and conducting various types of practice.

From this point of view, the region faces another challenge: developing a system of criteria for selecting leading companies, as well as mechanisms for motivating companies to participate in training and retraining programs.

Companies, relevant ministry and independent expert agencies evaluate education and the level of training of specialists. The Ministry of Education and Science of the Russian Federation annually makes adjustments to the scale of assessment of the activities of educational institutions. At the same time, the index of employment of graduates, the level of wages and promotion of a business career remain key, both during accreditation of educational institutions, and evaluation by independent expert agencies, also during certification of quality systems.

All approaches to assessing the quality of education can be divided into 5 groups:

1. Assessment of conditions for the provision of educational services.
2. Assessment of the result (knowledge, abilities, skills - competences of the student, graduate).
3. Assessment of satisfaction (student, graduate, industry).
4. Process approach.
5. Complex assessment including elements of the four previous groups.

The first approach assesses the possibility of implementing an educational standard in terms of the quality of the basic educational program, provision of educational facilities, training and research equipment, qualified teaching staff, educational materials. According to this scheme, assessment of universities is conducted by experts from Rosobrnadzor (Federal Education and Science Supervision Service) during accreditation examination, continuous online monitoring by the Ministry of Education and Science and independent expert organizations during certification and inspection of quality management systems (QMS).

The second group includes the recently introduced system of federal Internet exams, the system of state examinations which presumes the availability of qualified industrial workers in the State Examination Commissions (SEC).

Satisfaction is evaluated in several ways by different methods. The first group of methods is based on an expert survey of a specially selected group of students in the learning process and graduates - immediately after employment, in 5, 10, 15, etc. years after graduating from the educational institution (EI). The second group of methods is based on an expert survey of employers. This method was widely used in the period of planned economy.

The process approach is widely used to assess the focus on improving all EI processes (main and auxiliary). It is used for certification and inspection of the EI QMS for compliance with international quality standards ISO 9000.

Complex assessment of the training quality is a symbiosis of the four previous approaches in various combinations and ratios. It serves as the basis for making managerial decisions when accrediting EI, certifying QMS, evaluating rating agencies.

In the authors’ opinion, from the point of view of the rationale for regional policy, it is more appropriate to assess the joint activity of the university and interested companies in training, employment and career advancement of young professionals. At the same time, the effectiveness of the process of training a specialist in this context should be understood as ensuring maximum compliance of the graduate's competences with the industry requirements. Assessment of this effectiveness can involve a set of indicators, including: the duration of the adaptation period after employment of young specialist T_A in months; the probability of employment of graduate P_T; survival rate (job security) of young professionals V_{Too} over evaluation period T_{Oc} (in months) – the ratio of the number of young professionals who have worked at the enterprise for 5 years or more to the total number of young professionals employed over the estimated period.

It is ideal when a young specialist does not need an adaptation period or this period is combined with the process of obtaining higher education. In times of developed socialism, this period was 3 years. For this period the graduate of the university was given the status of a young specialist. This also involved a system of mentoring, restriction of the specialist's functions, reduction in the level of responsibility for unintentional shortcomings in the work. However, under conditions of competition and intensive innovation development of industries, it is necessary to find a compromise solution that is acceptable both for the education system with its limitations (requirements of the educational standard, the level of financing of the educational process) and the industry requirements, consisting in the desire to accelerate adaptation processes and ensure the fastest use of the university graduate to solve engineering problems.

The second assessment of effectiveness is "Ease ...", a high probability of employment. Dual education involves a significant contribution of the company to the training of a specialist. And in this key, a reasonable combination of the socio-cultural and competence-based approach is important (a person must share the values of the company). Why train if he or she goes elsewhere and causes damage to the enterprise? The real values of this indicator for motor specialties, according to the results of 10-year observations, are in the range of 0.5 to 0.85.

The survival rate (job security) of young professionals in the company depends on many factors, including industry, company personnel policy. For example, large oil companies are characterized by the availability of programs for young professionals, which involve a series of adaptation procedures that make it possible to shorten the period of adaptation as much as possible and reduce the probability of dismissal. The level of wages and prospects for promotion are essential. According to surveys of specialists of the personnel services of large companies engaged in the technical operation of vehicles, this indicator varies between 0.1 and 0.8.

For a comprehensive assessment of the effectiveness of the educational system, it is proposed to use the indicator characterizing the effectiveness of the educational system as the ratio of the effective work time of the graduate over evaluation period $T_{Оц}$ to the value of the evaluation period which is proposed to call the efficiency coefficient of the educational system - $K_Э$. Its physical meaning is the relative duration of the effective period of the graduate's work. One can understand the effective period of the graduate's work as the time from the moment of employment after graduation, except for the adaptation period.

It is related with indicators of the probability of employment, the adaptation period, the evaluation period, the survival rate by the ratio (1):

$$K_э = \frac{P_T V_{Too}(T_{Oc} - T_A)}{T_{Oc}}, \tag{1}$$

where T_A - assessment of the graduate's adaptation time;

P_T - assessment of the probability of employment of the graduate in the specialty within a year after graduation;

V_{Too} - assessment of the survival rate of young professionals in the company.

Depending on the objectives of the research, average values, mathematical expectation, or other valid values may serve as estimates of indicators.

In order to ensure visibility and form an integrated approach, it is customary to use radar charts in assessments. They show how the values of factors correlate in different companies Fig. 1.

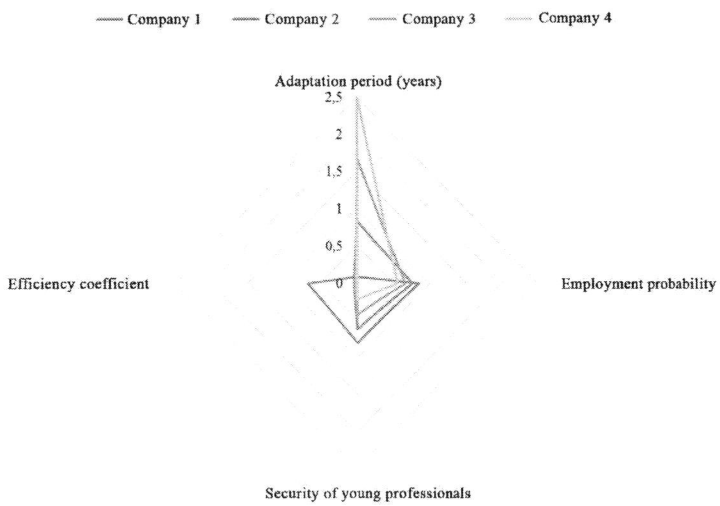

Figure 1. A radar chart of the efficiency of the company-university system

The proposed indicator assesses the joint activity of the company and the university in training. As can be seen from the above mentioned data, the greatest effect of the system is achieved where the company and the university carry out a complex of activities for transferring the adaptation period to the period of studying at the university (company 1). The worst result is when a young specialist begins to get acquainted with the company after graduating from the university. This acquaintance often involves obtaining a working profession, an internship, and a long-term job in a working position at once. The adaptation period is 2.5 years; low survival rate is typical for companies with such personnel policy (enterprise 4).

There are also companies on the market that implement a completely different personnel strategy. They do not work with young specialists, they do not recruit students from universities and colleges, and they do not spend resources on these "useless" activities. Often these enterprises are at a very good technological level. The essence of their personnel policy is the selection and re-purchase of specialists with the necessary core competencies in the labor market [3]. Such companies receive certain competitive advantages. From the point of view of the law - no complaints. From the point of view of state care for personnel training - there should be a system of preferences for companies that care about training young people, creating training centers, actively interacting with universities and colleges for training and retraining programs. And here the economic evaluation of the system becomes urgent, which will show how effective the personnel policy of the company is, where it gains, and where it loses. Issues of social and economic efficiency of investment in education have been discussed over many years in a huge number of publications by domestic and foreign authors [1, 2, 4, 5, 7, 8, 12]. According to studies conducted in the United States, every $35,000 invested in the training system brings $1,000,000 in profits [5]. What is the situation in Russia?

The economic model for assessing the investment of enterprises in education is given in [1]. The model is based on the rules for calculating the investment in the student, as well as calculating the additional profit and the payback period of investments in education. In the authors' opinion, this model describes well the closed corporate educational systems, not taking into account the options for involving young specialists from outside, which is typical for motor service and transport companies.

$$EFF = \frac{EF}{RC}, \tag{2}$$

where EFF – economic efficiency of investment in education;
EF – value of the received economic benefit;
RC – amount of resources, or costs on the part of the company.

To use this well-known formula for assessing the economic efficiency of the company-university system, it is necessary to modernize expression (2) by presenting the *EF* and *RC* indicators as functions of indicators T_A, P_T and V_{Too}, or the efficiency coefficient of educational system K_E.

Knowing the financial investments of the company in one student and the achieved efficiency coefficient, one can evaluate the efficiency of investments in educational programs.

3. Conclusion

The calculations conducted using the above mentioned approach for a number of companies in different sectors of the economy made it possible to estimate the payback period for company investments into educational projects (on tripartite contracts for training by the basic educational programs of higher education). They are from 5 to 0.5 years without taking into account K_E and from 10 years to 1 year, taking into account K_E.

Further research in this area can be aimed at typing educational systems, development of a system of criteria for selecting partner companies, as well as a system of preferences to motivate employers.

References

[1] Antropov V A, Parshina V S, Kudinov A V 2009 Socio-economic effectiveness of investment in personnel training. *Economy of the region* **1** 28-41
[2] Blinov A O 2015 Diagnostics of the Chinese miracle: what can we borrow. *Naukograd science production society* **1** 62-67
[3] Blinov A O 2012 Key competences - the subject of discussion around the world. *Alma mater Higher School Newsletter* **12** 15-19

[4] Blinov A O, Rudakova O S 2015 Innovations in education: concepts, problems, solutions. *Naukograd science production society* **2** 53-57
[5] Blinov A O, Rudakova O S 2015 New paradigms in the education system. *Naukograd science production society* **3** 69-73
[6] Voronkova G V, Zhadenova S V 2014 On forecasting the personnel potential of the construction industry on the basis of graduates of all levels of construction education. *Internet journal "Naukovedenie"* **5 (24)** 25
[7] Gavrilyuk V V 2012 Conceptualization of the practice of education of students of the XXI century. *Bulletin of the Tyumen State University. Socio-economic and legal research.* **8** 87-95
[8] *State educational standard of higher education in the area of training 23.03.03.* Operation of transport-technological machines and complexes (bachelor's degree). Approved by the order of the Ministry of Education and Science of the Russian Federation of December 14, 2015 № 1470
[9] Tkachenko E V 2013 New law on education and problems of vocational education for youth. *Vocational Education in Russia and Abroad* **2 (10)** 15-18
[10] *Federal Law of December 29, 2012 N 273-FZ "On Education in the Russian Federation" (with amendments and additions)* Information and Legal Portal "Garant". URL: http://base.garant.ru/70291362/#ixzz3OyWBRusP
[11] Fedorov M V, Cherepanov M G, Gredina O V, Kansafarova T A, Khramtsova I A 2011 Formation of a regional system for the development of qualifications. *Manager.* **3-4** 20-28

IOP Conf. Series: Materials Science and Engineering **357** (2018) 012010 doi:10.1088/1757-899X/357/1/012010

Destruction of linear part of pipeline under influence of high-frequency pressure change

K S Voronin and D A Cherentsov

Industrial University of Tyumen, 38 Volodarskogo St., Tyumen, 625000, Russia

E-mail: voronin_tsogu@mail.ru, cherentsovda@bk.ru

Abstract: The article deals with the vibrations of the pipe surface, connected with the change in pressure, which can lead to the destruction of the linear part.

1. Introduction

It follows from experience in the operation of main pipelines that most of the damage occurs suddenly, without apparent external causes. It is believed that this kind of destruction is associated with a variable load and has a fatigue nature [1-4]. Consequently, the problem of ensuring the reliability must be associated with dynamic processes in the gas pipeline.

2. Materials and methods

The fact is that the changing in the gas pressure along the length of the pipeline can cause considerable stresses. Deformations caused by these stresses negatively affect the reliability of the pipeline [14]. In this connection, the establishment of a relationship between the change in pressure and stresses arising in the pipes walls, caused by deformations as a result of dynamic processes, is an important task in solving problems of increasing the reliability [10-13].

Let us consider the oscillations of a cylindrical surface in a thin shell model.

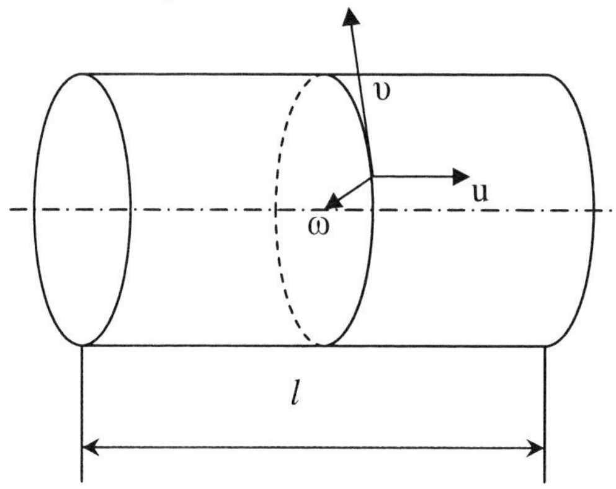

Figure 1. A model of a thin cylindrical shell

Let us introduce u, υ, ω - the variables denoting the displacement of the shell surface of the cylinder. The equations system of forced oscillations under the action of internal pressure will have the following form [2]:

$$\begin{cases} \frac{\partial^2 u}{\partial \xi^2} + \frac{1-\mu}{2}\frac{\partial^2 u}{\partial \varphi^2} + \frac{1+\mu}{2}\frac{\partial^2 \upsilon}{\partial \xi \partial \varphi} - \mu \frac{\partial \omega}{\partial \xi} = 0 \\ \frac{1+\mu}{2}\frac{\partial^2 u}{\partial \xi \partial \varphi} + \frac{\partial^2 u}{\partial \varphi^2} + \frac{1-\mu}{2}\frac{\partial^2 \upsilon}{\partial \xi^2} - \frac{\partial \omega}{\partial \xi} = 0 \\ \mu \frac{\partial u}{\partial \xi} + \frac{\partial \upsilon}{\partial \varphi} - \omega - k\nabla^4\omega = \frac{\rho r^2}{B} - \frac{\partial^2 \omega}{\partial t^2} + P(t) \end{cases} \quad (1)$$

where $\xi = \frac{x}{r}$, $k = \frac{h^2}{12r^2}$, $B = \frac{Eh}{1-\mu^2}$

x –distance from the cylinder edge to the considered point;
r – cylinder radius;
φ –angle between the axial plane and the considered point;
h – shell thickness;
E –modulus of elasticity of the shell material;
μ – Poisson's ratio of the shell material;
ρ – shell weight per area unit;
t – time.

Assuming the load as a periodic one: $P(t) = P \sin \omega t$, let us expand it to series:

$$P = \sum_{m=1}^{\infty}\sum_{n=1}^{\infty} P_{mn} \cos \alpha_m \xi \sin \varphi \quad (2)$$

where $\alpha_m = \frac{m\pi r}{l}$
P_{mn} - Fourier coefficients;
l – shell length;
m – number of half-waves in direction l;
n – number of waves around the circumference of the cross section.

After solving the equations system by the method described in [2], let us express the amplitude of forced oscillations:

$$A = A_\omega \sin \omega t$$

$$A_\omega = \frac{16P}{\pi^2 \rho h}\sum_{m=1}^{\infty}\sum_{n=1}^{\infty} \frac{1}{mn(\omega_{mn}^2 - \omega^2)} \cos \alpha_m \sin \varphi \quad (3)$$

where m , n = 1, 3, 5...
A_ω – maximum amplitude of forced oscillations;
ω_{mn} – natural oscillation frequency of the system.

However, in reality, the pressure amplitude is a random variable with an unknown distribution density.

The paper [3, 5] shows a histogram and a reconstructed function of the pressure distribution density in gas pipeline $f(P)$. To do this, daily fixed pressure values throughout the year were sampled. According to the results, the general form of the function can be approximated by the expression:

$$f(P) = P^\alpha e^{-\beta P} \quad (4)$$

where α, β – coefficients obtained by approximating the experimental results [3];
P – pressure in the gas pipeline.

Consequently, the amplitude of the forced oscillations taking into account the random distribution of the pressure amplitude will have the following form:

$$\overline{A_\omega} = \frac{16P}{\pi^2 \rho h}\int_0^\infty P^{\alpha+1}e^{-\beta P}\sum_{m=1}^{\infty}\sum_{n=1}^{\infty}\frac{1}{mn(\omega_{mn}^2-\omega^2)}\cos\alpha_m \sin\varphi \quad (5)$$

m, n = 1, 3, 5...

Taking m = 1, corresponding to the lowest frequency [3], the expression for $\omega_{mn}^2 = \omega^2$ can be written:

$$\omega_n^2 = \frac{P_{cr}}{\rho hr}\left(n^2 + \frac{1}{2}\frac{\pi^2 r^2}{l^2}\right)\left(1 - \frac{P^*}{P_{cr}}\right) \quad (6)$$

were P* - compression pressure of a cylindrical shell;

P_{cr} –the critical load of a cylindrical shell, determined by the formula of Mises [2]:

$$P_{cr} = \frac{Eh}{r}\frac{1}{n^2 + \frac{1}{2}\frac{\pi^2 r^2}{l^2}}\left[\left(n^2 + \frac{1}{2}\frac{\pi^2 r^2}{l^2}\right)^2\frac{h^2}{12(1-\mu^2)} + \frac{\pi^4 r^4}{l^4\left(n^2 + \frac{1}{2}\frac{\pi^2 r^2}{l^2}\right)^2}\right] \quad (7)$$

To carry out further analysis, let us rewrite expression (5) in a more convenient form:

$$\overline{A_\omega} = \frac{16}{\pi^2 \rho h}\sum_{n=1}^{\infty}\frac{\cos\alpha \sin n\varphi}{n}\int_0^\infty \frac{P^{\alpha+1}e^{-\beta P}dP}{[B(1-\delta P^*)-\omega^2]} \quad (8)$$

where $B = \frac{P_{cr}}{\rho hr}\left(n^2 + \frac{1}{2}\frac{\pi^2 r^2}{l^2}\right)$; $\delta = \frac{1}{P_{cr}}$.

During normal operation of the pipeline, the frequency range is far from resonant frequency $B(1-\delta P^*) \gg \omega^2$. Consequently, expression (8) can be simplified:

$$\overline{A_\omega} = \frac{16}{\pi^2 \rho h}\sum_{n=1}^{\infty}\frac{\cos\alpha \sin n\varphi}{n}\int_0^\infty \frac{P^{\alpha+1}e^{-\beta P}dP}{B(1-\delta P^*)} \quad (9)$$

Taking into account the fact that $\delta P^* \ll 1$, formula (9) can be written in the following form:

$$\overline{A_\omega} \approx \frac{16}{\pi^2 \rho h}\sum_{n=1}^{\infty}\frac{\cos\alpha \sin n\varphi}{n \cdot B(n)}\int_0^\infty P^{\alpha+1}e^{-\left(\beta-\delta\frac{P^*}{P}\right)P}dP \quad (10)$$

It follows from formula (10) that even if the given dynamical system is far from the resonance range, there is still pressure range $P \approx \frac{P^*}{P_{\kappa p}\beta}$, where the amplitude of the oscillations will increase with finite probability under condition $e^{-\left(\beta-\delta\frac{P^*}{P}\right)P} \approx 1$, as a power function of pressure. The pressure fluctuation limits usually reach 10% of the working pressure [5-7]. It is known that such fluctuations, being high frequency, cause small amplitudes of stresses and can not affect the process of appearance of short cracks.

The appearance of short cracks occurs during low-frequency action at large pressure drops during stopping and start-up of the gas pipeline. In this case, the resulting stresses caused by internal pressure drops can be compared with static stresses which are linearly dependent on the pressure [4]:

$$\sigma_{st} = \frac{P(D-h)}{2h} \quad (11)$$

In the case of high-frequency dynamic loads, as shown above, cases are possible in which the strain amplitude grows like a power function together with dynamic stress σ_ω.

$$\sigma_\omega = \frac{A_\omega}{r_0}\left[1 + \left(\frac{3\nu^2}{1-\nu^2}\right)^{0,5}\right]E \approx A_\omega^0\left[1 + \left(\frac{3\nu^2}{1-\nu^2}\right)^{0,5}E\right]P^{2+\alpha} \quad (12)$$

Consequently, a situation is possible where the stress caused by high-frequency pulsations will be comparable with low-frequency pulsations $\sigma_\omega \approx \sigma_{st}$.

In this case, the high-frequency load will cause the germination of short cracks. Crack length a_i, at which the avalanche failure mode begins to develop, will be determined by the expression [4, 8, 9]:

$$a_i = \frac{2}{\pi}\left(\frac{k}{\alpha\sigma_\omega}\right)^2 \cdot 10^3\text{mm} \qquad (13)$$

where k – stress intensity factor [4],
α – coefficient in the stress formula [4].

For $\overline{\sigma_\omega} \approx 7{,}5$ MPa, $k \approx 1$ MPa $\sqrt{\text{m}}$, crack length, after which achieving the avalanche destruction of the material begins at high-frequency oscillations, will be $a_i \approx 0{,}5$ mm by formula (13).

3. Conclusion

Thus, a random distribution of pressure inside the pipe can change the fracture mechanism. As it was shown, random pressure fluctuations do not lead to crisis situations under normal conditions. However, in a dynamical system, a critical condition can be realized, in which avalanche failure can be realized. In the theory of destruction, such situation is defined as self-organized criticality.

References:

[1] Bocharnikov V F, Pakharukov Yu.V *2005 Vibration and destruction in submerged and centrifugal pumps for oil production.* (Tyumen, Publishing house TyumGNGU)

[2] Oksogoev A A, Slepov B I 2003 *Applied Physics. Oscillation of structural elements.* (Tomsk, Publishing house of NTL)

[3] Syzrantsev V N, Nevelev Ya P, Golofast S L 2008 Probabilistic estimates of the permissible depth of corrosion of gas pipeline defects. *Training of personnel and modern technologies for the fuel and energy sector of Western Siberia* **1** 183-188

[4] Khazhinsky G M 2007 *Mechanics of small cracks and reliability of pipeline components.* (Moscow, Publishing house INEC)

[5] Voronin K S 2016 Forecasting and Evaluation of Gas Pipelines Geometric Forms Breach Hazard. Transport and Storage of Hydrocarbons. *IOP Conf. Series: Materials Science and Engineering* **154** 012020 doi:10.1088/1757-899X/154/1/012020

[6] Dudin S, Voronin K, Yakubovskaya S, Mutavaliev S 2016 Modeling Hydrodynamic State of Oil and Gas Condensate Mixture in a Pipeline. *MATEC Web of Conferences.* DOI: 10.1051/matecconf/20167302021

[7] Pirogov S P, Cherentsov D A, Gulyaev B A 2016 Prospects of applying vibration-resistant pressure gauges in the oil and gas industry. *IOP Conference Series: Materials Science and Engineering* 012013

[8] Dudin S M, Zemenkov Y D, Shabarov A B, et al. 2016 Modeling the Phase Composition of Gas Condensate in Pipelines. *IOP Conference Series: Materials Science and Engineering* **154** DOI: 10.1088/1757-899x/154/1/012010

[9] Zemenkova M, Zemenkov Y, Gladenko A, Podorozhnikov S 2016 *Estimation of Emissions During Monitoring of Pipelines in the Dynamic Mode of Operation.* 5th International Scientific Conference Integration, Partnership and Innovation in Construction Science and Education **86** DOI: 10.1051/matecconf/20168604053

[10] Shalay V, Zemenkova M, Zemenkov Y, et al. 2016 Modeling Parameters of Reliability of Technological Processes of Hydrocarbon Pipeline Transportation. *MATEC Web of Conferences* **73** 01029 DOI: 10.1051/matecconf/20167301029

[11] Zemenkova M, Shalay V, Zemenkov Y, Kurushina E 2016 Improving the Efficiency of Administrative Decision-Making when Monitoring Reliability and Safety of Oil and Gas Equipment *MATEC Web of Conferences* **73** DOI: 10.1051/matecconf/20167307001

[12] Kulikov A M 2016 On Choosing Multicomponent Multiphase System Separation Progress Optimization Criteria. *IOP Conference Series: Materials Science and Engineering* **154** DOI: 10.1088/1757-899X/154/1/012039
[13] Akulov K A, Petryakov V A, Mostovaya N A 2017 The Use of Foamed Concrete for Oil Spill Treatment and Carbon Capture. *IOP Conference Series: Earth and Environmental Science* **50** 17 DOI: 10.1088/1755-1315/50/1/012042
[14] Pirogov S P, Cherentsov D A, Chuba A Y 2016 Study of elastic sensing elements for vibration-resistant pressure gauges. *IOP Conference Series: Materials Science and Engineering* 012015 DOI: 10.1088/1757-899X/154/1/012015

Thermal interaction of underground pipeline with freezing heaving soil

S Y Podorozhnikov[1], P Mikhailov[2], L Puldas[1], A Shabarov[1,2]

[1]Industrial University of Tyumen, 38 Volodarskogo St., Tyumen, 625001, Russia
[2]Tyumen State University, 6 Volodarskogo St., Tyumen, 625003, Russia

E-mail: sergey_urevih@mail.ru, kaf_mms@utmn.ru

Abstract. A mathematical model and a method for calculating the stress-strain state of a pipeline describing the heat-power interaction in the "underground pipeline - soil" system in the conditions of negative temperatures in the soils of soils are offered. Some results of computational-parametric research are presented.

1. Introduction

Reliability of pipelines in the underground method of laying is determined to a large extent by the thermal and moisture state of the surrounding soils subjected to frost heaving. In conditions of seasonal temperature fluctuations, under certain conditions, frost heave forces are formed in the soil base and, as a consequence, additional stresses appear in the walls of the pipeline. The magnitude of the cramped forces depends on the soil type and, to a large extent, on the moisture state, therefore, to predict the thermophysical and strength parameters, a complex physico-mathematical modeling of heat and mass transfer in the soil and calculation of stresses in the walls of the pipeline are necessary [3, 8, 9, 10].

2. Materials and methods

In the offered model, the soil is considered as a substance consisting of three phases: water, ice and a dry soil skeleton. The mass balance equations for each time interval (t(n), t(n+1)=t(n)+Δt) in a fixed ij-th control volume (CV) V_{ij} are recorded separately for each phase [1,7]:

- for water

$$\frac{\rho_{ijw}^{(n+1)} - \rho_{ijw}^{(n)}}{\Delta t} \cdot V_{ij} = -\sum_{k=1}^{4} \rho_{ijkw} \cdot v_{ijkw}^{n} \cdot \Delta S_{ijk} + I_{iw} \cdot V_{ij} \quad (1)$$

- for ice

$$\frac{\rho_{ijI}^{(n+1)} - \rho_{ijI}^{(n)}}{\Delta t} \cdot V_{ij} = I_{wI} \cdot V_{ij} \quad (2)$$

IOP Conf. Series: Materials Science and Engineering **357** (2018) 012011 doi:10.1088/1757-899X/357/1/012011

where: $\rho_{ijw} = \frac{m_w}{V_{ij}}$ – reduced water density; $\rho_{ijI} = \frac{m_I}{V_{ij}}$ – reduced ice density; v_{ijk}^n – the velocity of the migrating moisture flux for the k-th bound CV ΔS_{ijk}; I_{Iw} – pore ice melting intensity; I_{wI} – pore water crystallization intensity.

The equation of internal energy in the assumption of the same temperature of the components of a substance, and taking into account the neglect of the power of internal forces for the corresponding time (t(n), t(n+1)=t(n)+Δt) in each ij-th CV, will be as follows:

$$\frac{\left(\rho_{ijs}\cdot U_{ijs}+\rho_{ijI}\cdot U_{ijI}+\rho_{ijw}\cdot U_{ijw}\right)^{(n+1)}-\left(\rho_{ijs}\cdot U_{ijs}+\rho_{ijI}\cdot U_{ijI}+\rho_{ijw}\cdot U_{ijw}\right)^{(n)}}{\Delta t}\cdot V_{ij}=$$

$$-\sum_{k=1}^{4}\rho_{ijkw}\cdot v_{ijkw}^{n}\cdot U_{ijkw}\cdot\Delta S_{ijk}-\sum_{k=1}^{4}q_{ijk}^{n}\cdot\Delta S_{ijk}, \tag{3}$$

where: $\rho_{ijc} = \frac{m_c}{V_{ij}}$ – reduced density of dry soil; U_{ijs}, U_{ijI}, U_{ijw} – specific internal dry skeleton of soil, ice, water; q_{ijk}^n – conductive specific heat flux through the *k*-th edge ΔS_{ijk} of the *ij*-th.

The specific internal energy of the substance component in the ij-th CV is defined as [6, 13]:

$$U_{ijs}=C_{ijs}\cdot T_{ij};\; U_{ijI}=C_I\cdot T_{ij};\; U_{ijw}=C_w\cdot(T_{ij}-T_o)+U_{wo},$$

where: T_{ij} - the temperature of the center of the *ij*-th control volume under consideration; C_{ijs}, C_I, C_w – specific heat for the dry skeleton of soil, ice, water; U_{wo} - a constant determined from the normalization condition of the phase transition: $U_{vo}=l_{lv}+C_L T_0+P_0/\rho_L^{(0)}-P_0/\rho_V^{(0)}$,

where: l_{lv} – specific heat of ice melting; P_0, T_0 – pressure and corresponding temperature of phase transition; $\rho_I^{(0)}, \rho_w^{(0)}$ – density of ice, water.

In view of the smallness of the temperature gradient, the process of transfer of liquid moisture is described by the law of isothermal moisture conductivity [5, 14]:

$$\vec{v}=-D(\theta,\theta_I)\cdot\nabla\theta, \tag{4}$$

where: $D(\theta,\theta_I)=D_T(\theta)/(1+n\cdot\theta_I)$ – the diffusion coefficient of moisture in the soil, determined by water content $\theta=\rho_w/\rho_w^{(0)}$ and the volume content of ice $\theta_I=\rho_I/\rho_I^{(0)}$; $D_T(\theta)$ – the diffusion coefficient of thawed soil; n - experimental constant, the value of which depends on the soil.

Specific densities of heat fluxes q_{ijk}^n are calculated according to the Fourier law. The coefficient of thermal conductivity of the soil is determined by taking into account the phase composition for each control volume at each time step. The thermophysical model of heat and mass transfer is closed by defining equations for transfer of liquid moisture and phase transitions, thermophysical and hydrodynamic characteristics of soils, and boundary conditions for heat and mass transfer.

The position of the pipeline in the vertical plane at an arbitrary time step can be found from the solution of the well-known longitudinal-transverse bending equation [4], written with regard to the boundary conditions corresponding to the rigid fixation at the ends of the section under consideration [0;*L*]:

$$\begin{cases} E\cdot J\cdot\dfrac{d^4W}{dz^4}+N\cdot\dfrac{d^2W}{dz^2}-q_н(z)=0 \\ W(0)=W(L)=0 \\ \dfrac{dW(0)}{dz}=\dfrac{dW(L)}{dz}=0 \\ z\in[0;L] \end{cases} \tag{5}$$

where $q_н(z)$ – the specific strength of the frost heaving of the soil acting per pipeline length unit; $W(z)$ – cross-motion of the pipeline section with coordinate z, function $W(z)$ determines the altitude of the pipeline section; L – length of the pipeline section subjected to the action of frost heaving; $E\cdot J$ – bending rigidity of the considered pipeline section; E – Young's modulus of the pipeline wall material; N – - longitudinal force at the ends of the pipeline section. The additional longitudinal stresses arising as a result of the frost heave action in the upper and lower generatrix of the pipeline are calculated as:

$$\sigma(z)=\pm E\cdot D_H\cdot\frac{d^2W}{dz^2}, \tag{6}$$

where: D_H – the outer diameter of the pipeline.

Function $q_н(z)$ is constructed according to the quadratic law of constrained heave [4]:

$$q_н(z)=D_H\cdot R\cdot\left(1-\frac{W(z)}{\Delta H^{(n)}}\right)^2, \tag{7}$$

where: R is the design resistance of the soil;

$\Delta H^{(n)}$ – displacement of the ground averaged over the pipeline diameter. For the n-time step, $\Delta H^{(n)}$ is determined by the increase in height $\Delta h_{ij}^{(n)}$ of the CV under the pipeline (Fig. 1), assuming that the increase in the volume of the ground during heaving is realized only in the vertical direction:

$$\Delta H^{(n)}=\frac{\sum_{i=1}^{N}\sum_{j=1}^{M(i)}\Delta h_{ij}^{(n)}}{N}, \tag{8}$$

Free piercing $\Delta h_{ij}^{(n)}$ of CV, at each time step according to the calculated fields of liquid moisture and ice content is calculated as:

$$\Delta h_{ij}^{(n)}=\begin{cases} 0, & m_{ij}\geq\dfrac{\rho_{ijB}^{(n)}}{\rho_B^0}+\dfrac{\rho_{ijL}^{(n)}}{\rho_L^0} \\ \Delta y_{ij}\cdot\left(\dfrac{\rho_{ijB}^{(n)}}{\rho_B^0}+\dfrac{\rho_{ijL}^{(n)}}{\rho_L^0}-m_{ij}\right), & m_{ij}<\dfrac{\rho_{ijB}^{(n)}}{\rho_B^0}+\dfrac{\rho_{ijL}^{(n)}}{\rho_L^0}, \end{cases} \tag{9}$$

where: m_{ij} - the porosity of the soil corresponding to the CV;

Δy_{ij} - vertical size of CV.

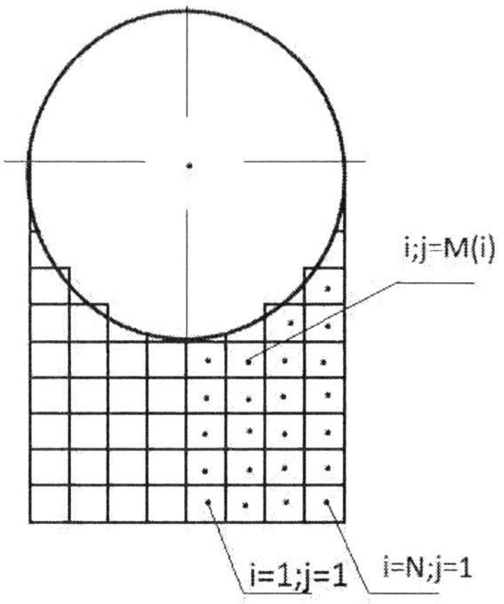

Figure 1. Determining the value of free soil heaving.

The calculation and parametric study of the effect of pipe and outdoor air temperatures on the pipeline's altitude position and on the value of additional longitudinal stresses in its walls was performed. The study was carried out for an underground pipeline with a diameter of 820 mm, and laying depth of 600-1000 mm [11, 12].

The ambient air temperature was set by a potential function:

$$T_B = A \cdot \sin\left(2 \cdot \pi \cdot \tau / 365\right) \tag{10}$$

where: A – maximum deviation of the temperature from 0°C;

τ - time in days ($\tau \in [0;180]$).

The calculated change in the vertical deflections of the pipeline over time is shown in Fig. 2.

Fig. 3 shows the calculated distributions of additional longitudinal stresses of the lower generatrix of the pipeline. Fig. 4, 5 show calculated dependences of the vertical displacement of the central section of the pipeline.

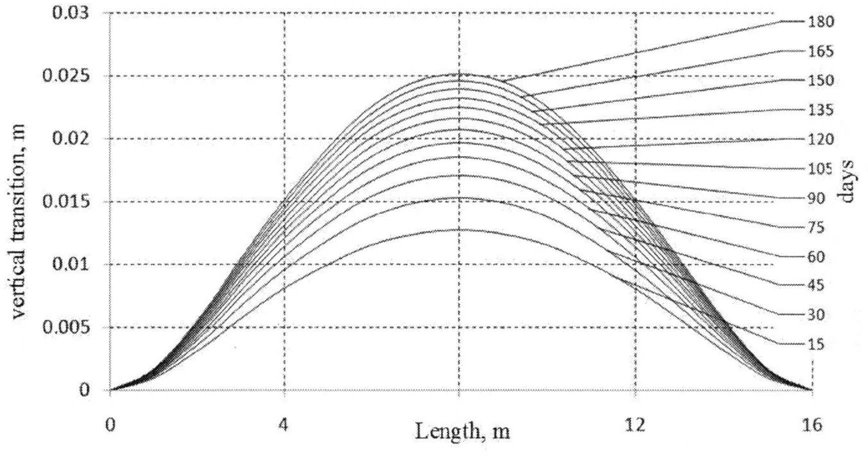

Figure 2. High-altitude position of the pipeline at different times. The pipe temperature is -10 °C; the amplitude value of the air temperature is -20 °C

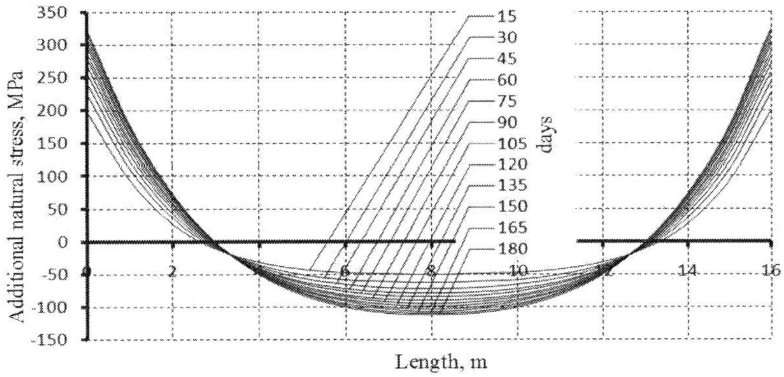

Figure 3. Distributions of additional longitudinal stresses under the pipeline over time. The wall temperature of the pipeline is -10 ^{0}C; the value of the parameter is A = -20 ^{0}C.

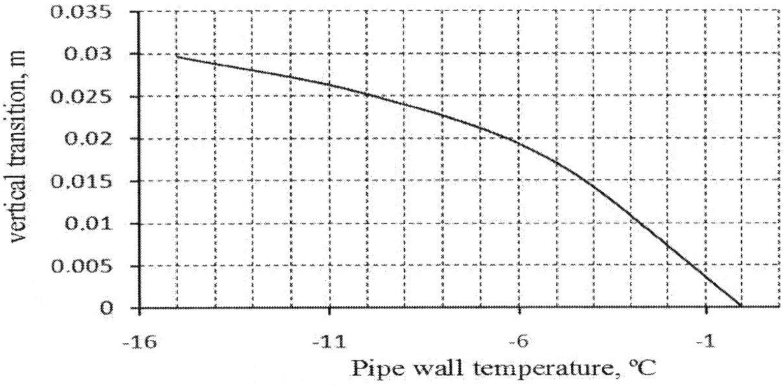

Figure 4. Dependence of the vertical displacement of the central section of the pipeline on its temperature.

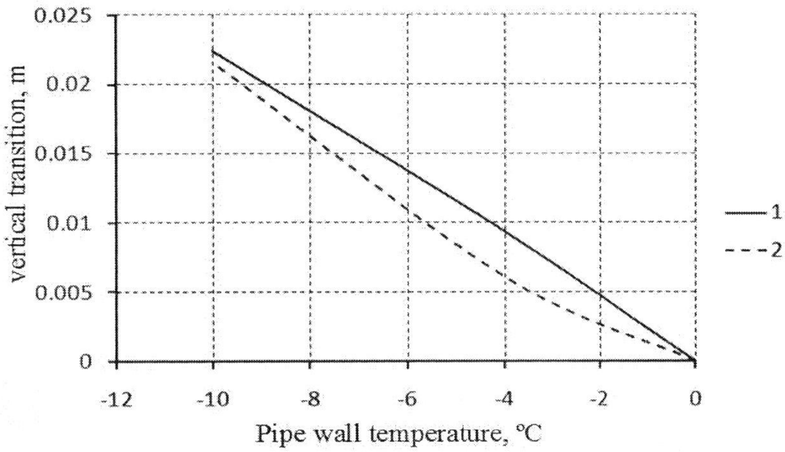

Figure 5. Dependence of the vertical displacement of the central section of the pipeline on its temperature, taking into account migration of moisture: 1 - taking into account migration of moisture; 2 - not taking into account the migration of moisture.

3. Conclusion

Based on the results of the computational and parametric study, a significant influence of the pipe wall and external air temperature on the additional longitudinal stresses arising in the walls was revealed. The maximum value of the additional voltage is formed at the ends of the section.

A significant effect of migration of moisture on the change in the altitude position of the pipeline is shown.

It has been established that at a pipe wall temperature of -4 °C, the deformation of the pipe in clayey soil increases 1.4-1.5 times in comparison with similar conditions in the absence of migration.

Based on the results of the study of the effect of pipe and outdoor air temperatures, a significant influence of the pipeline temperature on its altitude and the magnitude of the additional stresses arising in the walls of the pipe was established. The initial distribution of the reduced moisture density in depth was set at a constant value of 450 kg/m^3, which corresponded to the total moisture saturation of the soil. The greatest value of additional stress is observed at the ends of the pipeline section under consideration. It was noted that a decrease in the minimum air temperature from -10 °C to -20 °C with a given constant pipe temperature in the range from -5 °C to -15 °C leads to an increase in altitude position of only 1.01-1.03 times. Thus, under the conditions considered, there is a more noticeable effect of the pipe temperature than the ambient temperatures on the stresses.

The study of the influence of the moisture migration intensity made it possible to establish a significant role of migration in the pipeline deformation when the temperature of the pipe varies from -2 °C to -5 °C. At the initial moment, a linear increase in the reduced moisture density from 437 kg/m^3 on the surface to 450 kg/m^3 (full moisture saturation) was taken at a depth of 3 m or more. As the temperature of the pipe decreases, the role of migration decreases significantly, so at a constant pipe temperature of -10 °C, the ratio of vertical displacements calculated with and without migration is only 1.04, and at a temperature of -4 °C; the ratio is 1.54.

Investigations of the influence of snow cover thickness, carried out for a pipeline having a temperature equal to the local soil temperature, showed that under the considered conditions, a significant increase in the vertical pipeline movement to 0.012 m occurs with a decrease in the thickness of the snow cover from 0.2 m to zero. With a snow cover thickness more than 0.2 m, the pipeline deformations are insignificant.

Thus, the considered results of calculations show that the developed model and the calculation method are a convenient tool allowing determining the dynamics of changes in additional stresses in the pipe wall under real conditions of changing the parameters of heat and mass transfer in frozen soils.

References:

[1] Shabarov A B, Mikhailov P Yu, Puldas L A, Vakulin A A 2010 Physico-mathematical modeling of temperature fields and ice content in frozen soils around a buried pipeline. *Bulletin of Tyumen State University* **6** 14-19

[2] Grigoryev B V, Mikhailov P Yu, Shabarov A B 2010 *Experimental determination of the thermal conductivity of building materials, rocks and soils. Proceedings of the International Scientific and Practical Conference "Development and introduction of resource and energy-saving technologies and devices"* 117-121

[3] Aleksandrov P A, Kharionovsky V V 1986 *Calculation of underground pipelines in conditions of soil priming. Transport of natural gas.* (M.: VNIIGAZ)

[4] Gorkovenko A I 1999 Influence of frost heave forces on the high-altitude position of the pipeline. *Izvestiya Vuzov "Oil and Gas"* **3** 58-63

[5] Kulchitsky V A, Makagonov V A, Vasilyev N B, Chekov A N, Romankov N I 2002 Airfield pavements. A modern look. (M.: Physical and mathematical literature)

[6] Kutrunov V N, Mikhailov P Yu, Puldas L A, Vakulin A A, Vilkov M N 2012 Experimental research and physical-mathematical modeling of the process of oil cooling in an underground pipeline. *Vestnik TSU* **4** 67-73

[7] Shabarov A B, Kislitsyn A A, Grigoryev B V, Mikhailov P Yu, Puldas L A, Shastunova U Yu 2014 Heat and mass transfer in the oil and gas and construction techniques: manual. (Tyumen: Tyumen State University Publishing House)

[8] Shalay V, Zemenkova M, Zemenkov Y, et al. 2016 Modeling Parameters of Reliability of Technological Processes of Hydrocarbon Pipeline Transportation. *MATEC Web of Conferences* **73** 01029 DOI: 10.1051/matecconf/20167301029

[9] Zemenkov Y D, Shalay V V, Zemenkova M Y 2015 Expert systems of multivariable predictive control of oil and gas facilities reliability *Oil and Gas Engineering Procedia Engineering* **113** 312-315 DOI: 10.1016/j.proeng.2015.07.271

[10] Zemenkov Y D, Shalay V V, Zemenkova M Y 2015 Immediate analyses and calculation of saturated steam pressure of gas condensates for transportation conditions. *Oil and Gas Engineering* **113** 254-258 DOI: 10.1016/j.proeng.2015.07.330

[11] Dudin S, Bahmat G, Zemenkov Y, Voronin K, Shipovalov A 2017 Strategy for monitoring and ensuring safe operation of Russian gas transportation systems *MATEC Web of Conferences* **106** DOI: 10.1051/matecconf/201710606004

[12] Vlasenko V S, Slesarenko V V, Gulkov A N and Lapshin V D 2015 Preparation of arctic oil and gas condensate deposit formation fluid for storage and transport in the form of hydrocarbon hydrate-containing dispersed system Proceedings of the International Offshore and Polar Engineering Conference 1863

[13] Vlasenko V S, Slesarenko V V, Gulkov A N 2016 Using of triple-flow vortex tubes in associated petroleum gas preparation units. *Proceedings of the International Offshore and Polar Engineering Conference,* pp. 46-51

[14] Gulkov A N, Lapshin V D, Slesarenko V V, Morozov A A, Solomennik S F 2015 Adiabatic conversion method for the development of marine hydrate deposits. *Proceedings of the International Offshore and Polar Engineering Conference,* pp.73-76

Viability of using different types of main oil pipelines pump drives

A G Zakirzakov, Yu D Zemenkov and K A Akulov

Industrial University of Tyumen, 38 Volodarskogo St., Tyumen, 625000, Russia

E-mail: zakir1404@mail.ru, muzemenkova@mail.ru

Abstract. The choice of the pumping units' drive of main oil pipelines is of great importance both for design of pipelines and for modernization of existing ones. At the beginning of oil pipeline transport development, due to the limited number and types of energy sources, the choice was not difficult. The combustion energy of the pumped product was often the only available energy resource for its transportation. In this regard, the pipelines that had autonomous energy sources favorably differed from other energy consumers in the sector. With the passage of time, with the development of the country's electricity supply system, the electric drive for power-line equipment of oil pipelines becomes the dominant type of a pumping station drive. Nowadays, the traditional component is an essential factor when choosing some type of the drive. For many years, oil companies have been using electric drives for pumps, while gas transport enterprises prefer self-contained gas turbines.

1. Introduction

The development of transportation of the most important energy sources, such as gas, oil and electricity, makes it possible to design transport networks for the particular region, the one that is most expedient. However, when choosing a particular type of the drive for pumping units, analysis is often not conducted, because there is no alternative to using the electric drive [1, 2].

At the moment, there is no specific method for choosing the type of the pump drive. Important factors that may determine the decision-making in some cases can turn out to be unsustainable in others. Variants of technical solutions suitable for one region are inappropriate for another. The drive type accepted for the object under commissioning may be irrelevant for the reconstructed object. Detailed technical and economic analysis, conducted in two parallel directions: the ratio of calculable factors and the analysis of qualitative characteristics, should serve as a basis for rational choice [6-9]. To achieve the result, it is necessary to collect the initial data, which can be conditionally divided into several categories (technical, economic, security parameters, etc.).

2. Materials and methods

There are many factors influencing the choice of a particular type of drive for main pumps, and they are different for each unique case. Often, none of the factors is dominant, and then there is a need to compare several factors. Analysis of all factors seems to be quite time-consuming, so it is advisable to begin with preliminary comparative calculations to exclude obviously unsuitable options [3].

IOP Conf. Series: Materials Science and Engineering **357** (2018) 012012 doi:10.1088/1757-899X/357/1/012012

To determine the significant factors for selecting the pump drives, as well as their ranking, it was decided to use the expert evaluation method. In the case of application of the expert evaluation method, an analysis of their objectivity is topical. There are no guarantees that the obtained estimates are reliable. It is also possible to obtain similar results, if one takes into account the assessments of experts from one scientific community [4]. Therefore, there is a need to carefully select experts, rank them in detail, and also to analyze the results of the survey. There is also a problem of insufficient stability of the experts' evaluation, i.e. the same expert can assess the situation differently in another survey after a deeper discussion. In this case, the method of expert assessments "Delphi" helps to improve the effectiveness of the research. The method allows one to conduct a survey in several stages, providing feedback to experts, introducing them to the results of the first round of the survey. Consistency of opinions, as well as detailed group assessments in each of the survey rounds is achieved through a single format that preserves the composition of the expert group, the form of the questionnaires, and the methodology for processing the results [5].

Choosing the optimal parameters influencing the selection of the drive of the main pumps by the "Delphi" method (Fig. 1) was carried out in several stages.

At the first stage of the study, the problem was formulated - "The lack of a technique for choosing the drive of the pumping unit". Preliminary preparation was carried out: the existing types of drives of the main pumping units, the factors influencing the operation of different types of drives, the world experience of their use; the advantages and disadvantages of each of them were analyzed; the questions concerning the parameters having a significant influence on the choice of the optimal drive variant were indicated. Thus, possible options for solutions that could be adopted based on expert opinion were specified.

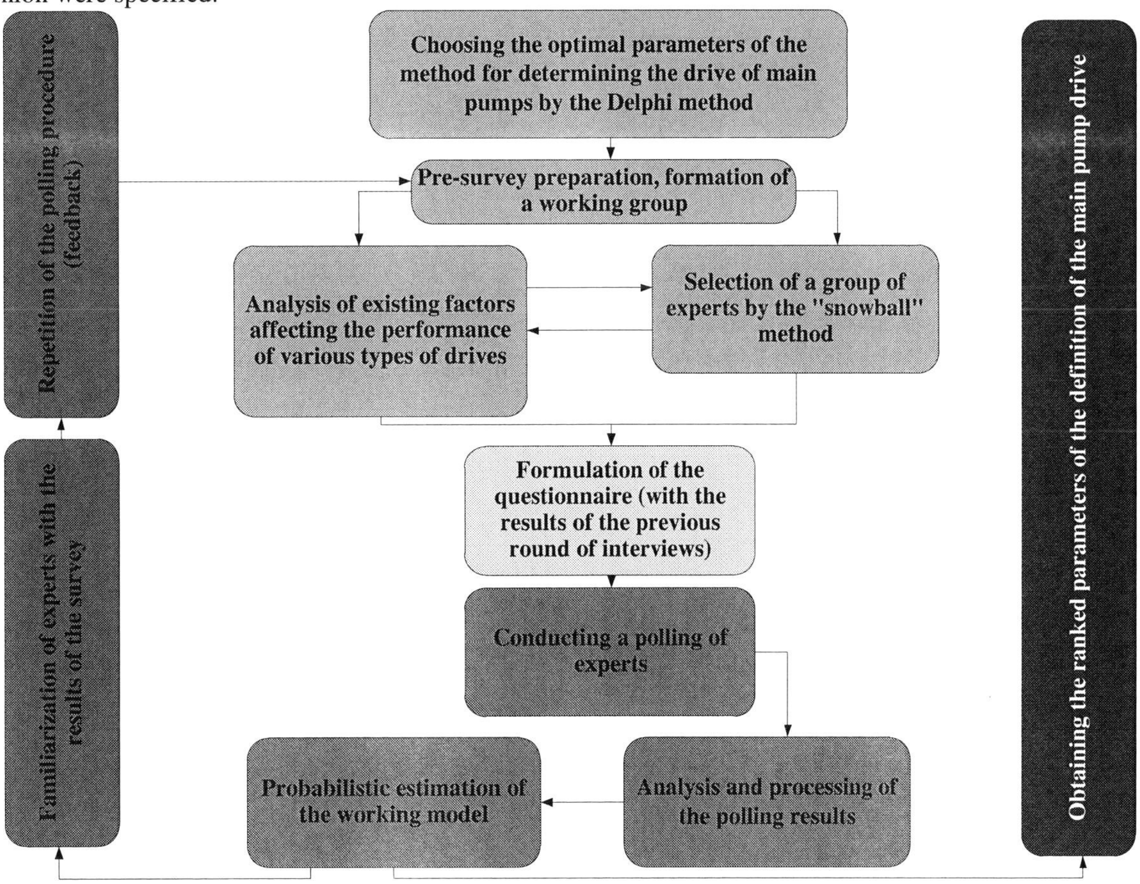

Figure 1. Diagram of the selection of optimal parameters by the "Delphi method"

The second stage was of an organizational nature. A group of specialists with professional polling and interviewing skills was formed. Questionnaires were prearranged, including the objectives and conditions of the survey, the procedure for processing results, evaluation criteria, and the use of survey results. At the same time, a group of experts was selected [10]. Specialists working in various oil transportation companies of the CIS (Russian Federation, Kazakhstan), specialists of design institutes, as well as the teaching staff of technical universities of the Russian Federation, were involved in the work. For the reliability of the results, specialists from various positions were involved in the work: from assistants to professors, from site masters to department heads. This allowed us to assess the situation "from different angles". The expert commission to obtain a qualitative result was formed by the method of co-nomination ("snowball method"). The essence of the method was that the first expert was asked to choose the three most competent specialists in the issue under consideration. The mentioned experts shared with the working group their list of three experts, etc. As a result, the commission selected 12 experts, whose names were often found in the questionnaire [11, 12].

At the third stage, the experts of the working group conducted a direct questionnaire. Then a statistical evaluation of the questionnaire was carried out, after which the experts were briefed on the result.

Expert analysis in the first round revealed 18 parameters (Table 1), having a significant influence on the choice of the optimal variant. The parameters were conventionally divided into four categories: functional, economic, safety parameters and infrastructure parameters.

Table 1. Pump unit drive choice parameters

parameters	letters of references
Functional parameters	
Capacity utilization	P1
Stability of modes	P2
Failure statistics	P3
Coefficient of extensive use	P4
Mean time between failures	P5
Coefficient of intensive use	P6
Coefficient of starting frequency	P7
Economic parameters	
Gas prices in the region	P8
Electricity prices in the region	P9
Payback period	P10
Gas consumption	P11
Electricity consumption	P12
Cost of installation and operation of equipment	P13
Safety and environmental parameters	
Equipment certification	P14
Impact on the human body	P15
Impact on the environment	P16
Infrastructure parameters	
Distance to the nearest power source	P17
Distance to the nearest gas supply source	P18

In the second round, it is proposed to rank the parameters for selecting the drive of the main pumping units according to the method proposed below.

Let x experts evaluate y objects with z parameters. The result of the evaluation will represent quantities $Y^{g}_{h\,i}$, where i is the number of the expert; h is the object number; g is the number of the comparison sign.

In the study, quantities $Y^g_{h\,i}$ (h = 1, 2... y ; i = 1, 2... x; g = 1, 2... z) are scores. They are set by each expert individually.

In order to obtain a group estimate of each object, one can use the comparison formula:

$$Y_h = \sum_{g=1}^{z}\sum_{w=1}^{x} q_{gi} Y^g_{h\,i} j_i \tag{1}$$

where q_{gi}– the weights of the drive selection parameters;

j_i – coefficient of the expert's competence.

Competence and weights coefficients are normalized values:

$$\sum_{g=1}^{z} q_g = 1; \quad \sum_{i=1}^{x} j_i = 1. \tag{2}$$

The coefficients of the weights of the parameters are determined expertly. If q_{gi} is the weight factor of the g-th parameter given by the i-th expert, then the average weight coefficient of the g-th index for all experts is equal to:

$$q_g = \sum_{i=1}^{x} q_{gi} j_i \,. \tag{3}$$

Expert competence factors can be calculated from the evaluation of the objects. Given that the expert competence should be evaluated by the degree of consistency of their assessments with the group assessment of the object, the algorithm for calculating the competence coefficients of experts will take the form:

$$Y^t_h = \sum_{i=1}^{x} Y_{h\,i} j^{t-1}_i \quad h = 1{,}2 \ldots y\ ; \tag{4}$$

$$\lambda^t = \sum_{h=1}^{y}\sum_{i=1}^{x} Y_{h\,i} Y^t_h \quad t = 1{,}2 \ldots\ ; \tag{5}$$

$$z^t_i = \frac{1}{\lambda^t} \sum_{h=1}^{y} Y_{h\,i} Y^t_h \quad i = 1{,}2 \ldots x\ . \tag{6}$$

The authors begin the calculation with t = 1. The initial values of the competence coefficients are assumed to be the same and equal to $j^0_i = \frac{1}{x}$.

At first approximation, the group estimate is:

$$Y^1_h = \frac{1}{x} \sum_{i=1}^{x} Y_{h\,i} \quad h = 1{,}2 \ldots y\ ; \tag{7}$$

normalizing coefficient is:

$$\lambda^1 = \sum_{h=1}^{y}\sum_{i=1}^{x} Y_{h\,i} Y^1_h\ ; \tag{8}$$

expert competency coefficient in the first approximation:

$$j^t_i = \frac{1}{\lambda^1} \sum_{h=1}^{y} Y_{h\,i} Y^1_h \,. \tag{9}$$

The ranking of experts is planned to be carried out in accordance with the weighting factors (Table 2).

Table 2. The weighting coefficients of experts

Indicators	weighting coefficient		
Academic degree	Without a degree	Candidate of Sciences (PhD)	Doctor of Science
	0	0,5	1
Work experience in the sphere of an electric drive use	X =n/N	where x – weighting coefficient n - expert's work experience N – experience of the most experienced expert in the sphere	
Work experience in the use of a gas turbine drive	X =n/N	where x – weighting coefficient n - expert's work experience N – experience of the most experienced expert in the sphere	
Availability of scientific works of Higher Attestation Commission (VAK), SCOPUS, WoS for the last 5 years	None	Up to 5 articles (inclusive)	Over 5 Articles
	0	0,5	1

3.Conclusion

Thus, the ranking of the parameters for selecting the drive of the main-unit pumps should form the basis of the methodology, which in turn will allow rational selection of the most economically, technically advantageous types of drives based on the initial parameters.

References

[1] Comprehensive program "Energy Saving and Improving Energy Efficiency in the Tyumen Region" for 2010-2020

[2] Gumerov A G, Gumerov R S, Akberdin A M 2001 *Operation of equipment of oil pumping stations Textbook* (M: Nedra-Business Center)

[3] Zakirzakov A G, Egorov A L 2015 *Analysis of the state of the main oil pipelines network of the Tyumen region on the basis of statistical data. Modern problems of science and education.*

[4] Tyrylgin I V, Shpilevoy V A, Zemenkov Yu D 2012 Energy saving and energy efficiency of the economy, production, transportation of oil and gas in Russia. *News of higher educational institutions. Oil and gas* 6

[5] Zemenkov Yu D, Shpilevoy V A, Podorozhnikov S Yu, Zakirzakov A G 2014 *Energy-technological complexes when designing and operating objects of transport and storage of hydrocarbon raw materials. Tutorial.* (Tyumen: Publishing house "Vector Buk")

[6] Bautin S P, Krutova I Y, Obukhov A G 2015 Twisting of a fire vortex subject to gravity and Coriolis forces. *High Temperature* 53 928-930 DOI: 10.1134/S0018151X1505003X

[7] Bautin S P, Obukhov A G 2013 Mathematical simulation of the near-bottom section of an ascending twisting flow *High Temperature* **51** 509-512 DOI: 10.1134/S0018151X1302003X

[8] Podorozhnikov S, Shabarov A, Shirshova A, Berdyshev A 2016 Dielectric Method for Diagnosing Formation of Gas Hydrates in Gas Pipelines *MATEC Web of Conferences* **73** DOI: 10.1051/matecconf/20167302025

[9] Kurushina V, Kurushina E, Shabarov A 2016 Technology Audit: Assessment of Innovative Portfolio *MATEC Web of Conferences* **73** DOI: 10.1051/matecconf/20167307028

[10] Nelyubov D, Shabarov A, Podorozhnikov S 2016 Complex-action reagent for oil-infield and oil-trunk pipelines *Solid State Phenomena* **871** 185 DOI: 10.4028/www.scientific.net/MSF.871.185

[11] Konev V, Polovnikov E, Krut O, Merdanov S, Zakirzakov G 2017 Investigation and

Development of the Thermal Preparation System of the Trailbuilder Machinery Hydraulic Actuator *IOP Conference Series: Materials Science and Engineering* **221** DOI: 10.1088/1755-1315/221/1/012001

[12] Polyakov D V, Merdanov Sh M 2005 Equipment for snow utilization using electric power Stroitel'nye i *Dorozhnye Mashiny* **12** 24-26

IOP Conf. Series: Materials Science and Engineering **357** (2018) 012013 doi:10.1088/1757-899X/357/1/012013

Snow as building material for construction of technological along-the-route roads of main pipelines

S M Merdanov, A L Egorov, V A Kostyrchenko and T M Madyarov

Industrial University of Tyumen, 38 Volodarskogo St., Tyumen, 625000, Russia.

E-mail: merdanovsm@tyuiu.ru, egoroval@tyuiu.ru, kostyrchenkova@tyuiu.ru, madjarovtm@tyuiu.ru.

Abstract. The article deals with the process of compacting snow in a closed volume with the use of vacuum processing for the construction of technological along-the-route roads of main pipelines. The relevance of the chosen study is substantiated; methods and designs for snow compaction are considered. The publication activity and defenses of doctoral and candidate dissertations on the research subject are analyzed. Patent analysis of existing methods and equipment for snow compaction with indication of their disadvantages is carried out. A design calculation was carried out using computer programs in which a strength calculation was performed to identify the most stressed places in the construction of a vibrating hydraulic tyre-type roller. A 3D model of the experimental setup was developed.

1. Introduction

During the construction and operation of main pipelines, it is necessary to provide transportation along the pipeline, as well as access to the nearest ports, motor roads and technological bases. The construction of capital roads is associated with significant financial costs and as a rule temporary roads are being built.

In regions with stable negative temperatures, temporary snow-ice roads are the most rational. The main road building materials for the construction of this type of roads are snow and ice. Snow is prepared before laying in the road bed, and ice is used in a crushed form as a solid filler or directly as a base.

Snow-ice roads must ensure uninterrupted traffic of the technological transport with the specified design loads and speeds, be built at a rapid pace with the maximum use of means of mechanization, as well as be strong and stable during the design life [1].

The undoubted advantage of temporary winter roads is their versatility, cheapness and the possibility of laying practically on any terrain, including waterways, swamps, etc.

With a small road traffic intensity, a simple construction of compacted snow is enough, but with high traffic intensities or high design loads, it is necessary to build a roadbed from pre-prepared snow with a given density and humidity.

To determine the optimum parameters of snow and to develop the working elements of snow-compacting and snow-wetting machines, the authors have been conducting theoretical and experimental studies for many years.

Purpose of the study is the process of compacting snow in a closed volume with the use of vacuum processing for the construction of technological along-the-route roads of main pipelines.

Objectives of the study:

- to analyze the publication activity and defenses of dissertations on the research topic;
- to develop and justify the design of an experimental installation for compaction of snow with the use of the vacuum processing;
- to study the effect of vacuum parameters and compaction regimes on the intensity of the snow compaction process;
- to develop a generalized mathematical model "Vacuum compaction- snow".

2. Materials and methods

This problem has actively been studied by the authors. Egorov A.L. in his thesis presented a specialized machine with a snow-compacting element. The main idea was introduction of the process of snow compaction during its loading into vehicles. To carry out experimental studies, an installation was developed and constructed that makes it possible to obtain dependencies of the applied load on the change in the final density, for different sizes of dies. On the basis of the proposed mathematical model, a formula is derived for finding the necessary force in working member F to create a block of required density p and size S.

In addition, in the course of the study, regularities in the variation of snow density were revealed, and also rational geometric and physical parameters of snow blocks were established.

The research was continued by Sharukha A.V. In his work by the example of the construction of snow-ice roads, he proposed supplementing the technology by introducing vibration into the snow-briquetting process. In the course of the study, the necessary load parameters were found to produce briquettes of the required density. A mathematical tool for calculating the parameters of the working element of the snow-removing machine was also developed. In the thesis, the processes occurring in the vibro-compaction of snow in a closed volume are considered [2,3].

To carry out the experiments, the authors used an installation based on a vibration platform and shown in Fig. 1.

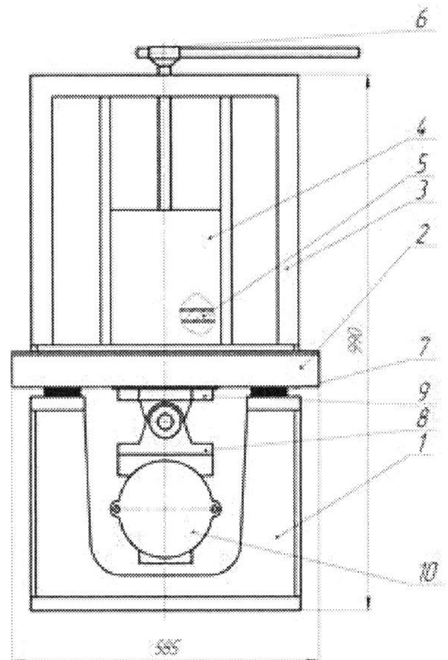

Figure 1. Installation for vibro-compaction of snow
where, 1–stand, 2–platform, 3–frame, 4–shape, 5– positive die, 6 – screw with handle, 7– elastic element (spring), 8 – lower support, 9– upper support, 10 – vibro-exciter.

After the production of snow briquettes, the values of their mass and geometric dimensions were recorded.

As a result of processing experimental data, numerical coefficients of mathematical models were obtained.

The result of these studies is the proof of the hypothesis of the rational use of vibration in the process of snow compaction, and it is also proved that when using vibration in a closed volume, an effect can be obtained in the form of saving costs for compaction.

Bykov V.Yu., a researcher from MADI (Moscow Automobile and Road Construction State Technical University), also carried out studies in this field. A feature of his scientific works is an increase in the efficiency of the process of pressing snow due to its vacuum processing. It was found that snow has such physical properties as porosity and the resulting air permeability of snow. It is noted that the nature of the interrelation between air permeability and porosity depends most strongly on the structure of snow and, in particular, on the grain sizes [4].

The experimental stand (Fig. 2) makes it possible to carry out a complex of studies of the snow compaction process under various conditions, including vacuum processing. The stand is equipped with automated systems for recording and processing a wide range of experimental data.

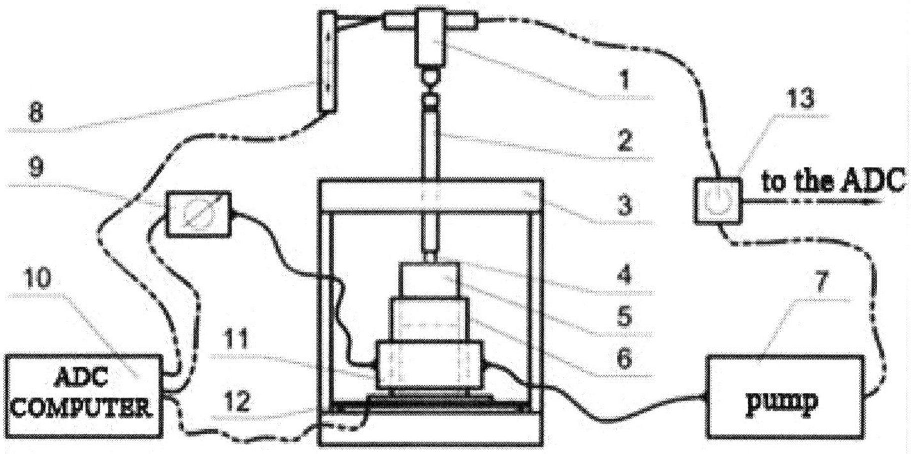

Figure 2. A scheme of the experimental stand
where, 1 – electric motor; 2 – screw; 3 – channel frame; 4 – ball end; 5 – piston; 6 – press chamber; 7 – vacuum pump; 8 – piston stroke gauge; 9 – pressure gauge; 10 – analog-to-digital converter connected to a PC; 11 – vacuum bandage of the press chamber; 12 – strain-gauge balance; 13 – start/end sensor.

Thus, the dependence of energy costs on snow pressing and its simultaneous vacuum processing was theoretically established and experimentally confirmed. Due to vacuum processing, it was possible to reduce the volume of compacted snow. The workflow was optimized, which made the compaction process more efficient. Energy consumption costs were also optimized and a methodology was developed for calculating the parameters of the vacuum snow-pressing equipment. It is proved that, depending on the intensity of gas tapping, it is possible to compress snow with the required characteristics.

All the works stated that the task of pressing snow is, first of all, the task of reducing the porosity of snow. Intensifying the gas tapping contained in the pores of snow makes it possible to achieve an increase in the degree of its compaction, while reducing the effort required for this. On the basis of experimental studies, the effect of the applied load on the change in the final density of snow was determined for static briquetting in a closed volume, with vibration and vacuum processing.

Researchers have developed and proposed options for attachments to road-building machines for snow compaction, for freezing snow and for making snow briquettes and ice blocks [4,12,15].

3. Overview of the methods of snow compaction

1. Static compaction.

Static compaction is provided by rolling. The compacting action of rollers depends on the maximum contact pressures, the distribution of pressures over the contact area, the contact surface dimensions, the rolling speed and the number of passes. The maximum contact pressures have a major influence on the strength of the material. An increase in the minimum size of the contact spot leads to an increase in the thickness of the compacted layer.

Increasing the length of the contact of the working element with the material and reducing the speed of movement lead to an increase in the loading time of the material being compacted. This positively affects the quality of compaction of materials, especially with viscous properties.

Among the various types of static rollers, rollers on pneumatic tires have proved to be the most universal and effective, so it makes sense to consider the rolling process exactly by these rollers.

When soils are compacted by vibrating, the vibrator mass is introduced into the state of oscillatory motions due to the energy brought to the exciter. After the vibro-exciter, due the kinetic energy of the latter, particles of soil are also introduced into the state of oscillatory motions. In the absence of shock pulses from the vibrator side, soil particles are only affected by inertial forces. The latter are proportional to the masses of these particles. In view of the fact that the masses of these particles are not the same, the arising inertial forces are also different. As a result of the difference in inertial forces, shear stresses arise at the contact points of the particles. Up to certain limits, these stresses are balanced by adhesion forces or the strength of the water-colloidal adhesive binders. After exceeding these limits, mutual displacements of these particles occur. Shear stresses are proportional to inertial forces; therefore, they are determined not only by the difference in the masses of the neighboring particles, but also by the accelerations that develop during vibrational motions.

The largest displacements of particles relative to each other are observed in disconnected soils, in which only frictional forces act. In cohesive materials, adhesion forces are added to frictional forces, which can not, to a large extent, be overcome by inertial forces during vibration; therefore, in cohesive materials, vibration causes a greater degree of elastic deformation.

The main parameters of particle oscillations are amplitude, frequency, velocity, acceleration.

3. Tamping.

For compaction by tamping, dropping plates on excavators and ramming machines are used.

Tamping is based on successive impacts of the plate on the material. Compaction happens due to the kinetic energy of the dropped weight, which is expended on the irreversible movement of the material particles and squeezing of the liquid phase films from the contact zones between the mineral particles, as well as the elastic compression of the material.

Compaction is most affected by the impulse of force, the time of impact, the speed of the dropped weight.

Specific impulse is the main characteristic of the tamping process.

Tamping is characterized by a short duration of stresses and a significant depth of propagation of the stress-strain state. Tamping machines are able to compact snow as layers of large thicknesses. Therefore, plates are used, as a rule, when due to the large thickness of the layer other machines are unsuitable.

4. Vacuum processing

Vacuum compaction can significantly reduce or eliminate the need for a loading embankment and ensure that strict requirements for subsidence sizes are met. Due to the lack of embankment and associated issues of soil stability under embankment, this technology is perfectly suitable for snow consolidation [5, 7, 8].

The main advantages of vacuum compaction technology are:

- significant reduction in the consolidation period compared to the static compaction time;

- reduction in the risk of shear deformations of snow.

4. Experimental

After the patent analysis, advantages and disadvantages of the methods used in devices and machines for snow compaction were taken into account, which allowed designing a laboratory installation for vibro-vacuum compaction of snow.

The experimental installation for the vibro-vacuum compaction (Fig. 3) is a camera mounted on a stand of a vibration platform, on which is a field vibrator of the type IV-98E. In the compacting chamber, a pneumatic cylinder is installed to create pressure of the output element on the compacted snow.

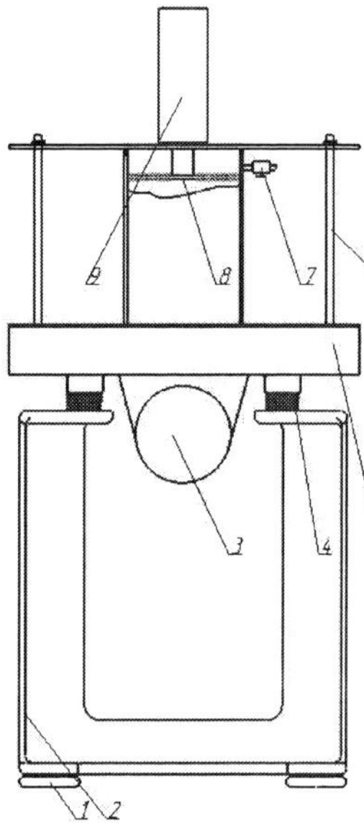

Figure 3. Experimental installation
where, 1 – anti-vibration spacers, 2 - stand, 3 – vibrator IV-98E, 4 – damping device, 5 - platform, 6 - stud, 7 – valve, 8 – compacting platform, 9 – pneumatic cylinder

A valve is installed in the chamber to connect a vacuum pump to vacuum process snow. The result of the installation operation is a compact snow briquette compressed by a combined action.

To create it, it is necessary to prepare a set of equipment, including:

1) welding inverter;
2) angle grinder;
3) drill;
4) drill kit;
5) set of keys;
6) measuring equipment (vernier calipers, tape measure, goniometer);
7) work wear

The complete set of the experimental installation includes:

1) Vibration stand, with vibration spacers.
2) Pneumatic cylinder.
3) Studs for fixing the cover top of the compaction chamber.
4) Nuts fastening to the studs of the stand, nuts fastening the studs to the compaction chamber.
5) Studs for the outside dimensions of the main structure, in our case 1m.
6) Vibrator IV-98E.
7) Bolts of fastening the vibrator to the stand.

Calculation of the experimental installation in the software Solidworks 16.

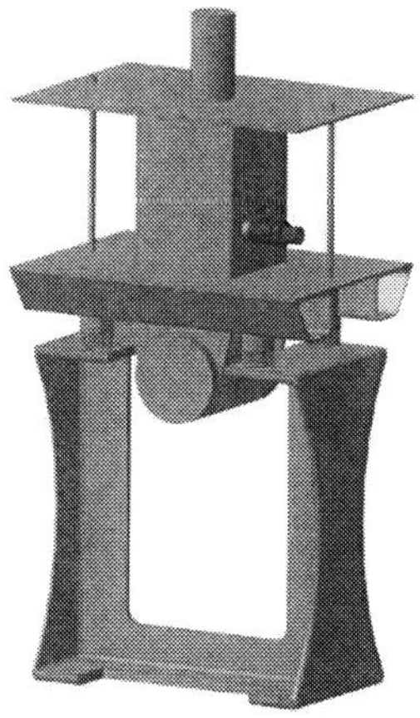

Figure 4. 3D Model in the software Compass 16

Figure 5. 3D Model in the software Solidworks 2016

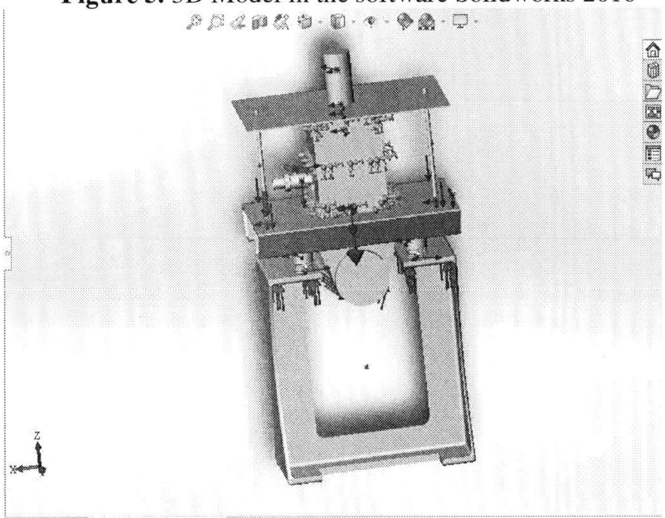

Figure 6. 3D Model with applied layers and material modulation, external factors (grid) for calculations in the software Solidworks 2016

5. Results

The methods used in the scientific work can be divided into theoretical and experimental. The method of ascension from the abstract to the concrete, the method of idealization, the method of formalization, the method of expert assessments, and others are referred to the theoretical method of research. Experiment as a method implies the study of a phenomenon in its pure form and allows you to explore the properties of objects of research in both natural and extreme conditions. An experiment can be repeated several times to check its results [9, 10, 11].

In the course of the research, the following basic working hypothesis was formulated: increasing the efficiency of snow compaction due to the combined effect of vibro-vacuum compaction in a closed volume, for utilization of snow from adjacent areas and roads in Tyumen, with a view to reducing the cost and timing of the utilization of snow.

The object of the research is the process of compacting snow with a vibro-vacuum method in a closed volume.

Table 1. Incomplete block plan and results of expert estimate

№	Parameter	Units	Experts								Total
			B1	B2	B3	B4	B5	B6	B7	B8	(Ti)
1	Thickness of the compacted snow layer	h (m)		14		12		5		8	39
2	Natural and climatic conditions (air temperature, humidity, atmospheric pressure)	$T_{ambient}$ (^0C); W (%); P (mmHg);	10		10	10		7			37
4	Frequency of vibration compacting action	n (Hz), A (mm)	14		11	14	6				45
5	Vacuum action	Y mmHg			12		8	9	8		37
3	Weight (load on snow)	m (ru)	14	10					11	12	47
6	Speed of static compaction	V (m/h)		14			11		9	8	42
	Results of expert estimate (B)		38	38	33	36	25	21	28	28	247
	B^2		1444	1444	1089	1296	625	441	784	784	7907

The subject of the research is the experimental installation.

Selection and justification of factors by the incomplete block plan method (expert estimates)

Planning the experiment includes selecting and justifying the factors that affect the process of snow compaction. During selection it is necessary to analyze the materials in scientific papers of researchers. When selecting the factors, it is necessary to take into account the significance of each of them. To select and justify the factors influencing the snow compaction in the construction of temporary winter roads, a questionnaire was formed to interview experts: Karnaukhov N.N., Merdanov Sh.M., Serebrennikov A.A., Zakirzakov G.G., Sharukha A.V., Egorov A.L., Konev V.V., Yarkin A.V. Eight experts evaluated the effect of six factors on the snow compaction on a 14-point system, with each expert being able to assess the quality of three factors, and each parameter was assessed by 4 experts.

Each expert evaluated the same number of objects. Each object is checked by the same number of experts; one expert should compare each pair of objects the same number of times. All these requirements are met when using a balanced incomplete block plan with the following parameters: b=8; v=6; q=3; r=4; N=vr=bq=24.

The objective of the expert estimate was to determine the factors of the best influence on the snow compaction (a greater influence factor is estimated by a larger number of points) and to establish significant differences between different parameters [6, 13, 14]. The incomplete block plan and the results of expert estimate y_{ij} are given in Table 1.

6. Recommendations

From the analysis of literature sources and the results of previous experiments conducted by Sharukha A.V., Egorov A.L., Merdanov Sh.M., etc., the following factors are identified for further investigation: number *n* of passes of the installation, units; installation weight - m, kg; frequency of the vibrator oscillations - v, Hz.

7. Conclusion

Materials that are given in the article are prerequisites for conducting a full-scale experiment to determine the dependence of snow compaction on vibro-vacuum compaction in a closed volume. After conducting the experimental studies, it is planned to form a generalized mathematical model "Vibro-vacuum compaction - snow".

References

[1] Dyunin A K 1983 In the realm of snow. (Novosibirsk: Science, Siberian Branch)

[2] Egorov A L, Kostyrchenko V A, Madyarov T M, Plokhov A A 2016 Designing of the Vibrating Hydraulic Tyre Roller in Order to Research the Optimal Regime Set Parameters for the Snow Mass Compacting. *International Journal of Applied Engineering Research* **11-19** 9956-9959

[3] Egorov A L, Merdanov S M 2016 Equipment for Construction of Reinforced Ice Crossings / *International Journal of Applied Engineering Research* **11-22** 10753-10756

[4] Grey D M 1986 *Snow Text: Directory* (JL: Gidrometeoizdat)

[5] Ivanov A A, Krivokhizha V N 2006 *Preparation of bases for snow-ice roads Text. Problems of the operation of transport systems: Materials of the regional NPK* (Tyumen: TyumGNGU)

[6] Karnaukhov N N, Ivanov A A, Van'tik V N 1989 *Device for compaction of snow Text. AU №1461815*

[7] Karnaukhov N N 2015 Technology and machinery for construction of winter roads International *Journal of Applied Engineering Research* **10 (15)** 35927-35935

[8] Kotelnikov V.V. Selection of optimal conditions for snow compaction during the construction of snow-covered roads. New technologies for the oil and gas region: theses of the reports of the XVII scientific and technical conference. - Tyumen: TSOGU, 1998. - Pp. 118-119;

[9] Madyarov T M, Kostyrchenko V A, Egorov A L, Plokhov A A 2016 Review of the Methods and the Constructions for the Waste Wood Recycling for the Machine Designing Based on Tractor Msn-10 for the Pellets Production. *International Journal of Applied Engineering Research* **11- 22** 10945-10951

[10] Merdanov S M, Egorov A L, Zakirzakov G G 2016 The Wireline Guide Assembly of Lifting Unit. *International Journal of Applied Engineering Research* **11 20** 10353-10357

[11] Obukhov A G, Sharukha A V, Shityy V P 2015 Sealing vibratory working body of the construction machine. *Bulletin of the Kurgan State Agricultural Academy* **4 (16)** 77-78

[12] Reznik L G, Karnaukhov V N, Evtin P V 2009 *Introduction to scientific research. Processing of experimental results: textbook* (Tyumen state oil and gas university. Tyumen)

[13] Shityi V, Sharukha A, Spirichev M 2015 *Temporary winter roads. Arctic dialogue in the global world The Proceedings of Joint Science and Education Conference,* pp. 319-321

[14] Sharuha A V 2009 *Substantiation of the parameters of a vibrating working body of the volumetric type of a snow briquetting machine. - The dissertation author's abstract on*

competition of a scientific degree of the candidate of technical sciences. (Tyumen state oil and gas university. Tyumen)
[15] Tikhomirov V B 1974 *Planning and analysis of the experiment* (Light Industry')

IOP Conf. Series: Materials Science and Engineering **357** (2018) 012014 doi:10.1088/1757-899X/357/1/012014

Optimization of geometric parameters of heat exchange pipes pin finning

K A Akulov, V V Golik, K S Voronin and A G Zakirzakov

Industrial University of Tyumen, 38 Volodarskogo St., Tyumen, 625000, Russia

E-mail: akukir@gmail.com

Abstract: The work is devoted to optimization of geometric parameters of the pin finning of heat-exchanging pipes. Pin fins were considered from the point of view of mechanics of a deformed solid body as overhang beams with a uniformly distributed load. It was found out under what geometric parameters of the nib (diameter and length); the stresses in it from the influence of the washer fluid will not exceed the yield strength of the material (aluminum). Optimal values of the geometric parameters of nibs were obtained for different velocities of the medium washed by them. As a flow medium, water and air were chosen, and the cross section of the nibs was round and square. Pin finning turned out to be more than 3 times more compact than circumferential finning, so its use makes it possible to increase the number of fins per meter of the heat-exchanging pipe. And it is well-known that this is the main method for increasing the heat transfer of a convective surface, giving them an indisputable advantage.

1. Introduction

One of the promising areas for improving heat exchangers is the use of pin finning of heat-exchange pipes [2] (Fig. 1). However, the use of this type of finning is constrained by its insufficient state of knowledge. One of the main issues facing engineers is the choice of geometric parameters of pin finning. This article is devoted to the answer to this question.

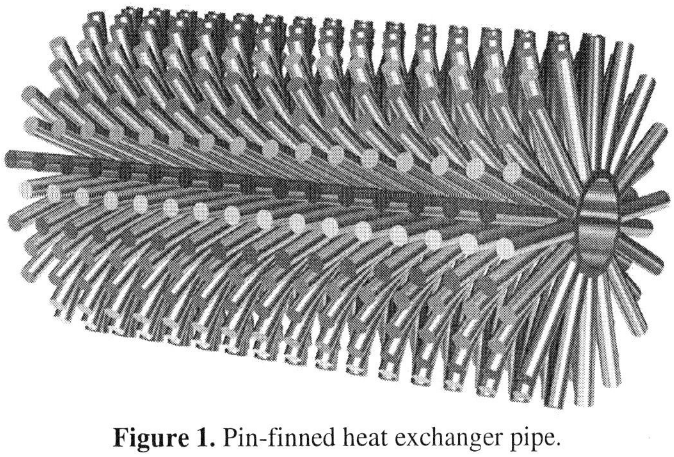

Figure 1. Pin-finned heat exchanger pipe.

2. Materials and methods

Pin-finned heat exchange pipes can be used in tubular heat exchangers - air cooling units and shell-and-tube heat exchangers. The flow of the washer fluid can move both parallel and perpendicular to the axis of the heat exchange pipe. In both cases, fin nibs located at an angle of 90 ° to the flow will have the maximum impact on them [1]. From the point of view of mechanics, the nib will be an overhang beam with length l with uniformly distributed load q (Fig. 2). It is known that maximum bending moment M from this load will be in the section of the rigid closing of the beam [11]. That is, the maximum stress in the beam will be observed at the base of the nib. This stress should not exceed the yield point for plastic materials (steel, aluminum, copper); otherwise deformation of the metal will be observed. It is necessary to find out at what geometric parameters of the nib (diameter and length), the stresses in it from the influence of the washer fluid will not exceed the yield strength of the material (aluminum). Let us solve this problem.

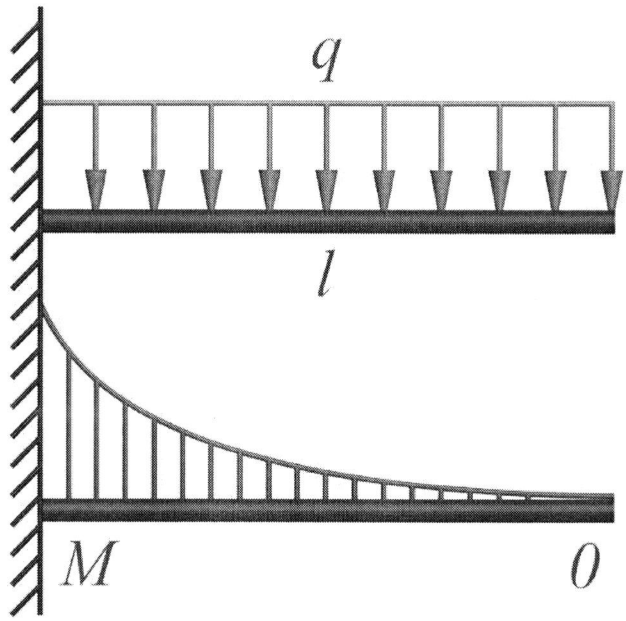

Figure 2. The design scheme of the nib as a cantilever beam with a uniformly distributed load and a bending moment diagram from this load

The maximum stress in the beam is given by formula [11]:

$$\sigma = \frac{M}{W}, Pa \quad (1)$$

where M – maximum bending moment, N/m; W – resistance moment, m^3.

The moment of resistance for a round-section beam [11]:

$$W = \frac{\pi d^3}{32}, m^3 \quad (2)$$

where d – diameter of a section of a circular beam, m.

The moment of resistance for beams of square section [11]:

$$W = \frac{a^3}{6}, m^3 \quad (3)$$

where a – cross-section side of a square beam, m.

Maximum bending moment for the cantilever beam [11]:

$$M = \frac{ql^2}{2}, \frac{N}{m} \quad (4)$$

where q – uniformly distributed load on the beam, N/m; l – length of beam, m.

Uniformly distributed load on a beam [12]:

$$q = S \cdot P \cdot c, \frac{N}{m} \tag{5}$$

where S – cross-sectional area of a beam, m^2; P – pressure on the beam, Pa; c - aerodynamic (hydrodynamic) factor (c=1,2 for round pipes, c=2,1 for square pipes [12]).

The cross-sectional area of a circular beam [11]:

$$S = \frac{\pi \cdot d \cdot l}{2}, m^2. \tag{6}$$

The cross-sectional area of a square beam [11]:

$$S = a \cdot l, m^2. \tag{7}$$

Pressure on the beam [11]:

$$P = \frac{\rho v^2}{2}, Pa \tag{8}$$

where ρ – density of the medium flowing around the beam, kg/m^3; υ - velocity of the medium flowing around the beam, m/s.

Substituting (8) and (6) in (5), let us obtain a uniformly distributed load on the beam of circular section:

$$q = 0{,}3\pi \cdot d \cdot l \cdot \rho \cdot v^2, \frac{N}{m}. \tag{9}$$

Substituting (8) and (7) in (5), one obtains a uniformly distributed load on the beam of the square cross-section:

$$q = 1{,}05 \cdot a \cdot l \cdot \rho \cdot v^2, \frac{N}{m}. \tag{10}$$

Substituting (9), (4) and (2) in (1), let us obtain the maximum stress in the beam of circular cross-section:

$$\sigma = \frac{4{,}8\rho v^2 l^3}{d^2}, Pa. \tag{11}$$

Substituting (10), (4) and (3) in (1), one obtains the maximum stress in the beam of the square cross-section:

$$\sigma = \frac{3{,}15\rho v^2 l^3}{a^2}, Pa. \tag{12}$$

From equation (11), it is possible to express the velocity of the washing stream and the diameter, at which there will be maximum stresses in the beam of the circular cross-section:

$$v = \sqrt{\frac{\sigma d^2}{4{,}8\rho l^3}}, \frac{M}{c}; \quad d = \sqrt{\frac{4{,}8\rho v^2 l^3}{\sigma}}, m. \tag{13}$$

From equation (12), it is possible to express the velocity of the washer flow and the side, at which there will be maximum stresses in the beam of the square cross-section:

$$v = \sqrt{\frac{\sigma a^2}{3{,}15\rho l^3}}, \frac{M}{c}; \quad a = \sqrt{\frac{3{,}15\rho v^2 l^3}{\sigma}}, m. \tag{14}$$

Using equations (13) and (14), substituting the yield strength values of the main grades of aluminum alloys $\sigma_{0,2}$ in them (Table 1), let us obtain the optimal values of the geometric parameters of the nibs for different velocities of their washed medium. For aluminum grades №1 and №6 (Table 1), these values are shown in the graphs of Fig. 3-6. As a flow medium, water and air were chosen, and the cross section of the nibs was round and square. The optimal geometric parameters of the pin fins of heat-exchange pipes for different flow rates of heat carriers are recommended to be chosen taking into account the safety factor, which is usually 10-20%.

One of the methods for improving circumferential-finned heat-exchange pipes is to increase the length and decrease the thickness of the fins. To date, these values have reached 16 mm and 0.3 mm. In the air-cooling apparatus, the airspeed reaches 13 m/s [8]. As calculations have shown, the geometric parameters of the pin finning will be much smaller than of the circumferential finning. Moreover, transverse fins are made only out of their alloys of the first group (Table 1), since the material is forced to stretch during the formation of the fin, and the pins can be made from the strongest aluminum alloy of the sixth group [4] (Table 1). If one sets the height of the pin fins and the airspeed similar to the pin-finned heat exchangers (height -16 mm, speed -13 m/s), then their thickness will be 0.09 mm for round nibs and 0.07 mm for square nibs. Reducing the thickness of the nib more

IOP Conf. Series: Materials Science and Engineering **357** (2018) 012014 doi:10.1088/1757-899X/357/1/012014

than 3 times allows one to increase the number of fins per one meter of the heat exchange pipe by the same amount. And as it is known, this is the main method for increasing the heat transfer of the convective surface. Thus, the optimal size of the pin fins is much smaller than the optimal size of the circumferential fins, which gives them an indisputable advantage.

Table 1. Guaranteed level of mechanical properties of Russian and American aluminum alloys

№	Alloys						Standard samples		
							Mechanical properties		
	System	Russia GOST 4784	USA ASTM B221	Russia	USA	Heat treatment	Strength limit $\sigma_в$, MPa	Yield strength $\sigma_{0,2}$, MPa	Relative extension ψ, %
1	Al>99%	A5, A6, AD, AD0, AD1	1050, 1060, 1070, 1100	AD1	1060	F	60	15	25
2	Al-Mn	MM, AMn, D12	3003, 3004, 3005, 3010, 3104	D12	3004	0	160	60	20
3	Al-Mg	AMG1, AMG2, AMG3, AMG4, AMG5, AMG6	5005, 5052, 5154, 5086, 5182, 5654	AMГ4	5083	0	270	110	14
4	Al-Mg-Si	AD31, AD33, AD35, AB	6063, 6262	AD31	6063	T6	205	170	10
5	Al-Cu-Mg	D1, D16, 1161, 1163, 1201	2024, 2117, 2219, 2419	D16	2024	T3511	450	315	9
6	Al-Zn-Cu-Mg	B93, B95, B96, 1915, 1935 1955	7050, 7075, 7070, 7108, 7004, 7475	B95	7075	T6511	560	490	6

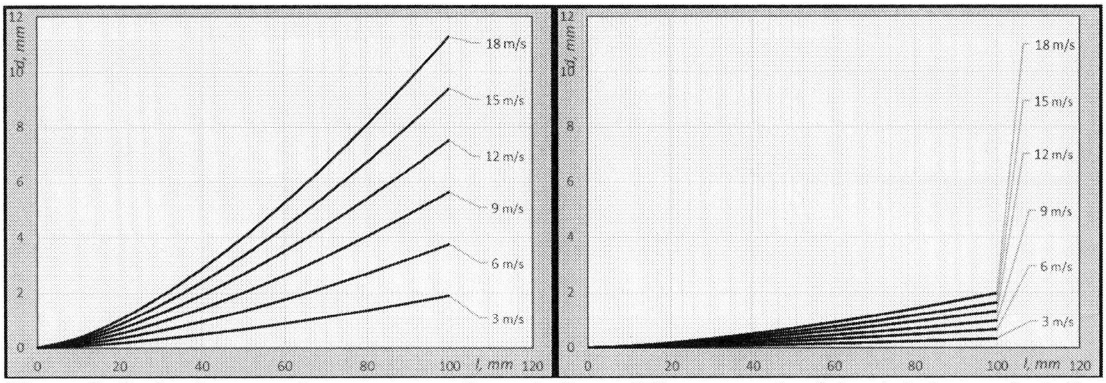

Figure 3. Optimal geometric parameters of round nibs at different speeds of the air being washed. On the left, there are alloys №1 from Table 1; on the right, there are alloys № 6 from Table 1.

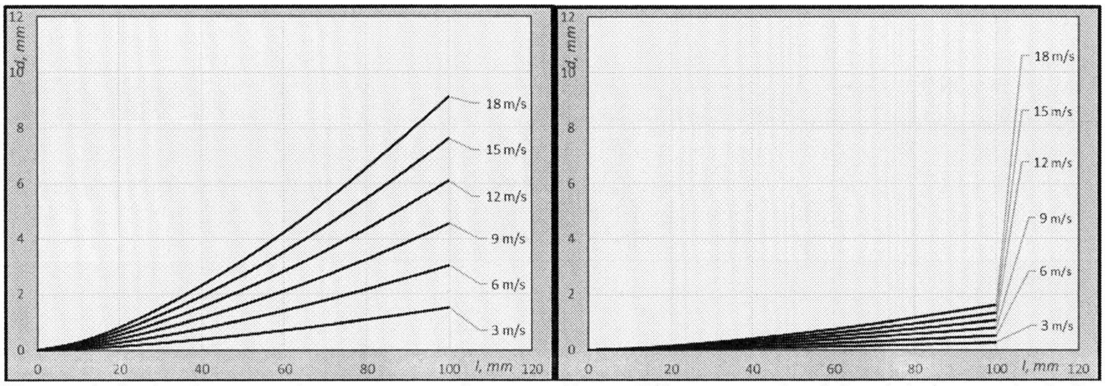

Figure 4. Optimal geometric parameters of square nibs at different speeds of the air being washed. On the left, there are alloys №1 from Table 1; on the right, there are alloys № 6 from Table 1.

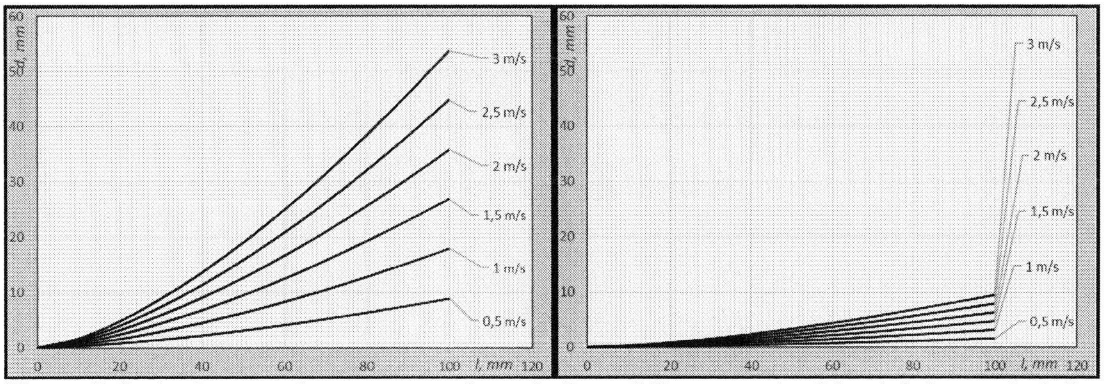

Figure 5. Optimal geometric parameters of round nibs at different speeds of the air being washed. On the left, there are alloys №1 from Table 1; on the right, there are alloys № 6 from Table 1.

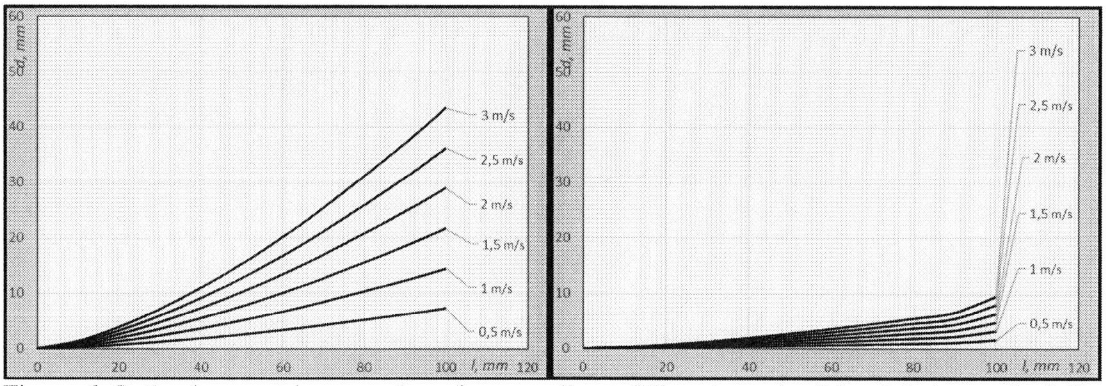

Figure 6. Optimal geometric parameters of square nibs at different speeds of the air being washed. On the left, there are alloys №1 from Table 1; on the right, there are alloys № 6 from Table 1.

3. Conclusion

Optimal values of the geometric parameters of nibs were obtained for different velocities of the medium washed by them. As a flow medium, water and air were chosen, and the cross section of the nibs was round and square. Pin finning turned out to be more than 3 times more compact than circumferential finning, so its use makes it possible to increase the number of fins per meter of heat-exchanging pipe. And it is well-known that this is the main method for increasing the heat transfer of a convective surface, giving them an indisputable advantage.

References:

[1] Akulov K A, Golik V V, Ponomareva T G 2015 Purification of units for air cooling of gas *Fundamental research* **12-3** 453-456

[2] Akulov K, Golik V, Chekardovsky M, Serebrennikov A 2016 Research of Spined Heat-Exchanging Pipes *MATEC Web of Conferences* **73**

[3] Akulov K A, Petryakov V A, Mostovaya N A 2017 The Use of Foamed Concrete for Oil Spill Treatment and Carbon Capture *IOP Conference Series: Earth and Environmental Science* **50**

[4] ASTM B221M-13 *Standard Specification for Aluminum and Aluminum-Alloy Extruded Bars, Rods, Wire, Profiles, and Tubes* (Metric)

[5] Brand A E, Chekardovskiy S M, Akulov K A 2017 The Analysis of the Development Dynamics and Structural Balance of Solar Energy in the World *IOP Conference Series: Earth and Environmental Science* **50**

[6] GOST 13726-97 *Tapes of aluminum and aluminum alloys*

[7] Kulikov A M 2016 On Choosing Multicomponent Multiphase System Separation Progress Optimization Criteria V *IOP Conference Series: Materials Science and Engineering* **154**

[8] Kuntysh V B, Sukhotsky A B 2013 To the problem of calculating the vibrational reliability of beams from bimetallic finned tubes of air cooling apparatuses *Bulletin of the International Academy of Refrigeration* **2** 46-50

[9] Pirogov S P, Cherentsov D A 2016 Theoretical foundations of the design of vibration-resistant manometers. *Measurement Techniques* 1-5

[10] Pirogov S P, Cherentsov D A, Chuba A Y 2016 Study of elastic sensing elements for vibration-resistant pressure gauges *IOP Conference Series: Materials Science and Engineering "International Scientific-Practical Conference of Students, Graduate Students and Young Scientists "Transport and Storage of Hydrocarbons"*

[11] Aleksandrov A V, Potapov V D, Derzhavin B P, Alexandrov A V 2003 *Resistance of materials* (Textbook for universities M.)

[12] SNiP 2.01.07-85 Loads and effects

[13] Zemenkov Y D, Shalay V V, Zemenkova M Y 2015 Immediate analyses and calculation of saturated steam pressure of gas condensates for transportation conditions. *Source: Oil and Gas Engineering* **113** 254-258

[14] Dudin S M, Zemenkov Y D, Shabarov A B, 2016 Modeling the Phase Composition of Gas Condensate in Pipelines *IOP Conference Series: Materials Science and Engineering* **154**

[15] Gusev A I, Kozyrev N A, Osetkovskiy I V, Kozyreva O A, Valuev D V 2017 Quality of metal deposited flux cored wire with the system Fe-C-Si-Mn-Cr-Mo-Ni-V-Co *Key Engineering Materials* **736** 23-28

[16] Valuev D V, Malushin N N, Valueva A V, Dariev R S, Mamadaliev R A 2016 Manufacturing component parts of mining equipment with application of hardening technologies *IOP Conf. Series: Materials Science and Engineering* **127**

[17] Gizatulin R A, Valuev D V, Trifonov V A, Valueva A V, Serikbol A 2015 Oxidation processes in blowing steel with inert gas into the ladle *IOP Conference Series: Materials Science and Engineering* **91** 1-7

[18] Saprykina N A, Saprykin A A, Matrunchik M S 2014 Formation of Surface Layer of Cobalt Chrome Molybdenum Powder Products with Differentiation of Laser Sintering Modes *AppliedMechanics and Materials* **682** 294-297

[19] Gizatulin R A, Kozyrev N A, Valuev D V, Valueva A V, Serikbol A 2016 research on operational characteristics of imported rails under conditions of east siberian railway *IOP Conf. Series: Materials Science and Engineering* **127**

[20] Malushin N N, Kovalev A P, Valuev D V, Shats E A, Borovikov I F 2016 Development and study of hard-facing materials on the base of heat-resisting high-hardness steels for plasma-jet hard- facing in shielding-doping nitrogen atmosphere *IOP Conference Series: Materials Science and Engineering* **142**

IOP Conf. Series: Materials Science and Engineering **357** (2018) 012015 doi:10.1088/1757-899X/357/1/012015

Features of realization of life cycle of commercial vehicles in western Siberia

E G Ishkina

Industrial University of Tyumen, 38 Volodarskogo St., Tyumen, 625000, Russia

E-mail: ishkinaeg@tyuiu.ru

Abstract. The management of a trucking enterprise is considered on the basis of reducing costs and increasing profitability during the life cycle operational period of the vehicles. Economic evaluation of the variants of the use of vehicles passed the normative operation period was carried out.

1. Introduction

Effective management of a trucking enterprise is not possible without planning and controlling the costs of rolling stock operation. Expenses for the cars operation are one of the components when pricing of transport services. In turn, the operating costs and cars profitability depend on their life cycle.

Sequentially arising periods: design, production, installation, storage, operation and disposal, are the car life cycle.

The car life cycle is realized by the car manufacturer and the owner buys the car. The car manufacturer realizes designing, manufacturing, installation and storage as well as can influence the car utilization. Influence involves the design of the car, taking into account its subsequent disposal, creation of conditions and organization of car reclaiming processes. The owner acquires the car and realizes the stages of its operation and utilization. The operational phase in turn may include vehicle modernization, repair and maintenance. The time of reclaiming of a car depends on the nature and conditions of its operation.

2. Materials and methods

For commercial vehicles, in terms of economic feasibility, it is important for the life cycle of the car, and at what time point it will be rational to stop the operation and reclaim it.

Let us consider the car operation process. In the operation process, every vehicle undergoes the maintenance, current and major repairs. For each car, the manufacturer enshrines a standard operating life; within this period there is a need for major repairs of the units and assemblies of the car. The major repair is intended for the regulated restoration of lost performance of cars and aggregates. Restoration of the main parts during major repair should ensure a level of quality close to or equal to the quality of new products [1].

Each owner of the car, especially commercial, solves the problem in the process of operation: what is better – to carry out a major overhaul of the car to restore its consumer properties or to buy a new vehicle with the reclaiming of the car that has passed its standard operating life.

Let us analyze the operation of commercial vehicles from the technical and economic point of view. What will be more expedient – to carry out a major repair of the car and continue to operate it or buy a new car and reclaim the old one.

Research was carried out at transport enterprises of the OAO NK Rosneft group in Western Siberia.

Let us compare the advantages and disadvantages of two options (projects) of car exploitation by conducting a SWOT-analysis based on a survey of the engineering and technical composition of enterprises (Table 1).

Table 1. Table SWOT-analysis

Options /evaluation	Acquisition of new cars	Cars major repair
Advantages	- high resource and reliability of new equipment; - improving the quality of services through the use of more productive cars (more complete in the design, as in the context of 10 years, it is significantly changing); - attraction of additional financial resources through the reclaiming of cars; - reduction of costs for current repairs and reduction of stocks of spare details in warehouses; - improving labor conditions for drivers and maintenance personnel	- less financial costs in comparison with the buying of new cars; - increase of professionalism of repair personnel; - development of invention and innovation; - low time for project implementation
Disadvantages	- high project costs; - possible reductions in stuff and related benefits	- low resource of overhauled automobiles; - increase in vehicle repair costs; - increase of stocks of spare details in warehouses; - lack of reserves to improve the quality of transport services; - relatively low major repair runs

On the basis of empirical studies, it was found that new cars can realize more customer orders by increasing the intensity of their use in comparison with the cars that have undergone major repairs. The increase in the operation intensity of new cars is due to greater consumer compatibility. This is connected most often with the fact that automakers regularly improve the consumer properties of cars. More comfortable car performance is more competitive in the market.

Let us consider the main differences between the operation of new cars and cars that have undergone major repairs. One of the operational indicators is the truck availability rate – $\alpha_т$; it determines the proportion of working time during which the truck fleet is in good repair and can be used in the transport process.

Coefficient of technical readiness ($\alpha_т$) is calculated by the formula:

$$\alpha_т = (AD_{hoz} - AD_{rem} - AD_{to}) / AD_{hoz}, \quad (1)$$

where AD_{hoz} – truck-days on the farm;

AD_{rem} –car-days are under repair;

AD_{to} – car-days in maintenance.

The coefficient of technical readiness is one of the indicators characterizing the performance of the car and truck fleet [2]. The coefficient of technical readiness is higher for new cars in comparison with cars after overhaul.

Fleet utilization of motor vehicles is the indicator characterizing the degree of use of the truck fleet for a calendar period. It is defined as the ratio of the car-days in operation to the value of car-days in the farm.

For any commercial enterprise, this coefficient should tend to a maximum, thus, let us obtain the objective function:

$$FLEET\ UTILIZATION\ (KIP)\ (\alpha_m, t) \rightarrow max\ . \tag{2}$$

With the increase in the operation life of the truck park, fleet utilization changes during time t. It is important that its decline occurs as slowly as possible.

The data on the change in fleet utilization in instrumentation (theoretically) for new machinery and overhauled equipment are given in Table 2, respectively.

Table 2. Dynamics of change in fleet utilization

Name / year	2015	2016	2017	2018	2019	2020	2021	2022	2023	2024	2025
Change in fleet utilization with new equipment											
Bulldozers B -170	0.77	0.75	0.73	0.69	0.67	0.65	0.63	0.6	0.57	0.54	0.51
Rippers T-25.01	0.76	0.74	0.72	0.69	0.67	0.65	0.63	0.6	0.57	0.54	0.51
Pipelayers ТГ-163 T-170	0.77	0.75	0.73	0.69	0.67	0.65	0.63	0.6	0.57	0.54	0.51
Truck crane KC-35714	0.77	0.75	0.73	0.69	0.67	0.65	0.63	0.6	0.57	0.54	0.51
KZKT-7428 with semitrailer KZKT-9101	0.72	0.7	0.68	0.65	0.64	0.63	0.62	0.6	0.57	0.54	0.51
KrAZ-260B with CHMZAP-99865	0.73	0.71	0.69	0.65	0.64	0.63	0.62	0.6	0.57	0.54	0.51
URAL-44202 with semitrailer CHMZAP - 9906	0.72	0.7	0.68	0.65	0.64	0.63	0.62	0.6	0.57	0.54	0.51
URAL-43204	0.76	0.74	0.72	0.69	0.67	0.65	0.63	0.6	0.57	0.54	0.51
Tatra-815 UDS-114	0.79	0.77	0.75	0.72	0.7	0.68	0.65	0.62	0.58	0.54	0.51
Change in fleet utilization of equipment after the major repair											
Bulldozers B -170	0.73	0.72	0.71	0.68	0.63	0.59	0.52	0.43	0.33	0.29	0.23
Rippers T-25.01	0.72	0.7	0.68	0.64	0.6	0.55	0.5	0.44	0.37	0.29	0.23
Pipelayers TG-163 T-170	0.74	0.72	0.68	0.64	0.6	0.55	0.51	0.45	0.38	0.3	0.23
Mobile crane KS-35714	0.74	0.72	0.68	0.65	0.6	0.55	0.49	0.43	0.36	0.29	0.23
KZKT-7428 with semitrailer KZKT-9101	0.67	0.65	0.63	0.6	0.58	0.56	0.5	0.44	0.37	0.3	0.23
KrAZ-260B with CHMZAP -99865	0.68	0.68	0.66	0.63	0.6	0.56	0.5	0.44	0.36	0.3	0.23
URAL-44202 with semi-trailer CHMZAP -9906	0.70	0.69	0.68	0.64	0.59	0.52	0.44	0.39	0.31	0.26	0.23
URAL - 43204	0.72	0.7	0.67	0.64	0.6	0.56	0.5	0.44	0.36	0.28	0.23
Tatra - 815 UDS-114	0.76	0.73	0.7	0.66	0.61	0.55	0.5	0.44	0.34	0.24	0.2

Based on the results of empirical and theoretical studies, the change in the fleet utilization instrumentation is as follows (Fig. 1). From the graph, it is obvious that in the initial period of the project realization, the fleet utilization is practically the same, however, with time t, the intensity of the decrease in the fleet utilization in the equipment after major repairs much larger, and the instrumentation of new and repaired equipment differs almost twofold in 10 years. That is, the resource of equipment after major repairs is much lower.

The figure shows the initial period of the project. For the new equipment during this period, there is acquisition, delivery, registration in the bodies of the State Traffic Safety Inspectorate and the

Rostekhnadzor as well as commissioning. For the old equipment, this is the period of the directly overhaul, and when replacing the number of units, registration in the bodies of State Traffic Safety Inspectorate and Rostekhnadzor.

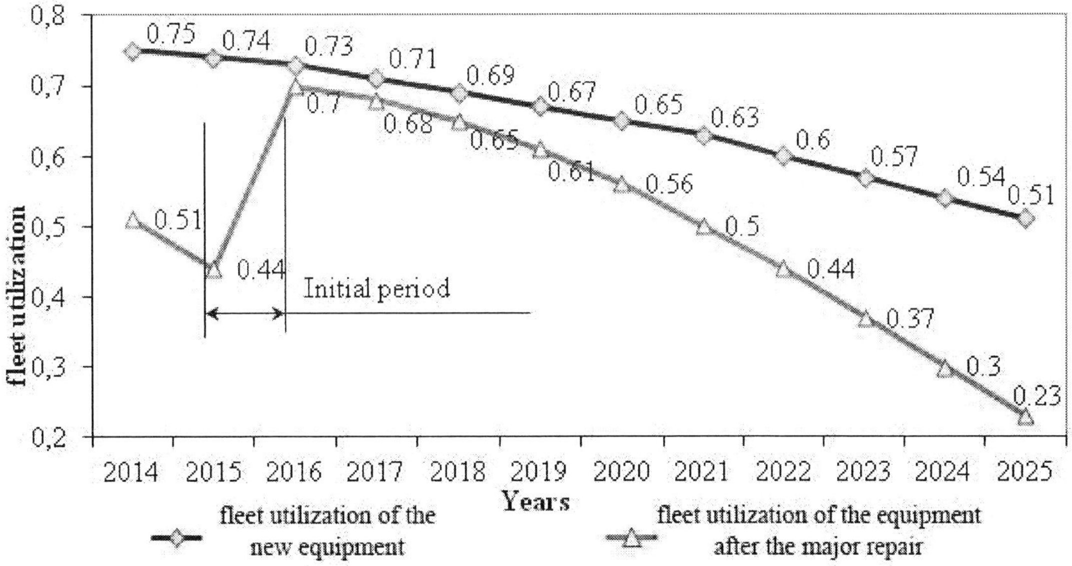

Figure 1. Changing in fleet utilization by years

It should be noted that according to statistics, the cost of overhaul is estimated at about 80% that of the cost of a new car or tractor.

The technical and economic calculation was based on the use of equipment for 10 years (the life-time recommended by the manufacturer). As a rule, the useable life expectancy of transport and tractor equipment lies, approximately, within these limits.

It is advisable to use the indicators of economic efficiency and time of recovery of outlay (project) taking into account the time factor, based on the concept of the flow of real money, in calculating the yield and the payback period of both options [3,9]. It involves measuring the inflow and outflow of real money for the entire life cycle of the car [4], i.e. for the entire investment period, starting from the investment of the first ruble and ending with the last year of the car's existence. Obviously, money has a temporary value, i.e. the value of the ruble invested in the project during the investment period changes, so the calculations were carried out taking into account the discounting.

The calculation of economic efficiency was carried out according to the specific indicator "costs per ruble of income", costs per 100 motor-hours (m/h) of work or costs per 1000 km of mileage. In the calculations, the operating revenue are compared, which is brought by the new and the overhauled equipment, a costs comparison required for the equipment operation and a comparison of the specific indicators per output unit [10].

Calculation of maintenance and repair costs consists of three parts: calculating the amount of maintenance, (performed by [5]), calculating the cost of labor and calculating the cost of materials. In these calculations, the costs for maintenance and repair per 1000 km of mileage (100 m/h of work) are determined, for new machinery and equipment after the major overhaul (Table 3).

Table 3. Costs for maintenance and repair for new machinery and equipment after the overhaul. Total repair costs, rub. Costs for maintenance and P per 1000 km (100 m / h of operation) Total repair costs, rub. The cost of maintenance and P per 1000 km (100 m / h of work)

Name	New machinery		Equipment after overhaul	
	Total repair costs, rubles	Maintenance and repair costs per 1000 km (100 m/h of work)	Total repair costs, rubles	Maintenance and repair costs per 1000 km (100 m/h of work)
Bulldozers B -170	80931.21	809.31	126820.27	1268.20
Rippers T-25.01	338218.06	3382.18	504035.70	5040.36
Pipelayers TG-163 T-170	104943.08	1049.43	168128.71	1681.29
Mobile crane KS-35714	497555.29	497.56	731881.32	731.88
KZKT-7428 with semitrailer KZKT-9101	945708.3	945.71	1445851	1066.50
KrAZ-260B with CHMZAP -99865	529579.7	529.58	775239	623.76
URAL -44202 with semitrailer CHMZAP -9906	345792.7	345.79	597779.1	399.64
URAL-43204	279479.78	279.48	484850.53	484.85
Tatra-815 UDS-114	969784.21	969.78	1314977.35	1314.98

Further, to determine the amount of operating costs, the following indicators were established: the volume of transport services, the total annual mileage and total fuel consumption.

In accordance with regulatory documents approved by the Ministry of Transport, the fuel consumption rate should be increased by 5% [6] for cars older than 8 years.

Knowing that the fall of the instrumentation for the overhauled equipment with the time occurs more intensively than the fall of the instrumentation of new equipment, it is quite enough to bring the indicators in one year of using the technology. Indicators of profitability of machinery for 2015 are presented in Table 4.

As can be seen from the table, when comparing the options by one unit the option of acquiring the new equipment certainly benefits, both in terms of profit and cost.

A comparison of the profit indicators (Figure 2) and the changes in costs (Figure 3) for the operating options are as follows:

Despite the longer payback period of the project for acquiring new equipment, the profit from this option is higher than that of the overhaul and maintenance of old equipment. The option of operating the equipment after the overhaul wins only for a shorter payback period of investment.

Table 4. Indicators of profitability of options for 2015

Name	new	old	new	old	new	old
	fleet utilization		Profitability, thous. rubles.		Costs for 1000 km of mileage (100 m/h of work), thous. rubles.	
Bulldozers B -170	**0.77**	**0.73**	481.34	399.0966	23.38	25.11
Rippers T-25.01	**0.76**	**0.72**	957.21	753.8751	44.78	48.85
Pipelayers TG-163 T-170	**0.77**	**0.74**	550.21	465.9078	20.69	22.34
Mobile crane KS-35714	**0.77**	**0.74**	652.17	585.0546	20.87	22.06
KZKT-7428 with semitrailer KZKT-9101	**0.72**	**0.67**	2142.71	1935.893	23.44	24.92
KrAZ -260B with CHMZAP -99865	**0.73**	**0.68**	2035.12	1885.402	13.15	13.94
URAL-44202 with semitrailer CHMZAP -9906	**0.72**	**0.70**	838.50	783.9671	11.29	11.78
URAL-43204	**0.76**	**0.72**	673.82	598.5667	10.70	11.49
Tatra-815 UDS-114	**0.79**	**0.76**	912.10	827.6743	26.18	27.71
			9243.17	**8235.44**	**194.47**	**208.21**

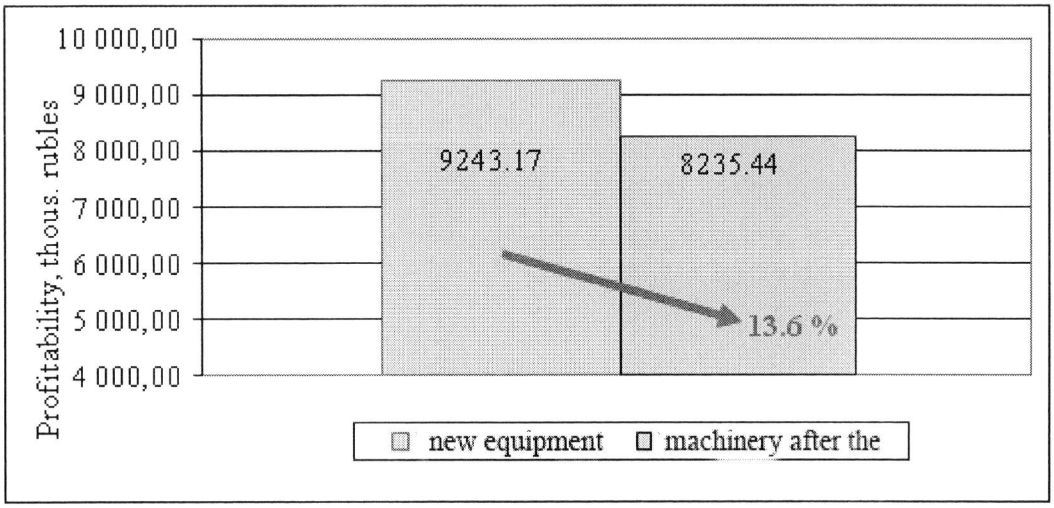

Figure 2. Comparison of profitability indicators

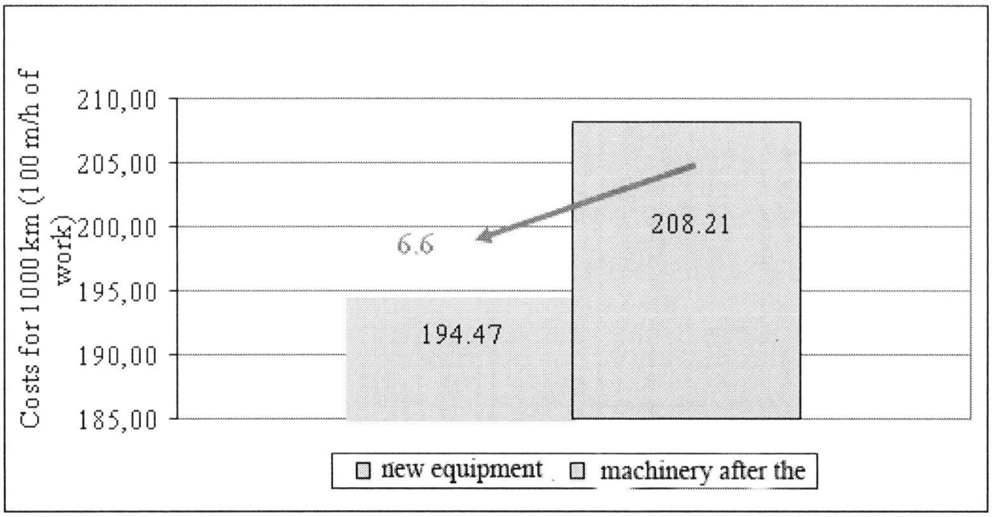

Figure 3. Comparison of costs indicators

The option of acquiring the new machinery wins both in terms of costs and profitability indicators per unit, as well as by the profitability of the project for the entire investment period. The new machinery has got a higher truck availability rate and fleet utilization, as, firstly, the efficiency of transport increases, and secondly, in connection with the reduction in the number of machinery, the number of car-days in the farm is reduced. This factor when calculating these coefficients is in the denominator and with its reduction the calculated factor is increased.

In addition, the old equipment is used for reclaiming, and the enterprise receives a material benefit from putting it into scrap (Table 5). The gain was determined based on the weight of the car and the approximate cost of a ton of car scrap 1700 rubles.

Considering the economic performance of the car operation before the onset of major repairs and after that, the authors conclude that it is more expedient to operate the automotive equipment before major repairs and then hand it over for reclaiming. Limitation of the life of cars may also be carried out at the legislative level. The Russian Ministry of Industry and Trade proposed setting a deadline for the cars operation. After this time, the machine will have to be disposed without fail [7, 8].

Table 5. Benefit from putting equipment into scrap metal

Name	Benefit, thous. rubles
Bulldozer B -170	27.03
Ripper T-25.01	65.17
Pipelayer TG-163 T-170	39.1
Mobile crane KS-35714	27.2
KZKT-7428	35.7
KrAZ-260B	21.72
URAL-44202	13.77
URAL-43204	14.53
Tatra-815 UDS-114	37.06
Total	281.28

3. Conclusion

The economic evaluation of the operational period of the vehicle life cycle confirms the proposed assumption that it is economically inexpedient to carry out major repairs after the recommended service life of vehicles due to the costs increase. This assumption is confirmed also by the decrease in the profitability of cars after major repairs.

References

[1] Kuznetsov E S, Boldin A P, Vlasov V M 2001 *Technical operation of cars* (textbook for high schools Moscow: Nauka)
[2] Khasanov R Kh 2003 *Fundamentals of technical operation of cars* (Orenburg: SEI OSU)
[3] Nepomnyashchiy E G 2003 *Investment design* (Taganrog: Publishing house TRTU)
[4] Ishkina E G 2016 Car reclaiming as a stage of its life cycle *Transport and transport-technological systems* 151-153
[5] Regulation on maintenance and repair of the rolling stock of road transport dated 20. September 1984
[6] Zakharov N S 2011 *Service of transport, technological machines and equipment in oil and gas production* (Tyumen State Oil and Gas University)
[7] Ishkina E G 2016 Management of a trucking enterprise according to economic indicators of the life cycle of cars *Economics and Management Systems Management* **22** 123-129.
[8] Ishkina E G, Ertman S A, Koryakina E A 2017 *Organization of the services of enterprises for the operation and repair of vehicles and special equipment* (Tyumen)
[9] Zakharov N S, Panfilov A A 2015 The influence of seasonal conditions on the heavy metals emissions during vehicle operation *Research Journal of Pharmaceutical, Biological and Chemical Sciences* **1** 1838-1851
[10] Zakharov N S, Elesin S V, Novoselov O A, Kichigin S J, Makarova A N 2016 Improving the efficiency of technical operation of the vehicles by optimizing the qualification structure of maintenance workers *International Journal of Applied Engineering Research* **3** 1998-2006

IOP Conf. Series: Materials Science and Engineering **357** (2018) 012016 doi:10.1088/1757-899X/357/1/012016

Methods of increasing efficiency and maintainability of pipeline systems

V A Ivanov, S M Sokolov and E V Ogudova

Industrial University of Tyumen, 38 Volodarskogo St, Tyumen, 625000, Russia

E-mail: ivanov_v_a@list.ru

Abstract. This study is dedicated to the issue of pipeline transportation system maintenance. The article identifies two classes of technical-and-economic indices, which are used to select an optimal pipeline transportation system structure. Further, the article determines various system maintenance strategies and strategy selection criteria. Meanwhile, the maintenance strategies turn out to be not sufficiently effective due to non-optimal values of maintenance intervals. This problem could be solved by running the adaptive maintenance system, which includes a pipeline transportation system reliability improvement algorithm, especially an equipment degradation computer model. In conclusion, three model building approaches for determining optimal technical systems verification inspections duration were considered.

1. Introduction

The possibility of the straightforward optimal pipeline transportation system (PTS) structure choice is eliminated by a typical for pipeline systems (oil, gas and other pipelines) tendency for recurrent process conditions to change, by various technical-and-economic indices and performance indicators for different trunk pipelines. All the mainly used technical and economic indices, playing a key role in determination of an optimal product supply system structure, can be divided into two classes. The first class includes indices, depending on the main properties of the environment, to which key PTS elements are directly related [1]. These include indices that feature the mode of conveyed fluid consumption: maximum capacity operating hours, delivery schedule density ratio, delivery schedule peaking ratio (ratio of minimum pipeline capacity to the longest maximum or minimum pipeline capacity operating duration), technical-and-economic indices of the resources used, pipeline environment conditions. The second class includes indices, displaying properties of a single pipeline as a PTS unit. The main pipeline properties can be defined variously, depending on the kind of connections (external or internal) they present. Therefore it would be convenient to divide the second class technical-and-economic indices into two groups: indices, determining external connections (connections between PTS and other national economy systems), and indices, mainly determining internal connections (connections between pipeline within PTS) [2].

According to the main principles of the systematic approach when selecting an optimal PTS structure, let us indicate these indices of each of the two second-class technical-and-economic indices groups that turn out to be necessary and possible to be considered in the current circumstances. It is obligatory to consider national economic costs to optimize the process of choosing optimal PTS structure models for the following reasons: firstly, to fully access all the national economic costs associated with PTS operating and, secondly, to choose the most appropriate pipeline system to internal and external connections [3]. On the one hand, first-class technical-and-economic indices (i.e.

indices that determined a mode of a conveyed fluid consumption) impose restrictions on the mathematical model of an optimal PTS structure and, on the other hand, directly or indirectly affect key management indicator – reduced costs. As reduced costs are influenced by the intensity of deterioration of equipment (which fully complies with the mode of a conveyed fluid consumption due to the specific characteristics of pipeline transportation), its direct impact on first-class technical-and-economic indices is mostly manifesting in changing the fixed and variable components of reduced costs. Second-class technical-and-economic indices determine reduced costs – management performance indicators of PTS operation. Pipeline transportation systems currently display certain traits consistent with huge cybernetic systems (a multitasking performance, complex associations, a hierarchy, probabilistic systems, a continuous development through time and a feedback mechanism) [4]. That is why these systems control has become more difficult. The key targets of PTS operation are rather connected with system control and environmental issues than with achieving high cost-effectiveness. This is because the achievement of mentioned targets is a mandatory requirement for an optimal and sustainable development of the national economy as pipeline transportation systems are an inseparable part of the national economy structure. The development of improved methods for optimizing the technical-and-economic structure of complex pipeline transportation systems is a critical task for its development management.

2. Methods

It is considered that PTS operating efficiency is provided when the following targets are achieved:

- maximum reliability of the fluid conveying;
- minimum capital investments;
- minimum annual costs;
- maximum pipeline adjustment range;
- minimum blackout probability;
- minimum consumption of scarce resources;
- minimum time required for raising capacity to a throughput capacity level;
- minimum time required for construction and putting into operation;
- minimum human labor costs while operating PTS.

The PTS maintenance strategy is determined by its specific purpose and structure. The selected maintenance strategy must fit in the existing PTS maintenance process.

In order to select the PTS maintenance strategy, the maintenance process is described further. Pipeline equipment and linear part are routinely inspected after a certain period of time. In the case of PTS failure, the emergency recovering repair is carried out; if not, the scheduled preventive maintenance is provided. After putting PTS into operation the scheduled current maintenance is carried out throughout period T; if PTS failed during period T, the emergency repair is carried out. This process continues until the major repair and then the cycle is repeated again [5, 6].

Let us consider stochastic process F(t), describing the state of the system at an arbitrary moment in time t.

Function F(t) in the case considered above will acquire the following values:

F(t)=
- F_0, if at moment t the system operates normally;
- F1, if at moment t a hidden failure occurs;
- F2, if at moment t an emergency preventive repair is conducted;
- F3, if at moment t scheduled preventive maintenance is conducted during storage;
- F4, if at working moment t the system is at a state of an emergency repair;
- F5, if at moment t a current maintenance is conducted;
- F6, if at moment t the plant control system is launched after repair.

IOP Conf. Series: Materials Science and Engineering 357 (2018) 012016 doi:10.1088/1757-899X/357/1/012016

The process of technical state changes can be displayed on a graph, which demonstrates the necessary PTS maintenance strategy. There are seven PTS maintenance strategies.

Strategy A: full system recovery is provided only after an evident failure.

Strategy B: full system recovery is provided at the moment of failure or at the scheduled time.

Strategy C (storage strategy): full system recovery is provided only at the scheduled time despite system failures.

Strategy D: full system recovery is provided after an evident failure or at the scheduled time.

Strategy E: after a separate failure demonstration only the failed component is replaced.

Strategy F: the system is fully restored when operating time reaches the set level; after a failure, only the failed component is replaced.

Strategy G: the system is fully restored after a failure or when the operating time reaches the set level.

The conditions of use of B and C strategies meet the requirements of PTS maintenance. Consequently, typical B and C strategies are suitable for creating a system for PTS maintenance. The combination of transient graphs forms a supergraph, reflecting a required strategy of PTS maintenance.

There are special purpose operational services which prevent and eliminate failures of pipeline equipment and linear part. The current system of PTS maintenance is based on a set of interrelated provisions given in reference documentation [7-9].

PTS maintenance is divided into two types:

– scheduled maintenance, which is provided periodically to exclude the increase in failure rate;
– unscheduled maintenance which is provided only after failures (emergencies) to restore PTS.

The first type of PTS maintenance, including routine inspections and preventive maintenance, is among important ways of maintaining PTS reliability while continuous system operation. However, due to non-optimal values of PTS components maintenance intervals, lack of a failure forecasting system and misallocation of backup components, this type of PTS maintenance turns out to be not sufficiently effective. In order to select the PTS maintenance strategy, the maintenance process is described further. The issue of PTS components failure impact assessment is considered further to explain the necessity of optimal PTS operating. Any PTS component failure leads to an optimal operational mode disturbance; meanwhile, the emerging productivity loss affects on fluid conveying efficiency. After a PTS component failure, the technological process comprises three stages:

– process shutdown, i.e. switch to the maintenance mode;
– the maintenance mode itself;
– process launch – process switch to the normal operational mode after PTS component maintenance.

Failure effects may be measured by an average PTS failure cost (the value of an average productivity loss due to the failure). The total duration of a shutdown and a maintenance mode composes the amount of time for PTS restoration in an emergency mode. Let us consider the components of a recovery time in detail. PTS component failure can be eliminated in two ways: with or without replacing the faulty component.

In the first case, failure can be eliminated by operational services on the spot. Stages of maintenance work include:

– administrative time τ_0;
– troubleshooting time τ_H;
– fault elimination time τ_y and launching time τ_p.

For the majority of pipeline transportation systems, a borderline case is typical: some failures are eliminated by operating staff and some by centralized repair stations if specific maintenance conditions are required. Replacement parts are installed instead of PTS components in need of centralized repair.

In order to operate PTS efficiently, the adaptive maintenance system, which includes PTS reliability improvement algorithm, should be used. The basis of the algorithm is PTS components failure stochastic process computer modeling. The modeling can be carried out when specific models (also called 'energy sources interfiled and trunk PTS components maintenance models') are used. The models are built to determine sustainable methods of the main reliability parameters improvement [10-12].

Maintenance models make consideration of failure elimination process possible; well-timed failure elimination creates opportunities for hydrocarbon PTS reliability improvement. Maintenance computer model is mainly determined by backup components availability and pattern of use, failure location and elimination methods, the ratio of fault-free operation time to maintenance duration. The selection of indices, used in PTS maintenance models, is important. Modern efficient methods of calculating PTS interrepair maintenance terms are based on usage of the dependence of predictive parameters on time. Due to arrange PTS maintenance, it is necessary to define the system and its components reliability parameters. In this case, the use of adaptive algorithms, providing combined operational reliability data capturing and system service, produces a large impact.

Failures can be located during specific PTS inspections named 'verification inspections' (VI). In addition, it is essential to solving inspection scheduling problem (i.e. the problem of inspection frequency determination) to minimize the expected value of costs associated with failures and inspections [13].

VI models describe degrading systems behavior. Because the degradation is a random process, the valid system state can only be identified during VI. Preventive maintenance of degrading PTS can be performed in two ways: regularly with fixed frequency or after measuring system parameters that are gradually changing while equipment degradation [14]. In the first case, only prior information on system state is used; the second method is based on the use of posterior information.

When applying the statistical modeling method (Monte Carlo method), system maintenance parameters are to be analyzed from certain probabilistic characteristics of PTS operation elementary process. For that purpose, the analyzed object probability prototype is built and then it is randomly and repeatedly implemented. The processed result serves as the problem-solving.

3. Conclusion

In summary, three model building approaches for determining optimal technical systems VI duration were considered. Each approach has both advantages and disadvantages. There are three ways to define an optimal PTS maintenance scheme.

In the first scheme, the maintenance is held after the fault-free operation. If a failure occurs before time, scheduled maintenance is held simultaneously with the emergency repair. After any kind of maintenance, PTS is considered as renewed. Then, the next schedule maintenance must be planned.

In the second scheme, scheduled maintenance is held after 25000 hours independently of a number of emergency failures happened between scheduled maintenance. After emergency failure only partial repair is held; after scheduled maintenance, PTS is considered as renewed.

The third scheme for PTS scheduled maintenance can be realized. In this case, the time period between scheduled maintenance will depend on the amount of PTS failures and on physical-chemical-mechanical PTS components condition. The third scheme is applicable to the so-termed degradation PTS failures. Degradation failures are characterized by a gradual decrease in the functionality of components in time. Degradation failures are the most frequently occurred failures. Based on fault inception statistical regularity, it is possible to prevent initiation of degradation failures and, partly, of random failures. PTS failures can be prevented if components predictor is known and if the gradual components structure change obeys a certain law.

A predictor variable is a variable that characterizes changes in component technological structure at any specific time and allows controlling component reliability margin.

Certain performance limits of PTS components allowable to control in operation are usually set. Components performance deteriorates over time, which means that component is close to breakdown.

That might give an opportunity to set a preliminary PTS predictor level at which the faulty component should be found and then replaced.

To change the faulty PTS component, it is necessary to have backup components. That is why, an optimal duration of schedule maintenance model should include a model for an optimal backup components quantity determination. These class models are considered under the question of redundancy optimization.

In this case, the cold redundancy is proposed to be used. This can be explained by the model assignment and usage: the model should determine the number of backup components necessary for maintenance.

References

[1] Ivanov V A, Lysianyi K K 2003 *Main pipelines structures reliability and performability* (SPb, Nauka)

[2] Ivanov V A, Yakovlev E I, Pushkin A A, Klyuk B A, Matrosov V I 1993 *Enhancement of operating efficiency of main pipelines* (M, All-Russian Scientific Research Institute of organization, management and economy of the oil and gas industry)

[3] Yakovlev E I, Ivanov V A, Klyuk B A, Matrosov V I 1993 *Pipeline transportation of liquid and gaseous energy carriers* (M, All-Russian Scientific Research Institute of organization, management and economy of the oil and gas industry)

[4] Chernyaev V D, Gallyamov A K, Yukin A F, Bondarenko P M 1990 *Oil pipeline transportation in difficult operating conditions* (M, Nedra)

[5] Chernyaev V D, Chernyaev K V, Berezin V L 1997 *Overall reliability of carbohydrate pipeline transport* (M, Nedra)

[6] Kurochkin B B 2000 *Service life and overhaul repair prediction of main oil pipeline* the Degree of Candidate of Engineering Sciences (Moscow)

[7] Gallyamov A K, Chernyaev K V, Shamazov A M 1997 *Assurance of operational reliability of oil pipeline structure based on technical diagnosis* (Ufa, Ufa State Petroleum Technological University)

[8] Voronin K S, Ogudova E V 2016 The effect of dynamic processes in the system "pipe-soil" on the pipeline deviation from design position *IOP Conference Series: Materials Science and Engineering "Transport and Storage of Hydrocarbons"*

[9] Shevchik V S 2005 *Dynamics of the development of the production infrastructure of oil production in the North of the Tyumen region Cand. tech. Sciences* (Tyumen State Oil and Gas University)

[10] Abramyan S G 2007 *Management of ecological compatibility of reconstruction and overhaul of main pipelines* (Volgograd: Volgograd state. architecturally-builds. University)

[11] Kharitonov V A 2008 *Construction of the main oil and gas pipeline* (Moscow: Association of Building Universities)

[12] Khaustov A P, Redina M M *2014 Emergency situations and environmental safety in the oil and gas sector Electronic fund of legal and normative technical documentation* (Electronic resource)

[13] Korshak A A, Nechval A M 2008 *Designing and operation of gas and oil pipelines* (St. Petersburg: Nedra)

[14] Zemenkova M Y, Zemenkov Y D, Shantarin V D 2016 Developing Mathematical Provisions for Assessment of Liquid Hydrocarbon Emissions in Emergency Situations *IOP Conference Series: Materials Science and Engineering* **154**

IOP Conf. Series: Materials Science and Engineering **357** (2018) 012017 doi:10.1088/1757-899X/357/1/012017

Force of resistance to pipeline pulling in plane and volumetrically curved wells

V S Toropov, S Yu Toropov and E S Toropov

Industrial University of Tyumen, 38 Volodarskogo St., Tyumen, 625000, Russia

E-mail: vladimir.s.toropov@gmail.com

Abstract. A method has been developed for calculating the component of the pulling force of a pipeline, arising from the well curvature in one or several planes, with the assumption that the pipeline is ballasted by filling with water or otherwise until zero buoyancy in the drilling mud is reached. This paper shows that when calculating this force, one can neglect the effect of sections with zero curvature. In the other case, if buoyancy of the pipeline is other than zero, the resistance force in the curvilinear sections should be calculated taking into account the difference between the normal components of the buoyancy force and weight. In the paper, it is proved that without taking into account resistance forces from the viscosity of the drilling mud, if buoyancy of the pipeline is zero, the total resistance force is independent of the length of the pipe and is determined by the angle equal to the sum of the entry angle and the exit angle of the pipeline to the day surface. For the case of the well curvature in several planes, it is proposed to perform the calculation of such volumetrically curved well by the central angle of the well profile. Analytical dependences are obtained that allow calculating the pulling force for well profiles with a variable curvature radius, i.e. at different angles of deviation between the drill pipes along the well profile.

1. Introduction

The technology of horizontal directional drilling, which is increasingly used in the construction of pipeline passages, includes three main stages [1]. The most energy-intensive and difficult from the technical point of view is the stage of pulling the working pipeline [2]. It is the parameters of the pulling process that are the criteria for choosing the equipment for construction.

During pulling, the pipeline gradually enters the well at the point where the drill string exits and moves along the well to the spudding point [3]. It is obvious that the pulling force will vary, depending on the length of the pipeline that has entered the well. As the pipeline enters the well, the pulling force will increase. The maximum force will be observed when the pipeline enters the well with its full length [4].

Since the maximum force required to pull the pipeline is of the main interest, let us limit ourselves to the case when the pipeline has completely entered the well. The force required at this stage of pulling will be the criterion for choosing the drilling equipment for the construction of this passage [5].

2. Materials and methods

The profile of the pipeline passage can have both rectilinear and curved sections [6]. In these sections, various pull-resistant forces will act.

IOP Conf. Series: Materials Science and Engineering **357** (2018) 012017 doi:10.1088/1757-899X/357/1/012017

In general, the resistance force to pulling a pipeline into a well depends on several components:

$$F_{res} = F_{res}\left(F_{cur}, F_{dm}, F_{gr}\right) \tag{1}$$

where F_{cur} - resistance force to pulling from the well curvature;

F_{dm} - resistance force from drilling mud weight;

F_{gr} – resistance force from friction against the ground.

At the curved sections, the pipeline will be pressed against the wall of the well due to the tension of the pipeline [7]. Figure 1 shows a section of the pipeline in a curved well with a constant curvature (shown in dotted lines). Let us study a pipeline moving in a well filled with drilling mud and ballasted until it reaches zero buoyancy.

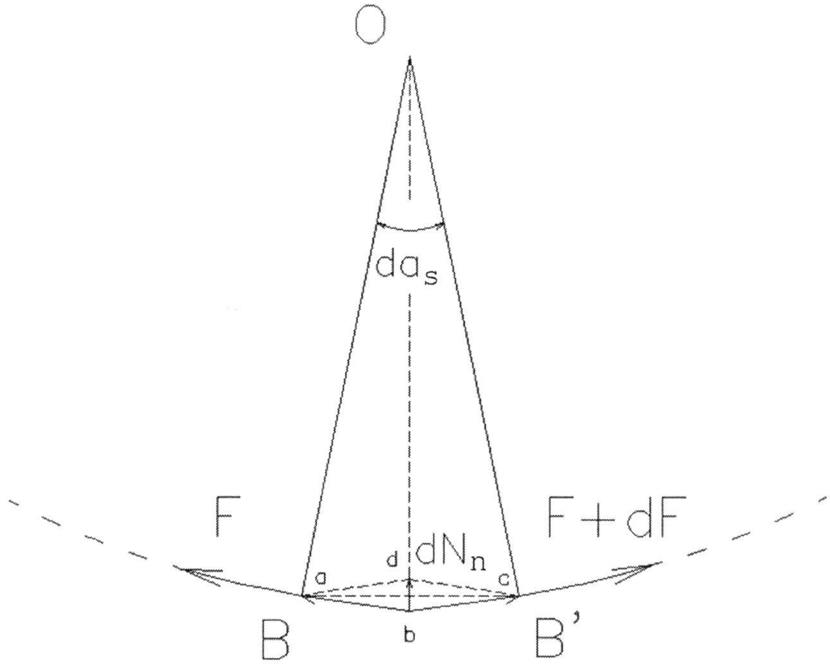

Figure 1. A design scheme for determining the force that presses the pipeline against the wall of the well in the process of pulling: *BB'* - infinitesimal element of the pipeline; dN_n – normal component of the force of pressing the pipeline against the wall of the well; *F* – force of pipeline tension; *dF* – force of tension of an infinitesimal pipeline element; da_s – central angle of the profile section.

If one considers the equilibrium of an infinitesimal element of pipeline *BB'* limited by central angle da_s, it is possible to obtain a component of total pressing force *dN* from pulling force *dF*:

$$dN_n = 2F\sin\frac{d\alpha_s}{2} + dF\sin\frac{d\alpha_s}{2}. \tag{2}$$

We discard infinitesimals of higher orders:

$$dN_n = Fd\alpha_s. \tag{3}$$

When deriving the equation for the component from the tension force of the pipeline without taking into account other forces of resistance, the solution is reduced to the Euler formula that is used in a number of standard documents for the design of trenchless pipeline passages by the method of horizontally directed drilling. With this approach, there is no need to divide the well profile into sections of different curvatures, as well as calculate for each section the component of the resistance force from the well curvature [8]. This approach only complicates calculations and introduces significant errors.

The following computations can serve as the basis for this provision.

Figure 2 shows a well of length L_s with the entry angle of the pipeline when pulling α_1 at point A and the exit angle to the day surface of α_2 at point C. The day surface is represented by straight line O_1O_2. Dotted lines indicate the tangents to the well profile at the points from which the radii of curvature are restored. Let us assume that the curvature of the well profile in the accepted coordinates is positive and does not change sign throughout its length.

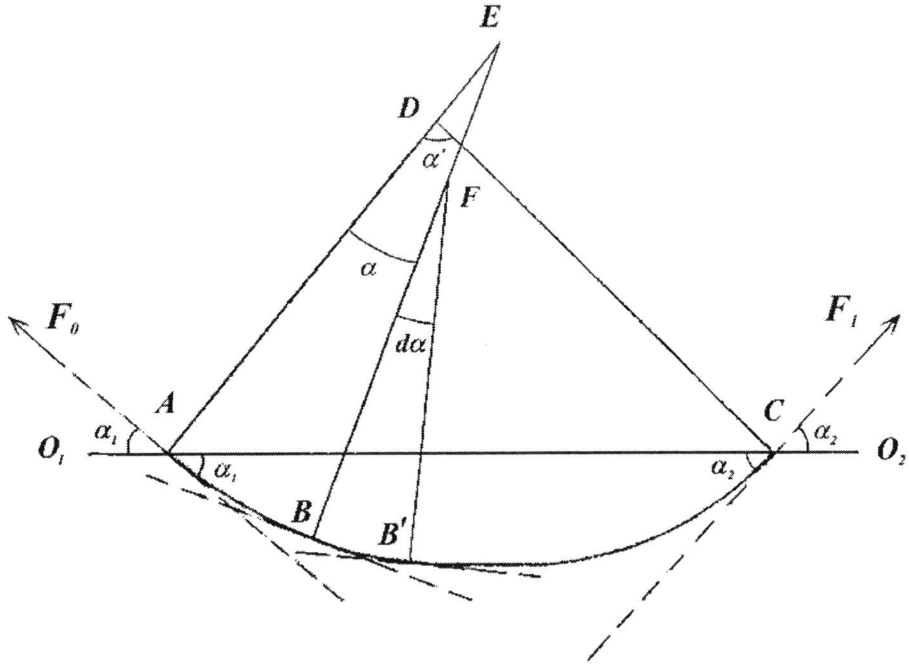

Figure 2. A design scheme for determining the component of the resistance force from the well curvature with a variable curvature radius: $\boldsymbol{F_1}$ – pipeline pulling force, $\boldsymbol{F_0}$ – pulling resistance force; O_1O_2 – earth's surface, A, C - entry and exit points of the pipeline.

At a well section, let us fix point B, the position of which is determined by current angle a between the radii of curvature recovered from the pipeline entry point A and point B. Moreover, well section AB can have variable curvature. In Figure 1, element BB' of the pipeline is presented in an enlarged form. The forces acting on element BB' of the pipeline corresponding to angle da are determined for point B as:

$$F = F(\alpha) \tag{4}$$

and for point B' as:

$$F + dF = F(\alpha + d\alpha). \tag{5}$$

Both forces are directed tangentially to the well trajectory. The normal force of pressure on wellbore roof $\boldsymbol{N}$ is defined as the total parallelogram force built on forces $\boldsymbol{F}$ and $\boldsymbol{F+dF}$ neglecting the change in the radius of curvature and the displacement of the center of curvature.

Since determining angle da is not connected with the change in the position of the center and the radius of curvature, the corresponding changes in the resultant force can be neglected, considering the parallelogram as a rhombus. Discarding infinitesimals of a higher order, let us obtain equation (3).

As the friction force of the pipeline against the wellbore roof is proportional to the normal force, one obtains:

$$dR = \mu \cdot dN_n = \mu \cdot F \cdot d\alpha. \tag{6}$$

where μ - coefficient of friction of the pipeline against the wall of the well.

Friction counteracts the beginning movement of the pipe, so that force *dR* together with pulling force *F* at point *B* must counterbalance force *F+dF* at point *B'*, whence:

$$dF = \mu \cdot F \cdot d\alpha . \quad (7)$$

Solving this differential equation for the entire borehole profile, which is determined by central angle *a'*, one obtains the well-known Euler equation:

$$F_0 = F_1 e^{-\mu\alpha'} . \quad (8)$$

Returning to Figure 2, let us note that from a simple geometric calculation, angle *a'* turns out to be equal to the sum of angles a_1 and a_2.

In actual pipeline laying conditions, especially in urban areas, the well profile can have a volumetric curvature. In this case, an additional component appears from the curvature of the profile in the horizontal plane.

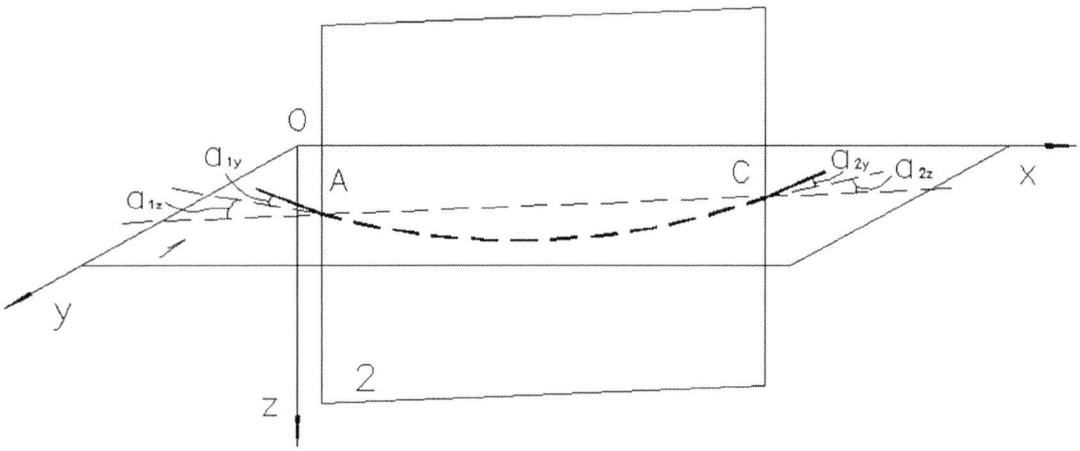

Figure 3. A design scheme of a well curved in two planes: *A, C* - entry and exit points of the well, *1* - plane of the earth's surface; 2 - plane perpendicular to the earth's surface; a_{1y}, a_{1z}, a_{2y}, a_{2z} – angles of entry and exit of the well relative to the corresponding planes.

Figure 3 shows a volumetrically curved well of length L_s. Straight line *AC* travels through the entry and exit points of the pipeline to the day surface, which is represented by plane *1*. Plane *2* intersects plane *1* along straight line *AC* and perpendicular to the day surface. At point *A*, the angle of the pipeline entry into the well with respect to plane *1* is denoted by a_{1y}, and relative to plane *2* - by a_{1z}, and at point C, the exit angle of the pipeline relative to *1* and *2* will be denoted by a_{2y} and a_{2z}, respectively.

In this case, the side force will have two components: dN_z - in plane *1* and dN_y - in plane *2*. They will give two components of the resistance forces, respectively, dF_z and dF_y, acting along the borehole axis. Adding these two components and solving the resulting differential equation, one obtains equation (8), in which:

$$\alpha' = \alpha_{1y} + \alpha_{2y} + \alpha_{1z} + \alpha_{2z} \quad (9)$$

where a_{1y}, a_{1z}, a_{2y}, a_{2z} – angles of entry and exit of the well relative to the corresponding planes.

Therefore, the volumetrically curved well can also be calculated from the central angle of the well profile.

Angle *a'*, which defines F_{cur} in (8), when the well is curved only in the vertical plane (the plane problem), is numerically equal to the sum of angles a_{1y} and a_{2y}. When the borehole is also curved in the horizontal plane, the sum of angles a_{1z} and a_{2z} is added to angle *a'*.

In fact, it is not always possible to drill a borehole section along one radius of curvature, i.e. ensure the same angle of deviation between drill pipes [9]. In particular, this can happen due to the fact that in order to strictly ensure the design radius of curvature in this section, it will be necessary to deflect drill pipes with respect to each other by fractional angles (in degrees), which cannot be realized in practice, because the drill head navigation system and the rig itself can only work with angles that are multiples of one degree [10]. The solutions obtained earlier make it possible to calculate the pulling force for well profiles with a variable radius of curvature, i.e. at different angles of deviation between the drill pipes along the well profile.

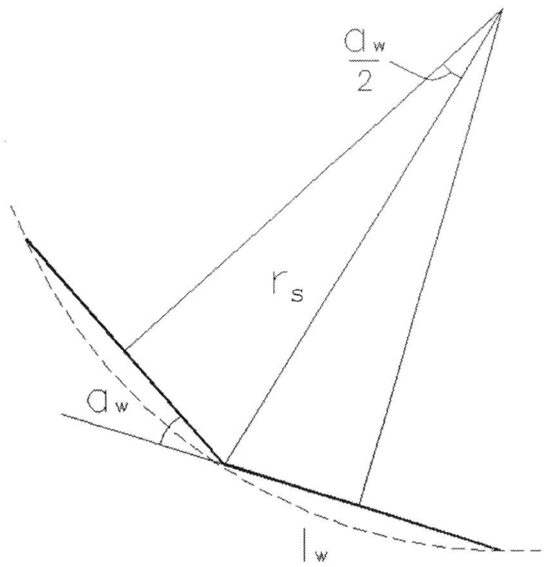

Figure 4. Finding the radius of curvature of the section of the profile from the angle of deviation of the drill pipes relative to each other in the general case: a_w - angle of deviation; l_w - drill pipe length; r_s - radius of curvature of the examined section of the well.

Figure 4 depicts a curved section of the well profile (shown in dotted lines), equal in length to two drill pipes. Perpendiculars are recovered from the centers of the drill pipes, the intersection of which gives the central angle of the profile section on the current drill pipe [11]. Radius r_s of curvature of the well profile in this section will depend on length l_w of the drill pipe and deviation angle a_w of this drill pipe, relative to the previous one.

Thus, if one uses the value of the current radius of curvature in this section when searching for the resistance force, one may find the pulling force in a section of one drill pipe length [12]. When summing these forces along the entire length of the profile, one gets a total pulling force for the entire profile. In this calculation, provided that a_w is zero, i.e. in a rectilinear section, it is necessary to neglect the curvature of the section in which this drill pipe is located.

3. Conclusion

Thus, the solutions obtained make it possible to draw the following conclusions:

1. When calculating the component of the frictional force from the curvature of the borehole, the effect of the sections having zero curvature can be neglected. In this case, this conclusion applies only to pipelines that have zero buoyancy [13]. With buoyancy other than zero, friction in curvilinear sections is calculated taking into account the difference between the normal components of the buoyancy force and weight.
2. With zero buoyancy of the pipeline, the total resistance to pulling, excluding the pulling forces from the viscosity of the drilling mud, is independent of the pipe length and is determined by the angle equal to the sum of the entry angle and the exit angle of the pipeline to the day surface.

Thus, finding the component of the resistance force from the curvature of the well can be reduced to finding the force of resistance according to the Euler formula (for the entire profile at once) using the angle numerically equal to the sum of the entry and exit angles of the pipeline.

Such calculation can be carried out for a profile of a well of any shape, since before drilling a well in this profile, a table is always compiled, in one form or another, indicating the deviation angles for each drill pipe relative to the previous one [14].

In addition, if the well is curved in several planes, one can calculate this volumetric curved well along the central angle of the well profile, taking into account the angles of the well entry and exit relative to the corresponding planes [15].

References

[1] Zemenkova M Yu, Shipovalov A N, Dudin S M and Zemenkov Yu D 2007 System analysis in oil and gas objects control processes *News of higher school. oil and gas* **5** 116-9

[2] Zemenkova M, Shalay V, Zemenkov Y, Kurushina E 2016 Improving the Efficiency of Administrative Decision-Making when Monitoring Reliability and Safety of Oil and Gas Equipment *MATEC Web of Conferences* **73** DOI: 10.1051/matecconf/20167307001

[3] Toropov V S, Ponomareva T G, Toropov S Yu, Toropov E S 2016 Quantitative assessment of changes in the pipeline - well system in the event of an emergency situation in horizontal directional drilling. *News of higher educational institutions. Oil and gas* **5** 71-77

[4] Zemenkov Yu D 2008 *Technique and technologic processes at transport of energy resources* **1** 380

[5] Kiryanov I E, Toropov V S and Toropov S Yu 2015 Determination of dependences of pipeline movement resistance forces in curve borehole from physical parameters and rheological model of using mud *Oil equipment* **8**

[6] Standard of the National Association of Builders STO NOSTROY 2.27.17 - 2011 "Development of underground space. Laying of engineering underground communications by the method of horizontal directional drilling". Official Edition

[7] Gorin A S, Spectrum Yu I, Kraskov V A 2003 *Pipeline Transportation of Oil Journal* **11**

[8] Willoughby D 2005 *Horizontal Directional Drilling Utility and Pipeline Applications* (New York; The McGraw-Hill Companies)

[9] Toropov V S, Tamer O S , Toropov S Yu and Nikiforov N A 2016 Dynamics of pipeline pulling process by horizontal directional drilling. International Scientific-Practical Conference of Students, Graduate Students and Young Scientists *"Transport and Storage of Hydrocarbons"*, **154**

[10] Toropov V S, Toropov E S, Podorozhnikov S Yu, Seroshtanov I V 2015 The use of additional equipment in horizontal directional drilling technology. *Mining information and analytical bulletin* **36** 36-44

[11] Pisarenko G S, Yakovlev A P and Matveev V V 1988 *Materials handbook* (Kiev; Naukova Dumka) 736

[12] Dudin S, Bahmat G, Zemenkov Y, Voronin K, Shipovalov A 2017 Strategy for monitoring and ensuring safe operation of Russian gas transportation systems *MATEC Web of Conferences* **106** DOI: 10.1051/matecconf/201710606004

[13] Shalay V, Zemenkova M, Zemenkov Y, et al. 2016 Modeling Parameters of Reliability of Technological Processes of Hydrocarbon Pipeline Transportation. *MATEC Web of Conferences* **73** 01029 DOI: 10.1051/matecconf/20167301029

[14] Hammerhead Mole pneumoshock machine usage as auxiliary equipment in horizontal directional drilling. Witch Company Information Center, http://www.ditchwitch.ru/articles/ispolzovanie - pnevmoudarnoi - ustanovki - hammerhead-mole – v – kachestve – vspomogatelnogo – oborudovaniya – v - gorizontalnom-napravlennom-burenii-39.htm#

[15] Zemenkov Y D, Shalay V V, Zemenkova M Y 2015 Expert systems of multivariable predictive control of oil and gas facilities *reliability Oil and Gas Engineering Procedia Engineering* **113** 312-315 DOI: 10.1016/j.proeng.2015.07.271

Dependence of energy characteristics of ascending swirling air flow on velocity of vertical blowing

R E Volkov[1], A G Obukhov[2] and V N Kutrunov [1]

[1] Tyumen State University, 6 Volodarskogo St., Tyumen, 625003, Russia
[2] Industrial University of Tyumen, 38 Volodarskogo St., Tyumen, 625000 Russia

E-mail:email@romanvolkov.ru, aobukhov@tsogu.ru, kvnkvnkvn@rambler.ru

Abstract. In the model of a compressible continuous medium, for the complete Navier-Stokes system of equations, an initial boundary problem is proposed that corresponds to the conducted and planned experiments and describes complex three-dimensional flows of a viscous compressible heat-conducting gas in ascending swirling flows that are initiated by a vertical cold blowing. Using parallelization methods, three-dimensional nonstationary flows of a polytropic viscous compressible heat-conducting gas are constructed numerically in different scaled ascending swirling flows under the condition when gravity and Coriolis forces act. With the help of explicit difference schemes and the proposed initial boundary conditions, approximate solutions of the complete system of Navier-Stokes equations are constructed as well as the velocity and energy characteristics of three-dimensional nonstationary gas flows in ascending swirling flows are determined.

1. Introduction

The proposed gas model, as a moving compressible continuous medium possessing dissipative properties of viscosity and thermal conductivity [1], in numerical simulation of complex swirling air flows arising during cold vertical blowing [2], gives the main gas dynamic characteristics that coincide with the results of field experiments [3]. Comparison of calculated numerical values of geometric characteristics and kinetic energies [4] led to the conclusion that the energy of rotational motion obtained from the energy of the Earth's rotation around its axis will exceed the half of the kinetic energy of the entire flow only for the diameter of the vertical part of the flow of not less than 5 meters and blowing velocity not less than 15 m/s. In this case, the main contribution of 97% to the total kinetic energy of the ascending swirling flow is given by the kinetic energy of rotational motion. It was this part of the energy that was obtained by an ascending swirling flow from the energy of the Earth's rotation.

In this paper, the developed method for parallelizing the numerical algorithm for solving the complete system of Navier-Stokes equations is used to numerically study the dependences of the energy characteristics of an ascending swirling air flow on the velocity of a vertical blowing. In particular, the aim of the paper is to establish by the numerical experiments the character of the change in the velocity and energy characteristics of the formed swirling air flow at a sudden step-like decrease in the velocity of the vertical blowing.

2. Materials and methods

To describe complex nonstationary three-dimensional flows of a compressible continuous medium with dissipative properties of viscosity and thermal conductivity, the paper uses the complete system of Navier-Stokes equations, which, being written in dimensionless variables taking into account the action of gravity and Coriolis forces in vector form, has the following form [4-10]:

$$\begin{cases} \rho_t + \vec{V}\cdot\nabla\rho + \rho\, div\,\vec{V} = 0, \\ \vec{V}_t + \left(\vec{V}\cdot\nabla\right)\vec{V} + \frac{T}{\gamma\rho}\nabla\rho + \frac{1}{\gamma}\nabla T = \vec{g} - 2\vec{\Omega}\times\vec{V} + \frac{\mu_0}{\rho}\left[\frac{1}{4}\nabla\left(div\,\vec{V}\right) + \frac{3}{4}\Delta\vec{V}\right], \\ T_t + \vec{V}\cdot\nabla T + (\gamma-1)\,T div\,\vec{V} = \frac{\kappa_0}{\rho}\Delta T + \frac{\mu_0\gamma(\gamma-1)}{2\rho}\left\{\left[\left(u_x - v_y\right)^2 + \right.\right. \\ \left.\left. + \left(u_x - w_z\right)^2 + \left(v_y - w_z\right)^2\right] + \frac{3}{2}\left[\left(u_y + v_x\right)^2 + \left(u_z + w_x\right)^2 + \left(v_z + w_y\right)^2\right]\right\}, \end{cases} \tag{1}$$

where the values of dimensionless coefficients of viscosity and thermal conductivity are as follows: $\mu_0 = 0.001$, $\kappa_0 \approx 1.46\mu_0$.

In system (1): t is time; x, y, z are Cartesian coordinates; ρ is gas density; $\vec{V} = (u, v, w)$ is gas velocity vector with projections onto the corresponding Cartesian axes; T is gas temperature; $\vec{g} = (0, 0, -g)$ is acceleration vector of force of gravity, $g = const > 0$; $-2\vec{\Omega}\times\vec{V} = (av - bw, -au, bu)$ is Coriolis forces' acceleration vector, $a = 2\Omega\, sin\,\psi$, $b = 2\Omega\, cos\psi$, $\Omega = \left|\vec{\Omega}\right|$; $\vec{\Omega}$ is angular velocity vector of Earth's rotation; ψ is latitude of point; O is the beginning of the Cartesian coordinate system rotating together with the Earth.

As initial conditions for describing the corresponding flows of a compressible viscous heat-conducting gas in the case of constant values of the coefficients of viscosity and thermal conductivity, functions determining the exact solution of [11] of the system (1) are taken as follows:

$$u = 0, \quad v = 0, \quad w = 0, \quad T_0(z) = 1 - kz, \quad \rho_0(z) = (1 - kz)^{\nu-1}, \tag{2}$$

where

$$k = \frac{lx_{00}}{T_{00}}, \quad l = 0.0065\ K/m, \quad x_{00} = 50m, \quad T_{00} = 288^o K, \quad \nu = \frac{\gamma g}{k} = const > 0. \tag{3}$$

Numerical experiments were carried out in the computational domain in the form of a rectangular parallelepiped (Fig. 1), which side lengths in a dimensionless form are 1 x 1 x 0.04 (in a dimension form they are 50m x 50m x 2m). Blowing the gas through a vertical pipe is modeled by setting vertical gas flow rate w as a time function in the form:

$$w(t) = M\cdot\left[1 - exp(-10t)\right], \tag{4}$$

through a square hole of size 0.1×0.1 (in dimensional form 5m x 5m) in the center of the upper bound of the computational domain; M is the maximum blowing velocity.

For the density on all six bounds of a parallelepiped, there is the condition of the flow continuity [12]: $x = 0$, $x = x^0$, $y = 0$, $y = y^0$, $z = 0$, $z = z^0$. The boundary conditions for the components of the gas velocity vector correspond to the non-flow conditions for the normal component of the velocity vector, and the symmetry conditions for the other two components of the velocity vector. For the temperature on all six faces, the thermal insulation conditions are set [12].

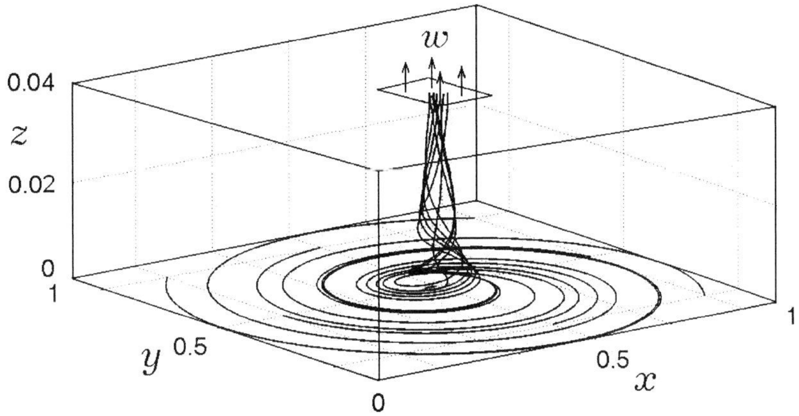

Figure 1. Computational domain

Calculations were carried out with the following input parameters: scale size values of density, velocity, distance and time were equal, respectively

$$\rho_{00} = 1.29\ kg/m^3,\quad u_{00} = 333\ m/s,\quad x_{00} = 50\ m,\quad t_{00} = x_{00} / u_{00} = 0.15\ s.$$

Difference steps in three spatial variables are $\Delta x = \Delta y = 0.005$ (dimensional value 0.25 m), $\Delta z = 0.004$ (dimensional value 0.2 m), and the time step was $\Delta t = 0.001$ (dimensional value 0.00015 s).

3. Results of calculations

The essence of the numerical experiments carried out is reduced to the following. Each separate calculation was started with the simulation of the gradual acceleration of the emerging ascending swirling flow with vertical air blowing through the upper opening at constant velocity $w = 20$ m/s.

With the passage of time, the airflow velocity in computational domain D, and, consequently, its total kinetic energy:

$$W = \frac{1}{2}\iiint_D \rho\left(u^2 + v^2 + w^2\right)dxdydz \approx \frac{1}{2}\sum_i\sum_j\sum_k \rho_{i,j,k}\left(u_{i,j,k}^2 + v_{i,j,k}^2 + w_{i,j,k}^2\right)\Delta x\Delta y\Delta z. \quad (5)$$

At the same time, the velocities of the circular (rotational) motion of the swirling flow also grows, as a result of which the kinetic energy of rotational motion increases

$$W_\varphi = \frac{1}{2}\iiint_D \rho\, v_\varphi^2 r dr d\varphi\, dz \approx \frac{1}{2}\sum_i\sum_j\sum_k \rho_{i,j,k}\, v_{\varphi i,j,k}^2\, r\,\Delta r\Delta\varphi\,\Delta z. \quad (6)$$

The increase in the velocities and energies of the ascending swirling flow occurs until it reaches the steady mode of its operation, in which all thermodynamic, velocity and energy characteristics stop to change. Since the parameters of all ascending swirling air flows, for which this series of computational experiments were carried out, were the same, the time to enter the steady mode was the same for them.

After reaching the steady-state regime, a different continuation of the numerical experiment was chosen for each variant of the calculation. Namely at the same time, the value of the blowing velocity suddenly decreased to a lower value, in comparison with the initial one. A further description of the computational experiments is devoted to the results of the changes in the functioning of the ascending swirling flow after such jump in the blowing velocity in each of the calculation options.

Fig. 2 shows the dependency plots of the total kinetic energy of the ascending swirling airflow versus time.

On the left-hand side of the figure, the plots of changing the total kinetic energy of the flow coinciding with each other, right up to reaching the steady mode (point $\dot{A}$ on the plots) at a dimensionless vertical blowing velocity of 0.06 (dimensional value $w = 20$ m/s). In the steady mode (point $\grave{A}$), the maximum value of the total kinetic energy is 0.001417 (dimensional value 25392640 J = 25.39 MJ), and the time to enter the steady mode is 17.5 minutes.

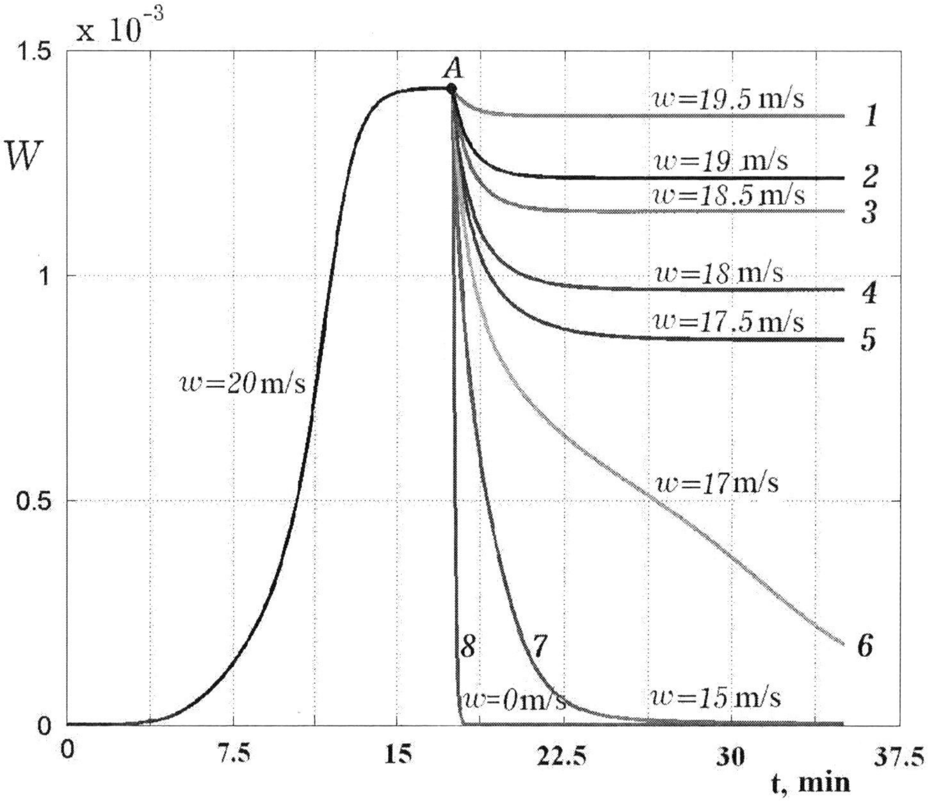

Figure 2. Dependences of the kinetic energy of a swirling flow on time

The right-hand side of the figure shows the dependency plots of the total kinetic energy on time after an step-like decrease in the velocity of the vertical blowing at point $\grave{A}$ up to velocities: $w_1 = 19.5$ m/s – plot 1; $w_2 = 19$ m/s – plot 2; $w_3 = 18.5$ m/s – plot 3; $w_4 = 18$ m/s – plot 4; $w_5 = 17.5$ m/s – plot 5; $w_6 = 17$ m/s – plot 6; $w_7 = 15$ m/s – plot 7; $w_8 = 0$ m/s – plot 8.

Plot 1 shows a gradual decrease in the kinetic energy of the flow and its entering the steady mode up to value 0.001354 (dimensional 24.26 MJ) during the time of $t_1 = 4.37$ minutes.

Plot 2 reflects a gradual decrease in the kinetic energy of the flow and its entering the steady mode to value 0.001217 (dimension value 21.81 MJ) during the time of $t_2 = 6.25$ minutes.

Plot 3 represents the decrease in the kinetic energy of the flow and its entering the steady mode to value 0.001142 (dimensional value 20.46 MJ) during the time of $t_3 = 8.51$ minutes.

Plot 4 shows a gradual decrease in the kinetic energy of the flow and its entering the steady mode to 0.0009686 (dimensional value 17.36 MJ) during the time of $t_4 = 11.25$ minutes.

Plot 5 reflects a gradual decrease in the kinetic energy of the flow and its entering the steady mode to value 0.0008578 (dimensional value 15.37 MJ) during the time of $t_5 = 13.75$ minutes.

Plot 6 represents a gradual decrease in the kinetic energy of the flow without its entering the stationary mode. Extrapolation of the dependence of the total kinetic energy on time shown by the plot makes it possible to estimate the time until the complete flow stop, which is $t_6 = 35$ minutes for a given blowing velocity.

Graph 7 visualizes the process of reducing the total kinetic energy of the flow to zero value, that is, the process of a complete flow stop occurs during time $t_7 = 12.5$ minute.

And, finally, plot 8 shows the process of sufficiently rapid reduction of the flow kinetic energy to zero value, or the flow moving stop during the time $t_8 = 3.5$ minute.

Further analysis of the calculation results presented in the figure can be carried out for two different ranges for reducing the blowing velocity. With a step-like decrease in the blowing velocity in the range from 20 m/s to 17.5 m/s, the swirling air flow does not collapse, but only smoothly enters the corresponding steady mode with a reduced value of the velocity characteristics, and, therefore, with a smaller margin of total kinetic energy. Moreover, with a decrease in the blowing velocity in this range, the time to reach the corresponding stationary mode increases almost linearly from 4.37 minutes to 13.75 minutes.

With regard to the step-like decrease in the blowing velocity in the range from 17 m/s to 0 m/s, it follows from the calculations that with such changes in the blowing velocity, the swirling air flow does not enter the steady mode, but gradually stops. At the same time, the velocity, and, consequently, the energy characteristics are smoothly reduced to zero values. The decay time of the flow reduces with a decrease in the blowing velocity from 35 minutes to 3.5 minutes.

A special case is a sharp decrease in the blowing velocity from 20 m/s to 0 m/s (plot 8). In fact, such numerical experiment corresponds to a sharp termination of the blowing velocity. The boundary conditions set in the numerical experiment do not allow air to escape from the computational domain. Therefore, the swirling flow quickly stops to exist. This numerical experiment simulates the destruction of a tornado by stopping the vertical gas flow.

4. Conclusion

Detailed three-dimensional nonstationary calculations simulated the flows arising in experiments on blowing air in the pipe from the bottom up. Numerical calculations of these flows set the boundary for the velocity, geometric and energy characteristics of the flow, when the Earth rotation begins to introduce an essential additive into the kinetic energy of the flow. Calculations of nonstationary three-dimensional flows made it possible to determine the time for these flows to enter the steady mode, taking into account energy characteristics. This justifies the obvious idea from the physics point of view: energy in tornadoes and tropical cyclones does not arise immediately and from nowhere. It accumulates over time from an understandable and well-grounded source – the kinetic energy of the Earth rotation around its axis.

References

[1] Abdubakova L V, Obukhov A G 2014 Numerical calculation of the velocity characteristics of a three-dimensional ascending swirling gas flow *Izvestiya Vuzov. Oil and gas* **3** 88-94

[2] Volkov R E, Obukhov A G 2016 Parallel computations in the studies of the dependence of the gas dynamic parameters of an ascending swirling gas flow on the blowing velocity *Izvestiya Vuzov. Oil and gas* **1** 92-97

[3] Volkov R E, Obukhov A G 2016 Method of parallelization of the algorithm for numerical solution of the complete system of Navier-Stokes equations *Izvestiya Vuzov. Oil and gas* **2** 92-98

[4] Bautin S P, Deryabin S L, Krutova I Yu, Obukhov A G 2017 *Destructive atmospheric vortices and Earth rotation around its axis* (Ekaterinburg)

[5] Bautin S P 1987 Representation of solutions of the Navier-Stokes equations system in the neighborhood of a contact characteristic *Applied Mathematics and Mechanics* **51** 574-584

[6] Bautin S P, Obukhov A G 2012 *Mathematical modeling of destructive atmospheric eddies* (Novosibirsk)

[7] Bautin S P, Obukhov A G 2012 Mathematical Modeling and Numerical Calculation of Flows in the Bottom of the Tropical Cyclone. *Bulletin of the Tyumen State University. Physics and mathematics. Computer science* **4** 175-183
[8] Obukhov A G 2012 Mathematical modeling and numerical calculations of flows in the bottom of the tornado *Bulletin of the Tyumen State University. Physics and mathematics. Computer science* **4** 183-189
[9] Bautin S P 2013 Obukhov A G Mathematical modeling of the bottom part of the ascending swirling flow *Thermophysics of high temperatures* **51** 567-570
[10] Bautin S P, Krutova I Yu, Obuhov A G, Bautin K V, 2013 *Destructive atmospheric vortices: theorems, calculations and experiments* (Novosibirsk; Ekaterinburg)
[11] Bautin S P, Obukhov A G 2013 One exact stationary solution of the equations system of gas dynamics *Izvestiya Vuzov. Oil and gas* **4** 81-86
[12] Bautin S P, Obukhov A G 2013 On one kind of boundary conditions in the calculation of three-dimensional unsteady flows of a compressible viscous heat-conducting gas *Izvestiya Vuzov. Oil and gas* **5** 55-63

IOP Conf. Series: Materials Science and Engineering **357** (2018) 012019 doi:10.1088/1757-899X/357/1/012019

Normalization of hydrocarbon emissions in Germany

R E Levitin

Industrial University of Tyumen, 38 Volodarskogo St., Tyumen, 625000, Russia

E-mail: 89028130230@mail.ru

Abstract. In connection with the integration of the Russian Federation into the European space, many technical regulations and methodologies are being corrected. The work deals with the German legislation in the field of determining of hydrocarbon emissions and the methodology for determining the emissions of oil products from vertical steel tanks. In German law, the Emission Protection Act establishes only basic requirements. Mainly technical details, which have importance for practice, are regulated in numerous *Orders on the Procedure for the Implementation of the Law* (German abbr. – *BimSchV*). Documents referred to by *the Technical Manual on the Maintenance of Clean Air* are a step below on the hierarchical ladder of legislative and regulatory documentation. This set of documents is represented by numerous DIN standards and VDI guidelines. The article considers the methodology from the guidance document VDI 3479. The shortcomings and problems of applying the given method in Russia are shown.

1. Introduction

The basis of environmental legislation of Germany is the Federal Law on Protection from Emissions (German *Bundes-Immissionsschutzgesetz*), adopted in 1974, which full name is the Law on Protection from Harmful Environmental Effects due to Air Pollution, Noise, Vibration and Similar Processes. The law was developed with the influence of its predecessor – the Clean Air Act, adopted in the United States in 1970. The Law on Protection against Emissions itself establishes only basic requirements. The technical details, which are of great importance for practice, are mainly regulated in numerous ***Resolutions on the Procedure for Enforcement of the Law*** (*German:* BImSchV), two of which, related to hydrocarbon emissions, can be singled out in relation to the regulation of hydrocarbon emissions from vertical steel tanks:

- **20. BImSchV** – *The regulation on the limitation of emissions of volatile organic compounds during transshipment and storage of gasoline* is directed to equipment used for storage, filling and empting operations and transport of gasoline. The main objective of the Resolution is to reduce emissions through the introduction of vapor recovery systems (Gaspendelsysteme). This resolution does not apply to refueling vehicles at gas stations. The specified area is regulated by the Resolution 21. BImSchV.
- **21. BImSchV** – *The regulation on restriction of hydrocarbon emissions when refueling the motor vehicles* prescribes the equipping of fuel dispensers with special devices (Gasrückführungseinrichtungen), returning gasoline vapor generated during the refueling operation, back to the filling station tank, and automatic control devices that interrupt the fuel supply if the vapor recovery system is not rectified within 72 hours.

IOP Conf. Series: Materials Science and Engineering **357** (2018) 012019 doi:10.1088/1757-899X/357/1/012019

The lower level of documentation is the level of normative and technical acts containing specific technical requirements for the implementation of the requirements of the above-mentioned Federal Law on Protection against Emissions. It is presented in the *Technical Manual on Air Cleanliness* (*German*: *Technische Anleitungzur Reinhaltungder Luft*[1], *abbr.*: ***TA Luft***). The first version of this Manual appeared in 1964 even before the adoption of the Federal Law on Protection against Emissions, and then the document was revised in 1974, 1983, 1988 and 2002. The technical regulation performs a low concretizing function, explaining the general and vague concepts that are used in the above mentioned Law, prescribing specific requirements for emissions, their measurements, etc. In judicial practice, guidance is recognized as being compulsory performed in most cases.

It should be noted that for liquid organic compounds, guidance ***TA Luft*** regulates the use of vertical tanks with a fixed roof, connected to a vapor collection system or to an installation for the purification of emissions. However, if the volume of the tank is more than 20,000 m^3 and if it is equipped with a floating roof and an effective sealing gate or pontoon under a fixed roof, such tanks are allowed to be used for oil storage provided that the reduction of emissions is at least 97 percent compared to similar tanks not equipped with a pontoon stationary roof.

A step below on the hierarchical ladder of legislative and regulatory documentation, the documents are referred to by the *Technical Manual on the Maintenance of Clean Air*. This set of documents is represented by numerous DIN standards and VDI guidance documents, integrated into the constantly updated and updated VDI/DIN Standards Guide for Air Purity (***VDI/DIN-Handbuch „ReinhaltungderLuft"***).

The methodology for calculating the loss caused by evaporation discussed in this paper was taken from the **Regulation Document of the German Association of Engineers VDI**[2] **3479** *Monitoring the emissions from tank farms that are not located in oil processing areas* (German ***VDI-Richtlinie 3479. Emissionsminderung – Raffenerieferne Mineralöltanklager***).

2. Application area of VDI 3479

Guidance document VDI 3479 is applied to new tank farms remote from oil refinery sites where the following products are stored and transshipped:

- raw oil,
- motor gasolines,
- diesel fuel,

---

[1]Technical Manual (German *Technische Anleitung*) is a document, that is so-called the General Administrative Order (Allgemeine Verwaltungsvorschrift), aimed at ensuring the uniform application of regulations by relevant institutions. For example, the receipt or non-receipt of a permit which is regulated by the Federal Emission Control Act (German *Bundes-Immissionsschutzgesetz*) may depend on compliance with the requirements of ***TALuft.***

[2]**German Engineers Association (VDI e. V),** founded in 1856, is the largest association of engineers and research scientists in Germany. One of the goals of the Association is to represent the interests of its engineers and researchers. In addition, the German Engineers Association (VDI) is a member ASIIN e. V. (Akkreditierungsagentur für Studiengänge der Ingenieurwissenschaften, der Informatik, der Naturwissenschaften und der Mathematik e. V. – Agency for the accreditation of educational programs, preparing students for engineering specialties, informatics, natural science disciplines and mathematics) and takes part in the accreditation of educational programs for engineering specialties in German universities. The Association of German Engineers, as well as other norm-setting organizations, the most famous of which is DIN (Deutsches Institut für Normung – German Institute for Standardization) is a commercial structure. Its main activity is development of standards and other normative documentation. At the moment, there are more than 1,700 standards created by the Association of German Engineers, the so-called VDI Guidelines (VDI-Richtlinien) covering different fields of methods and technology.

- fuel oil,
- aviation fuel,
- kerosene,
- special solvent gasolines,
- additives,
- fuel components.

When verifying the possibility of using the measures specified in VDI 3479 to reduce losses from the evaporation of oil and petroleum products in relation to old tank farms, it is necessary to take into account the existing technical conditions, as well as the environmental conditions and the proportionality of the proposed activities to these factors.

The document refers to the current legislative and regulatory acts, orders and regulations in the field of construction and operation of tank farms. It is emphasized that the rules and regulations for safety in VDI 3479 are not included. It is also mentioned that for important security elements, according to 12. BImschV[3], in some cases additional measures can be required not described in the document in question.

All values of the gas volumes in VDI 3479 are given for the state under normal conditions (273 K, 1013 hPa) after the deduction of the water vapor content.

Guidance document VDI 3479 is not applied to oil tank farms in oil refineries (for this, see VDI 2440) or petrochemical plants. It also can not be applied to petrol stations and reserve stores for end users. This document is not applied to fuels with a predominant share of the biological component, such as E85 and biofuels for diesel engines. In addition, the document is not applied to the storage and transshipment of liquid gases.

3. Calculation procedure for VDI 3479

Guidance document VDI 3479 presents a methodology for calculating hydrocarbon emissions from tanks with fixed and floating roofs.

The methodology for calculating for tanks with a fixed roof was developed based on the data obtained during the joint project BMI[4] and DGMK[5] 4590-01 – 4590-12 "Measurement and determination of hydrocarbon emissions during storage, transshipment and transportation of fuels and verification of methods for regulating these emissions"[6].

VDI 3479 states that the total emissions of the tank farm are formed not only of evaporation losses from tanks, but also of emissions from a vapor recovery unit and emissions from dispersed sources, which in turn are formed from the sum of losses in pumps, seals and etc., as well as vapor leakage due to leaks (see VDI 3479, Table 8 - 10).

Further, in this review, attention will be focused solely on evaporation losses from tanks.

Average annual evaporation losses L_{Tank} *from tank*

Average annual losses, $L_{Tank,\,a}$(kg/year), are calculated taking into account certain factors assigned

---

[3]Zwölfte Verordnung zur Durchführung des Bundes-Immissionsschutzgesetzes(Störfall-Verordnung – **12. BImSchV**) vom 8. Juni 2005 (BGBlI, 2005, Nr. 33, S. 1599–1620) / № 12 Resolution No. 12 on the execution of the Federal Emission Act dated 8 June 2005.

[4]BMI (Bundesministerium des Inneren) – Federal Ministry for Interior of Germany.

[5]DGMK (Deutsche Wissenschaftliche Gesellschaft für Erdöl, Erdgas und Kohle e. V – Scientific Society of Germany on oil, gas and coal.

[6]BMI-DGMK Gemeinschaftsprojekt 4590-01 bis 4590-12 „Messen und Ermittlung von Kohlenwasserstoff-Emissionen bei Lagerung, Umschlag und Transport von Ottokraftstoffen und Prüfen von Verfahren zur Beherrschung dieser Emissionen", Teil 1: Zusammenfassender Bericht des Gesamtprojekts Teil 2: Berichte zu den Teilobjekten.

IOP Conf. Series: Materials Science and Engineering **357** (2018) 012019 doi:10.1088/1757-899X/357/1/012019

to the time period of one year.

The following formula is used for the calculation:

$$L_{Tank,a} = (1-\eta_{SD})\cdot((1-\eta_{VD})\cdot f\cdot L_{A,a} + L_{B,a}) \quad (1)$$

where:

η_{SD} – pontoon efficiency (minimum 0.95);

η_{VD} – efficiency of the breathing valve;

f – color ratio (is determined according to Table 7 of VDI 3479 - see Table 1 below);

$L_{A,a}$ - losses as a result of evacuation of the product from the reservoir and respiration (kg/year);

$L_{B,a}$ - losses when filling the tank (kg/year).

Table 1. Coating ratios

Color	Coating ratio
White	1.0
Aluminum silver	1.1
Light gray	1.3
Shingle gray	1.4
Mouse gray	1.6
Green	1.6

Evaporation losses from atmospheric (freely ventilated) tanks with a fixed roof

Average daily loss from breathing, $L_{A,d}$ (kg/day), is calculated using the following formula:

$$L_{A,d} = f_A \cdot c_n \cdot V_n, \quad (2)$$

f_A – saturation degree (this is the ratio of the concentration reached to the saturation concentration),

c_n - saturation concentration of hydrocarbons in the air-vapor space above the liquid surface (kg/m^3),

V_n – average daily volume of evaporation (m^3/day).

$$V_n = \frac{T_n}{p_n}\cdot\left(\frac{p}{T_1} - \frac{p}{T_2}\right)\cdot V_G \cdot \frac{1}{t}, \quad (3)$$

T_n – normal temperature in Kelvin (T_n = 273 K);

p_n – normal pressure (in hectopascals) (p_n = 1013 hPa);

p – ambient pressure (hPa);

T_1 – the average minimum temperature in the gas space (in Kelvin) (averaging time: from 1 to 11 hours, average values are determined by Table 2);

T_2 – average maximum temperature in the gas space (in Kelvin) (averaging time: from 11 to 16 hours, average values (determined by Table 2);

t – application time (t = 1 day);

V_G – the volume of gas space above the product in the tank (in m^3) is determined by the formula from Table 2, depending on the type of use of the tank (for transshipment or storage). The storage tank, according to VDI 3479, is a tank with a turnover frequency of about 1 per year. All other tanks are transshipment ones.

To calculate the volume of gas space above the product in the tank in Table 12 of the standard VDI 3479 (see Table 2 below), it is proposed to use the following formulas:

for a transshipment tank, m^3, $V_G = 0{,}5\cdot V$

for a storage tank, m^3,

$V_G = 0{,}075\cdot V$

where V – tank volume.

Average annual losses when breathing and pumping out the product from the reservoir,

$L_{A,\,a}$ *(kg/year).*

The calculation takes into account the summer and winter period and uses the following formula:

$$L_{A,a} = L_{A,S} \cdot d_S + L_{A,W} \cdot d_W \quad (4)$$

or:

$$L_{A,a} = f_{A,S} \cdot c_n \cdot V_{n,S} \cdot d_S + f_{A,W} \cdot c_n \cdot V_{n,W} \cdot d_W \quad (5)$$

where:

$L_{A,S}$ – average daily loss in breathing in summer (kg/year),

$L_{A,W}$ – average daily loss in breathing in winter (kg / year),

$f_{A,S}$ - the saturation degree in summer; it is determined by the formula from Table 2,

$f_{A,W}$ - the saturation degree in winter; it is determined by the formula from Table 2,

c_n – concentration of hydrocarbons saturation in the air-vapor space above the liquid surface (kg/m^3),

d_s – number of summer days in a year,

d_w – the number of winter days per year (in the note it is pointed out that the summer period is from May to September for the latitudes of Central Europe: d_s = 153 153 days per year, and the winter one is from October to April: d_w= 212 days per year)

$V_{n,S}$ – average daily volume of evaporations in summer (m^3/day), ·

$V_{n,W}$ – average daily volume of evaporations in winter (m^3/day).

Table 2. Data for tanks used for transshipment and storage.

	The average temperature of the product in the tank, in K		The temperature of the gas space, in K		Saturation degree f_A		Volume of gas space V_G, in m^3
	in summer	in winter	in summer	in winter	in summer $f_{A,S}$	in winter $f_{A,W}$	
Tank for transshipment	287		T1=286.5 T2=303.5[a)]	278 286	0.63	0.57	0.5*V[c)]
Storage tank	291	280	b)	b)	1	1	0.075*V[c)]

a) The average maximum temperature of the gas space measured in summer can be at the upper limit of the possible values, if it is taken into account that this temperature was determined at a distance of only approximately 1 m under the dome, with a gas space height of 5.7 m.

b) There are no data for storage tanks, therefore, the gas temperature values given for the transshipment tanks can be applied.

c) Tank volume is measured in m^3.

Note: Due to the fact that procedures for measuring the temperature of gas space T1 and T2, the average product temperature in tank T and the degree of saturation f_A are very labor-intensive, the averaged values based on real measurements can be used for approximate calculations. It should be noted that these values were obtained as a result of measurements on the fuel storage tank, the color ratio of which was 1.1. For tanks with a different color factor, it should be substituted in the formula for calculating evaporation L_A in relation to 1.1.

Evaporation losses from breathing and pumping from tanks with a fixed roof equipped with breathing valves

To determine the emissions from a fixed roof tank equipped with breathing valves, the losses for total breath L_A of a freely ventilated tank are multiplied by the coefficient of the breathing valves (1 – ηVD) (see formula (1)).

4. Conclusion

As can be seen from the description, the methodology for calculating oil losses from evaporation from tanks with fixed roofs from the Guideline Document of the German Engineers Association VDI 3479 (Germany), and the Russian methodologies equally include the main components of the total evaporation losses: "large" and "small breaths" (while in Germany they operate with the concepts: *loss at work, loss during storage*), but differently evaluate the importance of factors affecting the process of loss occurrence. First of all, this is reflected in the initial data, which are necessary for calculation.

In particular, the German technique operates with three basic formulas and offers most of the data relating to the product temperature, the gas space of the reservoir, the saturation degree of the gas space, etc., based on the data in the table, which gives the average values obtained on test tanks in climatic conditions of Germany. And if this can be permissible for Germany, and even, to some extent, for Europe, then, for example, for Russia such approach in determining the initial data due to the large extent of the territory and the significant difference in the climate between its individual parts would be absolutely unacceptable.

References

[1] Richtlinie 2010/75/EU des Europäischen Parlaments und des Rates vom 17. Dezember 2010 über Industrieemissionen (integrierte Vermeidung und Verminderung der Umweltverschmutzung).
[2] Richtlinie 2003/87/EG des Europäischen Parlaments und des Rates vom 13. Oktober 2003 über ein System für den Handel mit Treibhausgasemissionszertifikaten in der Gemeinschaft und zur Änderung der Richtlinie 96/61/EG des Rates-
[3] Risk Engineering Position Paper: Atmospheric Storage Tanks, Marsh, February, 2011. URL: http://uk.marsh.com/NewsInsights/Articles/ID/5015/Risk-Engineering-Position-Paper-Atmospheric-Storage-Tanks.aspx
[4] Zwölfte Verordnung zur Durchführung des Bundes-Immissionsschutzgesetzes (Störfall-Verordnung – 12. BImSchV) vom 8. Juni 2005 (BGBl I, 2005, Nr. 33, S. 1599–1620)
[5] VDI 1000. VDI-Richtlinienarbeit - Grundsätze und Anleitungen, 2010
[6] VDI 3479: Emissionsminderung. Raffinieriefeme Mineralöltanklager, 2010
[7] Nekrasov V O, Levitin R E, Tyrylgin I V, Zemenkov Yu D 2012 *Pat. 2500486 RU, IPC B05C7/04, B65D88/74. Device for improving the performance of vertical steel tanks/ (RU)-No. 2012125478/05*
[8] Levitin R E 2007 Change in the quality of oil products in the process of delivering them to consumer. *Electronic scientific journal Oil and gas.* **1**
[9] Levitin R E 2008 *Underground storage of petroleum products in horizontal steel tanks using inert gases* (Tyumen State Oil and Gas University)
[10] Lubin E A, Korshak A A 2010 Criterial equations of mass transfer in operations with oils in vertical cylindrical tanks *Electronic scientific journal Oil and gas* **2**

IOP Conf. Series: Materials Science and Engineering **357** (2018) 012020 doi:10.1088/1757-899X/357/1/012020

Effect of elemental composition on structure and corrosion resistance of electrolytic iron-based alloys

E V Koreshkova and A A Kulemina

Industrial University of Tyumen, 38 Volodarskogo St., Tyumen, 625000, Russia

E-mail: elenavk0276@mail.ru, newalena@mail.ru

Abstract. The article deals with the formation of electrolytic coatings taking into account the participation of basic components and elements of nonmetallic nature by the example of iron-phosphorus and iron-molybdenum alloys. The data of X-ray photoelectron spectroscopy of surface layers are presented, which confirm the presence of nonmetallic inclusions in the coatings under study. The influence of the structure being formed on the corrosion resistance of coatings is also studied.

1. Introduction

At present, the solution of the issues of saving metal associated with corrosion processes and wear of parts of oilfield equipment is often realized by changing the properties of the surface layers of products by applying functional metal coatings. Among the various coating methods, electrolytic precipitation of alloys is the most common one. This is due to the comparative simplicity of the deposition process, the availability of control and automation, the wide variation possibilities of the properties of deposited coatings.

To strengthen and restore worn surfaces of equipment parts, electrolytic iron-based alloys can be effectively used. Recently, special interest is caused by alloys, which tend to be amorphous in the course of electrodeposition. The formation of an amorphous-crystalline and amorphous structure, associated with the influence of amorphizing elements introduced into the alloy, leads to fundamental changes in the physicomechanical and operational properties. For the practical use of such coatings, it is necessary to have a clear understanding of the structure of electrolytic alloyed amorphous alloys, which is the link between the given deposition conditions and the properties obtained.

The aim of the work was to study the influence of nonmetallic impurities on the formation of the structure and process of amorphization of the coatings obtained.

2. Material and methods

The object of the study was electrolytic alloys of iron alloyed with molybdenum and phosphorus, since these elements are effective amorphizers. The elemental composition was determined by an X-ray photoelectron magnetic spectrometer. To research the distribution of elements along the depth of the sediment, etching of the surface with an argon ion beam was used; the rate of bleeding of the surface was 5 nm/min. The change in the content of an element was judged by the relative intensity of its lines of the X-ray electron spectrum, calculated with allowance for the relative sensitivity coefficient.

A quantitative gravimetric method was used to evaluate the corrosion resistance of coatings.

IOP Conf. Series: Materials Science and Engineering **357** (2018) 012020 doi:10.1088/1757-899X/357/1/012020

3. Results and discussion

In the process of electrodeposition of iron-phosphorus and iron-molybdenum alloys, not only the basic component but also the elements of a nonmetallic nature participate, which is confirmed by the conducted studies.

The formation of iron-phosphorus coatings occurs as a result of the simultaneous discharge of ions of iron, hydrogen and hypophosphite ions. [1].

According to X-ray photoelectron spectroscopy data, the surface layers of the coatings studied are enriched with impurity elements (oxygen) (Fig. 1, Table 1).

Table 1. Composition and structure of electrolytic iron-phosphorus alloys

Phosphorus content in the sediment, %	Oxygen content in the sediment, %	Structure	Perfection of texture [111], %
0	-	BCC	100
2	-	BCC	70
4	-	BCC	55
5	0,6	BCC	30
8	-	BCC	10
9	-	BCC	5
10	0,8	BCC +AP	-
14	-	BCC +AP	-
15	-	AP	-
18	-	AP	-
22	-	AP	-
24	1,5	AP	-

AF – amorphous phase

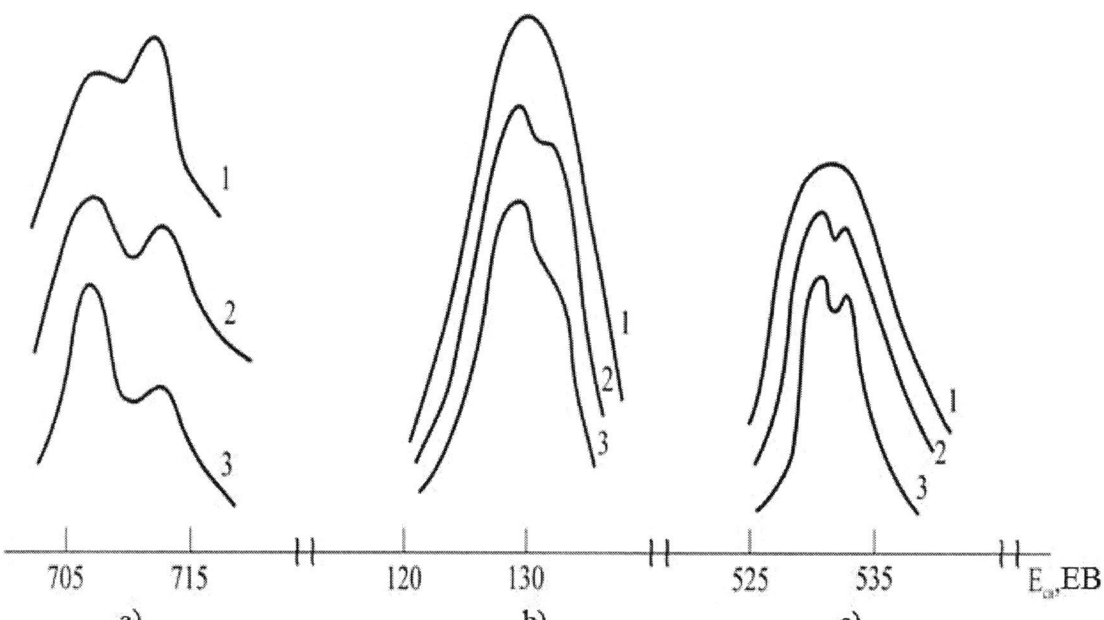

Figure 1. X-ray photoelectron spectra of Fe2p (a), P2p (b), O1s (c) amorphous (1), amorphous-crystalline (2) and crystalline (3) iron-phosphorus alloys: 1 - without ionic etching of the surface; 2, 3 - after 8 minutes of etching

As can be seen from spectrum Fe2p (Fig. 1a), the iron in the coatings is in the metallic state (binding energy 707.8 eV) and in the form of hydroxide compounds $Fe(OH)_2$ and $Fe(OH)_3$ (710.2 - 712.3 eV). Moreover, in amorphous alloys with a high content of phosphorus, iron predominates in the oxidized state.

For the $P2_p$ spectrum, two peaks with binding energies of 129.7 and 132.5 eV are observed on the surface of the coatings (Fig. 1, b). The first peak is in good agreement with the binding energy for red phosphorus (130.1 eV), and the second peak with similar values for phosphorus in hypophosphite (132.4 eV), phosphite (132.9 eV) and phosphate ions 132.1 eV) [2]. Taking into account the data on the adsorption mechanism, the main reaction proceeding on the catalytically active surface of sample $H_2PO_2^- + OH^- - e \rightarrow H_2PO_3^- + H$, it can be considered that the peak of oxidized phosphorus reflects the presence of hypophosphite and phosphite ions in the coatings. The state of phosphorus in the deeper layers of the sediment, in contrast to iron, almost does not change. All spectrograms, after the ion etching of the surface, fix peaks of oxidized and reduced form (Fig. 1, curves 2, 3).

Additional information on the composition of the surface layers was obtained by studying the spectra of the O1s level. For all the samples, one peak with a binding energy of electrons in the region of 530.5-531.5 eV is fixed in the authors' experiments (Fig. 1, c). The considerable width of this peak (3.5-3.8 eV) and, in some cases, its asymmetry allow one to assume the complex structure of this signal. The peak decomposition shows that it is the result of superposition of two peaks with an electron binding energy ~ 530,7±0,3 and 532,6±0,2 eV. The first peak is due to oxygen in the iron oxides, and the second one, with a higher binding energy, is due to oxygen in the oxide compounds of phosphorus (532.4 eV).

In the deposition of iron-molybdenum alloys, as in the case of iron-phosphorus alloys, three parallel processes also occur: the discharge of ions of iron, hydrogen, and molybdate ions.

However, in a number of cases, the discharge of molybdate ions occurs partially and a film is formed on the surface of the growing sediment containing the products of their incomplete reduction:

$$xMoO_4^- + 2xH_{ad} = Mo_xO_{2x} + 2xOH^-$$

Along with the molybdenum oxides in the solid-phase film, iron hydroxides occur in the near-cathode space due to the alkalization of the electrolyte. The solid-phase film blocks the surface of the growing sediment, and thus, it should reduce the deposition rate of the alloy. Indeed, as the concentration of molybdate ions increases, the rate of deposition of coatings decreases, their quality and appearance change.

As it is known [3], in the electrodeposition of alloys of the subgroup of iron with molybdenum, a film of insoluble oxide and hydroxide compounds of lower valence is formed on the surface of the growing sediment. Moreover, the resulting film, even at a significant thickness (about 50 nm), is permeable to metal ions of a subgroup of iron, although it reduces the deposition rate of coatings. The introduction of these ions into the film leads to a change in its composition and structure: the Mo-O bonds are weakened, the bonds between molybdenum and the metal of the iron subgroup are manifested, as a result of which the molybdate ion is reduced to a metal with the formation of an alloy. Oxygen-containing compounds present in the film (Mo_xO_{2x}, $Fe(OH)_x$) are also usually included in the resulting coatings, dramatically changing their structure and properties. Therefore, electrolytic alloys with molybdenum in their composition can contain a significant amount of oxygen.

The results of the experiment showed that in iron-molybdenum alloys, in addition to the base elements, there are impurity elements, oxygen and carbon with the highest concentration of nonmetallic inclusions found in the surface layer of sediments (Fig. 2, Table 2).

IOP Conf. Series: Materials Science and Engineering 357 (2018) 012020 doi:10.1088/1757-899X/357/1/012020

Table 2. Composition and structure of electrolytic iron-molybdenum alloys

Molybdenum content in the sediment, %	Oxygen content in the sediment, %	Carbon content in the sediment, %	Structure	Perfection of texture [111], %
0	-	-	BCC	100
2	-	-	BCC	80
5	-	-	BCC	70
7	0,8	0,5	BCC	60
10	-	-	BCC	30
13	-	-	BCC	10
15	-	-	BCC	5
17	-	-	BCC	2
18	2,0	1,6	BCC +AP	-
21	-	-	BCC +AP	-
22	-	-	AP	-
24	2,4	1,8	AP	-
27	-	-	AP	-
30	-	-	AP	-

AF – amorphous phase

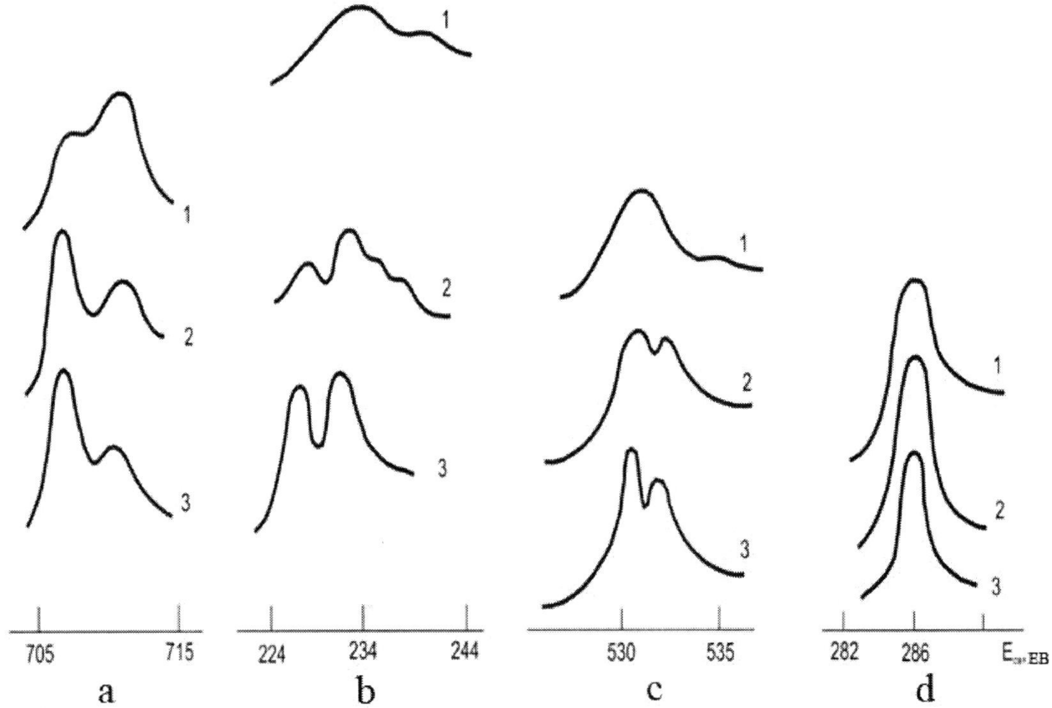

Figure 2. X-ray photoelectron spectra of Fe2p (a), Mo3d (b), O1s (c) and C1s (d) amorphous (1), amorphous-crystalline (2) and crystalline (3) iron-molybdenum alloys: 1 - without ion etching surface; 2, 3 - after 8 minutes of etching

As can be seen from the Fe2p spectrum (Fig. 2a), iron predominates in the form of Fe (II) and Fe (III) hydroxide compounds (~ 710.2 - 712.3 eV). One broad peak is fixed in the Mo3d spectra in the region of 227.8±234.1 eV (Fig. 2, b). The decomposition of this signal gives rise to the assertion that it is a superposition of several peaks conditioned by the presence of molybdenum in metallic (binding energy 228.0 eV) and oxidized (oxidation state from 2+ to 6+) states.

On the O1s spectrum, the main peak is observed (Fig. 2, c) in the area of 530.5 - 531.7 eV. The considerable width of the peak and its asymmetry indicate the complex structure of this signal. Oxygen, which formed a chemical bond with metals, oxygen in solid solution, chemisorbed oxygen, as well as various forms of chemically and physically adsorbed water on the surface of the growing sediment can contribute to this peak. The bonds of O1s in hydroxide and in water bound inside the sediment are almost the same (~ 531 eV) and their identification is practically impossible.

One peak with an energy of 285.0 eV is recorded in the C1s spectrum, which indicates the presence of a C-H bond in the citrate-ion composition (Fig. 2, d).

The corrosion resistance of galvanic coatings is influenced by various factors: the nature of the deposited and base metal, the adhesion between them, the uniformity of the distribution of the coating over the surface, its porosity, nonmetallic inclusions, internal stresses in the coating. The influence of these factors is studied in detail in [4]. In addition, the structure and texture of the coatings, which can be controlled by alloying, have a significant influence on the corrosion process.

Iron-phosphorous electrolytic alloys differ significantly in their corrosion resistance from pure iron coatings having a pronounced columnar structure, so that passive films are not formed on their surfaces under the influence of corrosive media. At the same time, the corrosion resistance of alloys increases with the increase in the phosphorus content, which determines the different structural state of the coatings, in the following sequence: crystalline, amorphous-crystalline and amorphous. It is known that phosphorus in carbonaceous and stainless steels is an undesirable impurity, which worsens their corrosion resistance. However, in nonequilibrium systems, such as electrodeposited iron-phosphorus alloys, the inhibitory effect of phosphorus on the corrosion process is observed. Obviously, in the process of sedimentation, the possibility of forming groups of atoms with covalent bond of phosphide type, capable of imparting high corrosion resistance to coatings, is realized in the coating structure [5]. Moreover, with an increase in the phosphorus content in the alloys, the internal tensile stresses decrease, which contributes to the development of the corrosion process.

Amorphous coatings have the greatest protective properties, since they lack structural defects (grain boundaries, dislocations, packaging defects) that are active centers of corrosion. In addition, amorphous coatings are deposited with internal compressive stresses that impede the corrosion process [5].

The protective ability of iron-molybdenum alloys is somewhat higher than that of iron-phosphorus (Fig. 3). With increasing molybdenum content in the alloy, the corrosion resistance of coatings increases, which is explained by the greater stability of molybdenum in acid media than phosphorus [6].

As in the case of iron-phosphorus alloys, amorphous iron-molybdenum coatings have a greater protective ability than crystalline coatings. The reason for the high corrosion resistance of these alloys is associated with the formation of a protective oxide film with a high concentration of molybdenum, responsible for self-passivation [5].

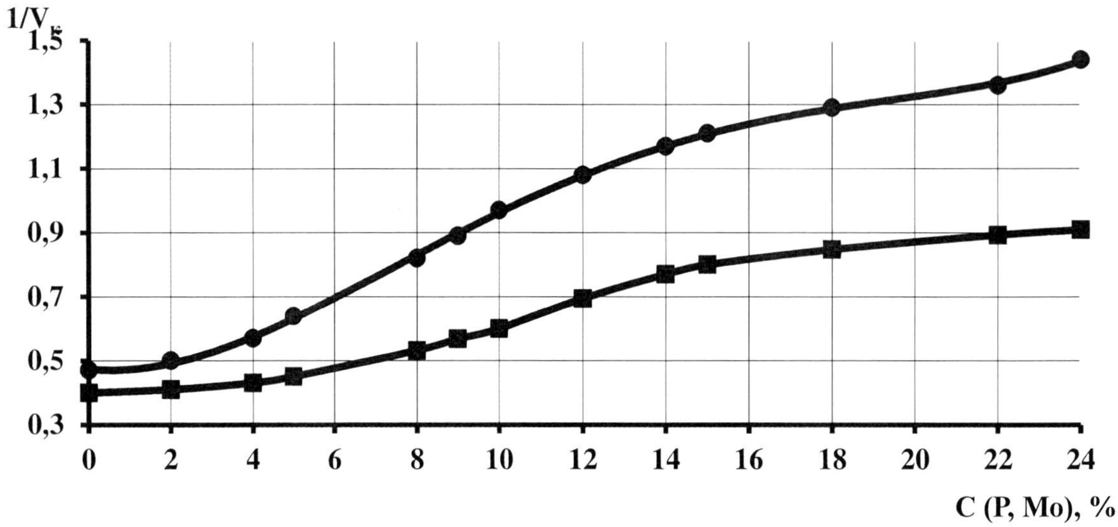

Figure 3. Dependence of the corrosion resistance of coatings with iron-phosphorus (■) and iron-molybdenum (●) alloys on the content of the alloying element

4. Conclusion

The data obtained, as well as the results of [7], show that a solid-phase oxide-hydroxide film (~ 50 nm thick) of variable composition is formed on the surface of the coatings during the electrolysis, consisting of products of an intermediate reduction of molybdate ions, citrate ions and hydroxides of the coprecipitated metals. Such film inhibits the growth of the precipitate, and the greater the concentration of molybdate ions in the electrolyte is, the more nonmetallic particles are introduced into the coatings and the more rapidly the rate of their deposition decreases.

X-ray photoelectron spectroscopic analysis shows that the surface layers of the coatings studied (thickness ~ 50 nm) are enriched with impurity elements. For iron-phosphorus alloys, it is oxygen, and for iron-molybdenum alloys it is carbon and oxygen, which in the course of electrodeposition form compounds with elements of the system that impede the diffusion mobility of metal atoms and inhibit the growth of crystals. This leads to a dispersion of the structure of the resulting coatings, the appearance of a developed surface of the boundaries between highly disoriented grains and facilitates the transition of the alloy to an amorphous state.

The increase in the corrosion resistance of the coatings studied is due to the absence of structural defects (active corrosion centers) and the appearance of internal compressive stresses that impede the corrosion process, which is a consequence of the transition of the resulting alloys to an amorphous state.

References

[1] Kovensky I M, Povetkin V V, Koreshkova E V, Plekhanov V I 2008 *Nanocrystalline and amorphous electrolytic iron-phosphorus alloys* (Moscow: MGIU)

[2] G Y Hunger 1985 *Selected methods of research in metallurgy* (Moscow: Metallurgy)

[3] Kovensky I M, Povetkin V V 2003 *Electrolytic alloys* (Moscow: Intermet Engineering)

[4] Ginberg A M 1984 *Increase of anticorrosive properties of metal coatings* (Moscow: Metallurgy)

[5] Arzamasov B N, Krasheninnikov A I, Pastukhova Zh P, Rakhshtadt A G 1994 *Scientific foundations of materials science* (Moscow: MSTU)

[6] Koreshkova, E V, Kulemina, A A Influence of Molybdenum on Structure and Properties of Iron-Based Alloys *Key Engineering Materials* **736** 3-7

[7] Povetkin V V, Kovensky I M, Koreshkova E V, Neupokoeva A A, Zolotareva Ye V 2008

IOP Conf. Series: Materials Science and Engineering **357** (2018) 012020 doi:10.1088/1757-899X/357/1/012020

Electrolytic alloying of iron with phosphorus and molybdenum *Izvestiya Vuzov. Mechanical engineering* **4** 55-58

[8] Saprykin A A, Saprykina N A, Dudikhin D V, Emelyanenko S M 2014 Influence of Layer-by-Layer Laser Sintering Modes on the Thickness of Sintered Layer of Cobalt-Chromium-Molybdenum Powder *Advanced Materials Research* **1040** 805-808

[9] Fedoseev S N, Valuev D V, Mamadaliev R A, Sokolov P N 2017 Application of integrated methods to improve the technological properties of steel *Key Engineering Materials* **736** 110-115

[10] Gusev A I, Kozyrev N A, Osetkovskiy I V, Kozyreva O A, Valuev D V 2017 Quality of metal deposited flux cored wire with the system Fe-C-Si-Mn-Cr-Mo-Ni-V-Co *Key Engineering Materials* **736** 23-28

[11] Gizatulin R A, Valuev D V, Trifonov V A, Valueva A V, Serikbol A 2015 Oxidation processes in blowing steel with inert gas into the ladle *IOP Conference Series: Materials Science and Engineering* **91** 1-7

[12] Mamadaliev R A, Kuskov V N, Popova A A, Valuev D V 2016 Alloying Elements Transition into the Weld Metal When Using an Inventor Power Source *IOP Conference Series: Materials Science and Engineering* **127**

[13] Gizatulin R A, Kozyrev N A, Valuev D V, Valueva A V, Serikbol A 2016 Research on operational characteristics of imported rails under conditions of east siberian railway *IOP Conf. Series: Materials Science and Engineering* **127**

[14] Malushin N N , Kovalev A P, Valuev D V, Shats E A, Borovikov I F 2016 Development and study of hard-facing materials on the base of heat-resisting high-hardness steels for plasma-jet hard- facing in shielding-doping nitrogen atmosphere *IOP Conference Series: Materials Science and Engineering* **142**

[15] Malushin N N, Valuev D V, Valueva A V, Serikbol A 2014 Kinetic study of the effect of plasticity and its role in stress relaxation in the weld speed steel during the martensitic transformation *Applied Mechanics and Materials* **682** 58-63

Ballasting pipeline moving in horizontal well as method of control sticking phenomenon

V S Toropov and E S Toropov

Industrial University of Tyumen, 38 Volodarskogo Str., Tyumen, 625000, Russia

E-mail: vladimir.s.toropov@gmail.com

Abstract. The mechanism of the phenomenon of sticking a pipeline moving in a well while pulled by the facility horizontal directional drilling is investigated. A quantitative evaluation of the force arising from sticking is given. At the same time, the working hypothesis takes a view of the combined effect of adhesion and friction interactions as the reasons that cause this phenomenon. As a measure to control the occurrence of sticking and to reduce the resistance force to movement of the pipeline in the well, several methods of ballasting the working pipeline are proposed, depending on the profile of the well and the ratio of the length of the curved sections of the inlet and outlet and the straight horizontal sections of the profile. It is shown that for crossings, which profile contains an extended horizontal section, it is possible to partially fill the pipeline with water to achieve zero buoyancy, and for crossings with curvature along the entire profile, the ballasting efficiency will be minimal.

1. Introduction

The phenomenon of sticking the pipeline is possible at the stage of pulling on sections of the profile of a subsoil crossing with permeable soils. Its reason is the interaction of the pipeline with the filter cake, which is formed on the borehole walls [1].

To form a filter cake, it is necessary that the hydrostatic pressure of the solution in the considered section of the well is greater than the pressure in the rock, provided that it is permeable.

During the flow of a mixture of slurry and mud particles into the soil through the borehole wall, the borehole wall itself delays the penetration of solid particles through itself. Thus, particles suspended in the drilling mud accumulate, forming a filter cake.

An increase in the thickness of the filter cake leads to a decrease in the amount of liquid penetrating into the soil from the well, and accordingly slows down and then stops the process of formation of the filter cake.

2. Materials and methods

The factors determining the rate of formation and the size of the filter cake are:

- differential pressure quantity;
- proportion of solids in the drilling mud;
- filtering degree of the water solution base in the soil.

The dependence of the main parameters of the formation process of the filter cake on the factors determining this process is shown in Table 1.

Sticking caused by a high differential pressure value occurs, as a rule, in soils with a high degree of permeability, for examples sandstones [2]. This can happen because the pipeline is adjacent to the filter cake. The contact surface, in this case, is exposed to the difference between the internal pore pressure of the soil and the column pressure of the drilling fluid (Fig. 1).

Table 1. Dependence of the parameters of the formation process of the filter cake on the determining factors

Factor	Value	
	High	Low
The value of the differential pressure.	The speed of crust formation increases, its thickness increases.	The speed of crust formation decreases, its thickness decreases.
The proportion of solids in cuttings in the drilling mud.	The porosity and permeability of the crust as well as the rate of its formation and thickness increase.	The crust becomes thinner and more solid.
Degree of filtration of the water solution base in the soil.	The thickness of the crust increases; the crust is softer.	The thickness of the crust is small; the hardness is high.

With the use of horizontal directional drilling technology, there are no serious limitations on the crossing depth, and therefore the pressure difference can be significant [3]. The force arising from this interaction hinders the movement of the pipeline in the well [4].

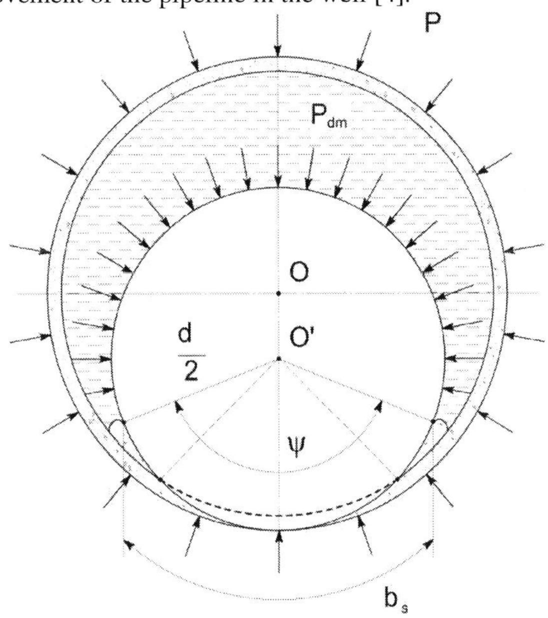

Figure 1. A design model for determining the sticking force of a pipeline moving in a well

At present, there is no single approach to the interpretation of the mechanism for the appearance of the sticking effect. There are researchers who suggest that the effect of sticking the pipeline is due to the filter cake, which becomes stronger during the growth of the pressure difference as the main cause of sticking.

If the soil, where the well is drilled, is composed of almost impenetrable clay soils, the sticking is caused not by the pressure difference but directly by the hydrostatic pressure of the drilling fluid column, causing the pipeline to adhere to the filter cake.

There is also another opinion according to which when sticking, not only the process of adhesion to the borehole wall takes place, but the frictional resistance arising between the pipeline and the clay cake, which increases with the growing pressure drop of the drilling mud column and pore pressure of the soil.

In order to take into account as many factors as possible, negatively affecting the process of pulling the pipeline, let us consider the combined effect of adhesion and friction interactions on the working hypothesis. Considering this, the sticking force will be equal to:

$$F_s = \left[\left(D_1 + D_{st}\right)\gamma_{dm} - P\right]\frac{d}{2}\cdot\beta\cdot l \tag{1}$$

where D_l - penetration of the pilot well from the spudding site;
D_{st} - the depth of the spudding site in the starting pit from the day surface;
d – outer diameter of the pipeline;
γ_{dm} –specific gravity of drilling fluid, N/m^3;
P – reservoir pressure, Pas;
ψ –the angle of the contact zone of the pipeline with the filter cake, radians (Figure 1);
l – length of the sticking segment.

And the force of the resistance to the movement of the pipeline from the effect of picking up the filter cake is defined as follows:

$$F = \left[\left(D_1 + D_{st}\right)\gamma_{dm} - P\right]\frac{d}{2}\cdot\beta\cdot l\cdot f \tag{2}$$

where f – friction coefficient, taking into account adhesion and frictional resistance to pipeline movement.

As can be seen from Figure 1, with increasing the filter cake thickness, the contact area of the pipeline with the borehole wall also increases. The thicker the filter cake becomes, the more the area of contact with the pipeline increases, and the more significant is the resultant force of differential sticking.

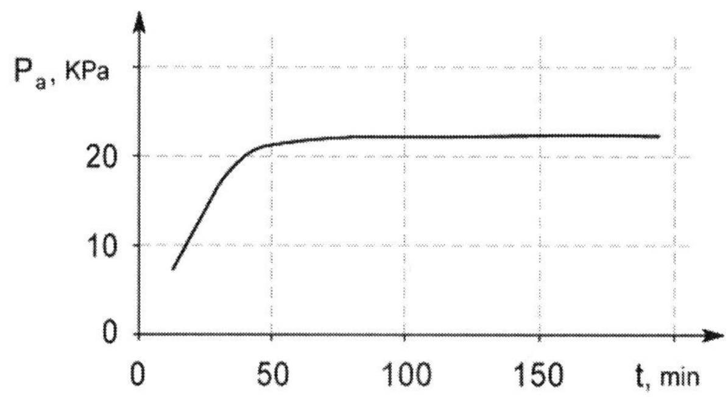

Figure 2. A value of adhesion of the filter cake to the steel wall of the pipeline depending on the contact time at a pressure difference of 0.1 mPa.

The risk of the sticking effect is significantly increased in case of a long stay of the pipeline without movement inside the borehole drilled in the sandy ground, as the process of covering the pipeline with a filter cake gradually increases with an increase in the contact surface area [5].

Because of the increase in the area, the value of frictional forces grows, the frictional interaction effect appears, since the drilling fluid begins to be displaced from the contact zone, and an area is formed in which there is already a sliding friction of the pipeline material over the soil. As a result, the efforts necessary for ultimate strength of the pipeline increase significantly.

To reduce the friction force of the pipeline against the soil surface of the well, the pipeline to be pulled is ballasted with water [6]. The method of ballasting depends on the well profile and the ratio of the curved sections of the inlet and the outlet and the straight horizontal sections [7].

For crossings, which profile contains an extended horizontal section, it is possible to fill partially the pipeline with water until it reaches zero buoyancy. This makes it possible to minimize the frictional forces on the soil when pulling the pipeline.

For a well without a horizontal section, i.e. with curvature along the entire profile, water will collect in the lower part of the crossover [8]. In this case, the lower part of the pipe will have the excess negative buoyancy, and the peripheral sections will have the maximum positive buoyancy, i.e. the efficiency of ballasting will be minimal.

It should be noted that positive buoyancy has other negative factors. When pulling the pipeline with insufficient ballasting under the influence of tension forces, it is possible to lift the pipeline to the borehole high. This is especially evident in the curved sections of the inlet and outlet of the pipeline from the well [9]. In this case, the pipeline destroys the soil of the upper part of the borehole high, which can lead to the high collapse and sticking of the pipeline.

From this it follows that the "excessive" ballasting of the pipeline is less harmful from the point of view of protection against sticking than the preservation of "excessive" buoyancy. This is especially true for profiles that do not have a horizontal section, i.e. the variant of the pipeline movement with the pressure against the lower forming wall of the well is more preferable [10].

The floating force acting on the pipeline depends on its diameter and wall thickness. With increasing the diameter, there is an intensive growth of the floating force.

The condition for floating force on the pipeline is an equality:

$$q = \pi \cdot d \cdot \gamma_t \cdot \delta - \frac{\pi \cdot d^2}{4} \cdot \gamma_{dm} \tag{3}$$

where q – load from the weight per unit length of the pipeline, taking into account the floating force occurred in the drilling mud, N/m;

γ_t – specific gravity of the material of the pipeline being pulled into the well, N/m^3;

δ – wall thickness of the pulled pipeline, m.

Figure 3 shows the alignment chart of the dependence of the load on the weight per unit length of the pipeline, taking into account the floating force occurred in the drilling mud, and on the wall thickness for different pipeline diameters.

From the condition that the upper well height does not collapse (the pipeline lies at the bottom of the well), the previous equation can be represented in the form of an inequality:

$$q = \pi \cdot d \cdot \gamma_t \cdot \delta - \frac{\pi \cdot d^2}{4} \cdot \gamma_{dm} \geq 0, \tag{4}$$

or if one expresses the value of the wall thickness:

$$\delta \geq \frac{d}{4} \cdot \frac{\gamma_{dm}}{\gamma_t} \tag{5}$$

where δ is the thickness of the pipeline wall at which the pipeline's zero buoyancy is ensured. The inequality sign shows that for large values of δ, the pipeline will lie on the lower generating line of the well and will not touch the upper high.

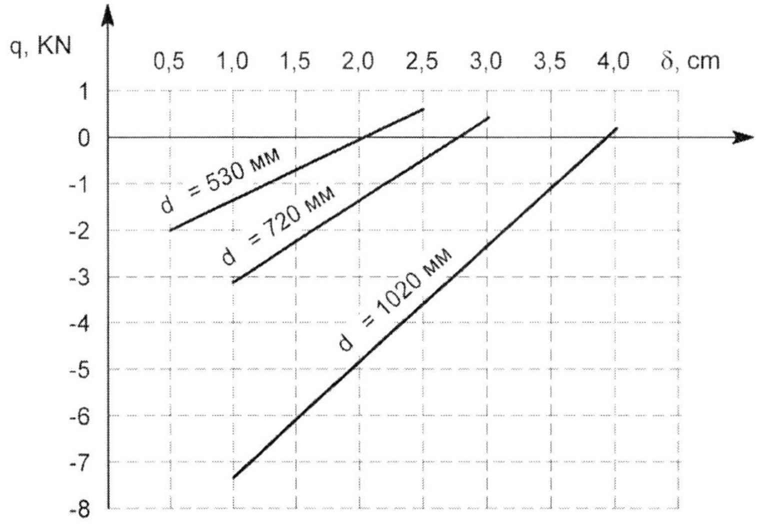

Figure 3. The dependence of the load on the weight per unit length of the pipeline, taking into account the floating force occurred in the drilling mud, on the wall thickness of the pipeline.

In expression (5), the value of q will be numerically equal to the weight of water that must be poured into the pipeline to create zero buoyancy. When pulling, water is periodically poured into the cavity of the pipe, thereby unloading the pipeline to be pulled (4).

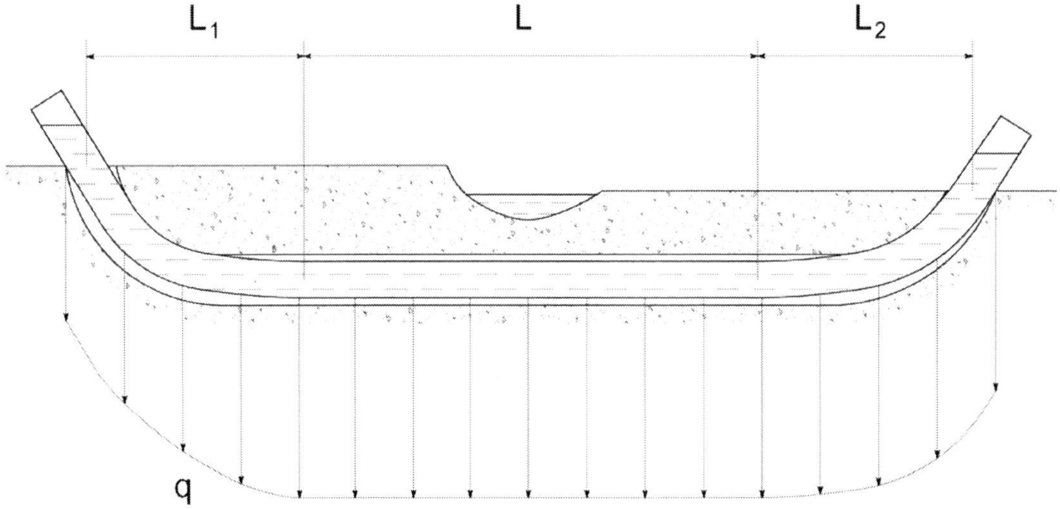

Figure 4. A flow chart of ballasting of the pipeline by completely filling it with water: *L1, L2* - curved sections of the inlet and outlet; *L* - straight section of the profile.

When the pipeline is completely filled with water, too much effort may be exerted to hold down the pipeline to the lower generating line of the well [11, 12]. They can be adjusted by using an internal pipe or by partial filling with water (Fig. 5, 6).

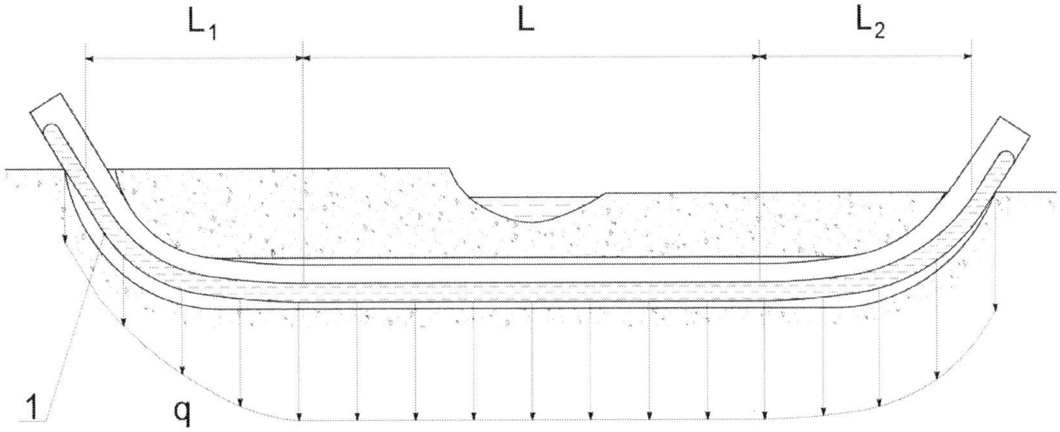

Figure 5. A flow chart of ballasting of the pipeline with the help of a water filled pipe placed in the inner cavity: *1* – a pipe of a smaller diameter in the internal cavity of the working pipeline.

In the last two cases, the loading flow charts will differ.

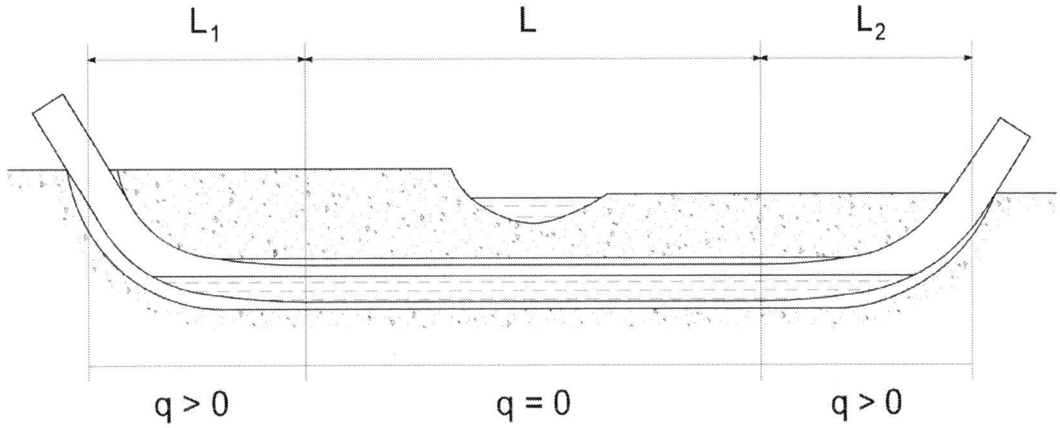

Figure 6. A flow chart of ballasting the pipeline by partially filling it with water: q - the resulting force of the pipeline effect on the upper generating line of the borehole wall in different sections.

The pulling process should go without stopping and interruptions (excluding those justified by the horizontal directional drilling technology itself) in order to prevent sticking of the pipe in the well [13].

3. Conclusion

In order to avoid sticking the drill string and the pipeline to be pulled to the borehole walls, in cases when the pulling process is stopped because of technical reasons and breaks during which the pipeline remains stationary in the borehole, it is necessary to rotate the drill pipes and flush the well with drilling mud after certain periods [14].

In addition, several different flow charts for ballasting the pipeline with water are proposed, which use will not only reduce the load on the well wall from the pipeline side, thereby reducing the resistance to pipeline movement in the well, but will reduce significantly the danger of the pipeline sticking to the well walls [15]. These schemes of full and partial ballasting in the process of pulling are relevant for crossings, which profile contains an extended horizontal section.

References

[1] **Shilov A G, Glichev V A 2013 Prevention of packing during drilling in clayey sediments. Quality management in the oil and gas sector. 4 60 - 62**

[2] **Fateev V S 2006 Key-seating during drilling. Magazine "Vestnik" of the Institute of Geology, Komi Science Center, Ural Branch of the Russian Academy of Sciences 11 11 - 12**

[3] Zemenkov Yu D 2008 *Technique and technologic processes at transport of energy resources*: two volume instructional aid. (Tyumen; Vector Book)

[4] Zemenkov Yu D 2004 *Transporting and storage of oil and gas in examples and sums* (St. Petersburg; Nedra)

[5] Toropov V S, Temirbaev R M, Toropov E S, Toropov S Yu 2014 *Interaction Between the Pipeline and Additional Equipment for Trenchless Technologies.* International Scientific-Practical Conference of Students, Graduate Students and Young Scientists *"Transport and Storage of Hydrocarbons"*, **154**

[6] Kiryanov I E, Toropov V S, Toropov S Yu 2016 Determination of dependences of pipeline movement resistance forces in curve borehole from physical parameters and rheological model of using mud. Oil equipment 8

[7] Toropov V S, Tamer O S, Toropov S Yu, Nikiforov N A 2015 *Dynamics of pipeline pulling process by horizontal directional drilling.* International Scientific-Practical Conference of Students, Graduate Students and Young Scientists *"Transport and Storage of Hydrocarbons"*, **154**

[8] Toropov V S, Osyka A V, Toropov E S 2001 Pulling the pipeline into volumetric curved boreholes. Proc. of the interregional youth scientific conference "Severgeoekotech-2001" pp. 70-71

[9] Zemenkova M Yu, Shipovalov A N, Dudin S M and Zemenkov Yu D 2007 System analysis in oil and gas objects control processes *News of higher educational institutions. Oil and gas* **5** 116-9

[10] Antipiev V N, Zemenkov Yu D and Shabarov A B 2002 *Technical and parametrical diagnostics in pipeline systems* (Tyumen; Vector Book)

[11] Shalay V, Zemenkova M, Zemenkov Y, et al. 2016 Modeling Parameters of Reliability of Technological Processes of Hydrocarbon Pipeline Transportation. *MATEC Web of Conferences* **73**, 01029 DOI: 10.1051/matecconf/20167301029

[12] Smirnov V V and Zemenkov Yu D 2013 Using the finite element method for calculation of intense deformed condition of above ground pipelines constructed on permafrost soil *Pipeline transport: theory and practice* **4** 18-23

[13] 2008 *Monitoring of hydrodynamic and technical characteristics of pipeline systems* ed by Zemenkov Yu D (Tyumen; "Vector Book")

[14] Ibragimov A A, Podorozhnikov S Yu, Shabarov A B, Medvedev M V and Zemenkova M Yu 2014 Calculation model and algorithm for determining residual resource of the pipeline under the periodic variations of stress and corrosion *Mining informational and analytical bulletin Scientific and technical journal* (Moscow; Gornaya kniga) **S4** 199-206

[15] Voronin K S, Ogudova E V 2016 The Effect of Dynamic Processes in the System "Pipe-Soil" on the Pipeline Deviation from Design Position. Transport and Storage of Hydrocarbons. *IOP Conf. Series: Materials Science and Engineering* **154** 012019 doi:10.1088/1757-899X/154/1/012019

IOP Conf. Series: Materials Science and Engineering **357** (2018) 012022 doi:10.1088/1757-899X/357/1/012022

Development of layout algorithm for pipeline, considering topographic features

P V Grigorieva, K S Voronin1 and D A Cherentsov

Industrial University of Tyumen, 38 Volodarskogo St., Tyumen, 625000, Russia

E-mail: polino4ka.g@yandex.ru, voronin_tsogu@mail.ru, cherentsovda@bk.ru

Abstract: In the article algorithms of path-tracing are considered; the developed algorithm for searching the optimal pipeline route is presented, taking into account topographical features of site.

1. Introduction

Drawing up the route plan is an obligatory stage in the pipeline design. This issue is considered at the stage of route location, which consists of a complex of engineering and geodesic activities to find the path. Route location includes two main elements [1-3].

The first is a route plan which is an exact projection of the route on the horizontal plane.

The second is a longitudinal profile of the route, which is its vertical section along the projection line [4, 5]. This plan consists of segments of straight lines with different slope ratio, which can be connected by circular curves if necessary.

An ideal route should be straightforward, without deviations and bends, which can lead in practice to a significant increase in the cost of construction and its operation. However, the direct and longitudinal plans most often do not correspond to each other, and subsequently these discrepancies are solved by curving the general plan of the path (curves of constant and variable radius of curvature). So, the plan should reflect the trajectories of bypassing sites with negative geological conditions, large slopes and other unfavorable obstacles [9-12].

It is not always possible to take into account the structure of topographic features at the initial stages of compiling the path of the main pipeline route, carried out on small-scale topographic maps.

At present, designing of pipelines is a difficult task. There are many options for passing the pipeline route between the start and end points. The choice of the route is more often based on the expert opinion of the technicians developing the project documentation. However, automating the search for a route with the lowest capital costs is an urgent task [7].

2. Materials and methods

There are many different algorithms to solve the problem of finding the path between the start and end points. Their comparison is presented in Table 1. These algorithms differ in the following characteristics: the accuracy of the trace performing, the performing speed of the algorithm, the number of limits taken into account and the requirements for pipeline design. The following trend is clearly traced in these algorithms: the more accurately the algorithm performs the tracing, and the more rules it takes into account, the slower is its performance, and vice versa, the faster the algorithm performs, the less precise decisions it gets. The task is to create an algorithm that will most quickly and accurately route the pipeline on a map, taking into account the topographic features of the site [8]. The wave algorithm was chosen as the basis of the created algorithm, as it can find a path under any conditions and is fairly simple in its use.

Table 1. Comparison of algorithms for pipeline route location

Algorithm	Algorithm A*	Dijkstra's algorithm	Wave algorithm
Advantages	1.Combines the length of the previous path. 2.Takes into account the cost to the target point.	1. High speed of performance. 2. Calculation of ribs with a negative weight 3. High accuracy of result.	1. The best path will be always found. 2. The possibility of introducing the cost of the cell. 3. The possibility of constructing a path not to one but directly to several goals, that is, the search for the nearest target is in progress. 4. The clarity and simplicity of the algorithm.
Disadvantages	1. Memory limitations. 2. It takes a lot of process time.	1.Complexity of understanding. 2.On each pass, it takes into account the length of only one rib.	1. A large amount of memory is required.

The process of building a route can be divided into 2 stages [8].

1. The wave transmits in the 4 directions from the initial element. The element in which the wave came is the front of the wave. In Fig. 1, numbers indicate the wave front numbers.

Each element of the first wave front is the source of the secondary wave. Elements of the second wave front generate a wave of the third front and so on. The process continues until the end element is reached or until it becomes clear that it can not be achieved.

2. The route itself is constructed. Its construction is carried out from the final element to the initial one.

The initial task of finding the path for laying the pipeline layout is to apply a grid and "weight" of cells to the site of the map. The "weight" of cells is selected in accordance with the criteria of optimality: the reduced costs; the pipeline length; labor costs; reliability of pipeline operation; time of construction as well as the cost of laying the pipeline taking into account the site features.

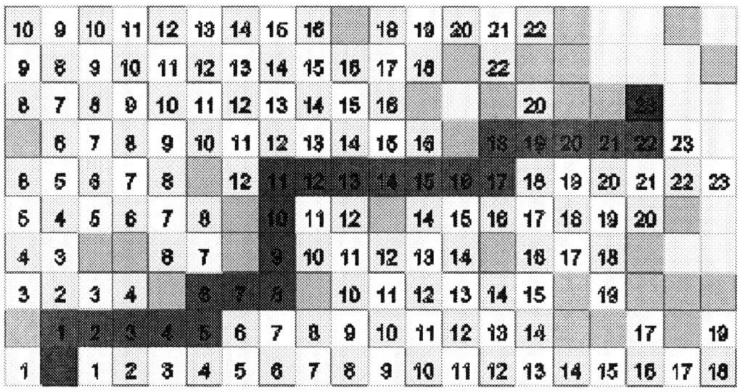

Figure 1. The result of the wave algorithm

Then the matrix with the "weights" of the grid is transferred to the developed program, where the laid path with the optimal costs is shown by simple pressing the button.

To implement the wave algorithm, a flow chart has been developed (Fig. 2a), which looks like this:

the initial (input) parameters are given. This is the dimension of the grid (matrix), the initial and final points of the trace and the "weight" applied to the grid, taking into account topographical features. The essence of the developed algorithm is to perform the task: to lay the optimal path from the initial to the final element. For this, a recursive function will be used, i.e. if the condition, where the start point is equal to the final one, is not fulfilled, let us move to the other cell along the X and Y axes and add the "weight" of the cell to which the existing sum is added. That should be done until the extreme point is reached. At the end, the minimum amount of "weights" is displayed [8].

The algorithm was implemented in the VBA language. The choice of this particular language was based on the ease of transferring the "matrix" of the terrain map to Excel electronic work sheets.

Fig. 2b shows the grid with the applied "weights" and the results of the shortest distance found on the map using the implemented wave algorithm. For a more detailed study of the route, it is necessary to share the map into smaller cells, i.e. create a matrix containing more elements.

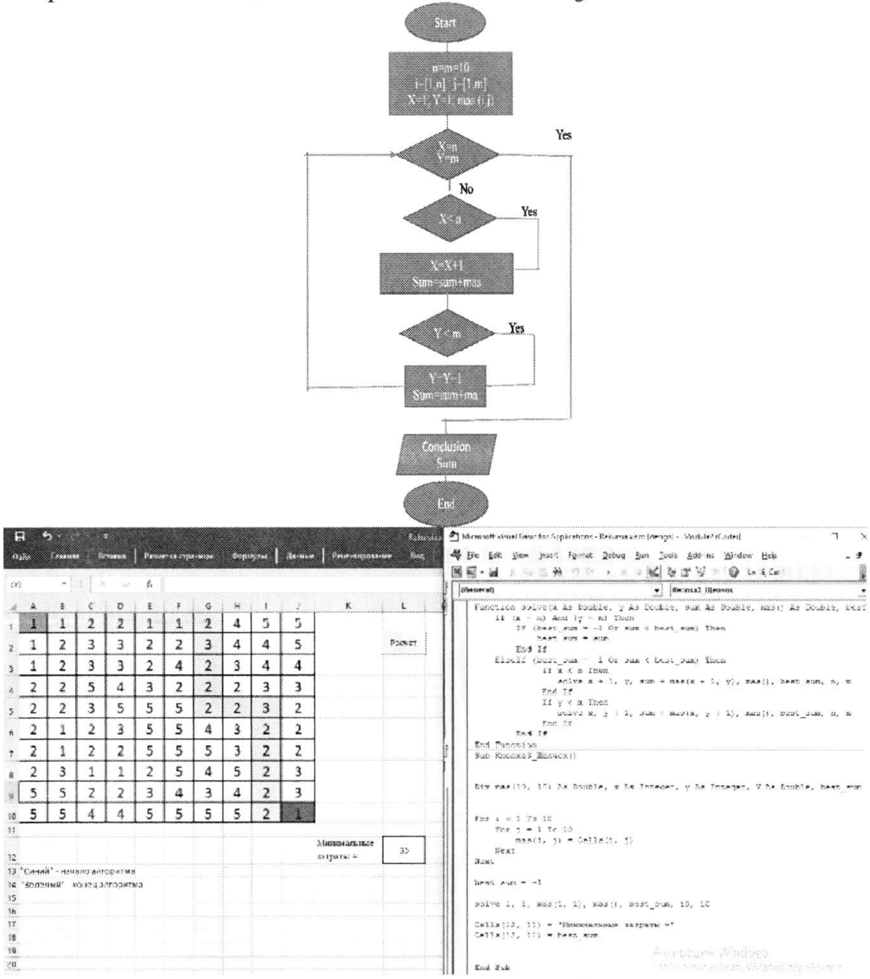

a) b)

Figure 2. A flow chart and implementation of the wave algorithm in the VBA language

The resulting pipeline route, using the developed algorithm, is angular, because 4 directions were used. To correct this, one needs to divide the grid into smaller cells, or use the interpolation function, i.e. one finds approximate values of the obtained trace for values of variable x that are different from the interpolation nodes [6]. To do this, let us use interpolation with splines: one of the most common cases in practice is a cubic spline.

On each of segments $[x_{i-1},x_i]$, $i=1, 2, \ldots, n$ let us search for spline function $s(x)=s_i(x)$ in the form of a polynomial of the third degree:

$$s_i(x) = a_i + b_i(x - x_i) + \frac{c_i}{2}(x - x_i)^2 + \frac{d_i}{6}(x - x_i)^3, \quad (1)$$
$$x_{i-1} \le x \le x_i, i = 1,2,\ldots,n$$

where a_i, b_i, c_i, d_i are the required coefficients, which are found by differentiating function $s(x)$ on x $a_i=s_i(x_i)$, $b_i=s_i'(x_i)$, $c_i=s_i''(x_i)$, $d_i=s_i'''(x_i)$:

$$a_i = a_{i+1} + b_{i+1}(x_i - x_{i+1}) + \frac{c_{i+1}}{2}(x_i - x_{i+1})^2 + \frac{d_{i+1}}{6}(x_i - x_{i+1})^3, i = 0,1,2,\ldots,n-1. \quad (2)$$

Denoting $h_i=x_i-x_{i-1}$ and omitting the intermediate operations, let us finally obtain a system of equations for determining coefficients c_i:

$$hc_{i-1} + 2(h_i + h_{i+1})c_i + h_{i+1}c_{i+1} = 6(\frac{y_{i+1} - y_i}{h_{i+1}} - \frac{y_i - y_{i-1}}{h_i}), i = 0,1,2,\ldots,n-1. \quad (3)$$

$c_0= c_n=0$.

Since the coefficient matrix is tridiagonal, system (3) has a unique solution. Having found coefficients c_i, let us define the remaining coefficients by the explicit formulas:

$$d_i = \frac{c_i - c_{i-1}}{h_i}, b_i = \frac{h_i}{2}c_i - \frac{h_i^2}{6}d_i + \frac{y_i - y_{i-1}}{h_i}, i = 1,2,\ldots,n. \quad (4)$$

3. Conclusion

Thus, the developed algorithm for finding the optimal pipeline route according to the proposed criteria, taking into account topographic features, possible unforeseen emergency situations and potential economic losses will allow selecting the most rational option as well as significantly improving the quality of the taken decisions.

References:

[1] Babin L A et al. 1970 *Selection of main pipeline routes.* (Leningrad, "Nedra")

[2] Borodavkin P P, Berezin V L, Ruderman S Yu 1974 *Choice of optimal routes of main pipelines.* (Moscow, "Nedra")

[3] Zemenkova M, Zemenkov Y, Gladenko A, Podorozhnikov S 2016 Estimation of Emissions during Monitoring of Pipelines in the Dynamic Mode of Operation. 5th International Scientific *Conference Integration, Partnership and Innovation in Construction Science and Education* **86** DOI: 10.1051/matecconf/20168604053

[4] Spasskaya K A 2014 Criteria for choosing a variant of the direction of the railway route, taking into account the development of landslide processes. *Internet-journal "Naukovedenie"* **3**

[5] Shalay V, Zemenkova M, Zemenkov Y, et al. 2016 Modeling Parameters of Reliability of Technological Processes of Hydrocarbon Pipeline Transportation. *MATEC Web of Conferences* **73** 01029 DOI: 10.1051/matecconf/20167301029

[6] Voronin K.S. Forecasting and Evaluation of Gas Pipelines Geometric Forms Breach Hazard. Transport and Storage of Hydrocarbons. *IOP Conf. Series: Materials Science and Engineering* **154** 012020 doi:10.1088/1757-899X/154/1/012020

[7] Dudin S, Voronin K, Yakubovskaya S, Mutavaliev S 2016 Modeling Hydrodynamic State of Oil and Gas Condensate Mixture in a Pipeline. *MATEC Web of Conferences.* DOI: 10.1051/matecconf/20167302021

[8] Pirogov S P, Cherentsov D A, Gulyaev B A 2016 Prospects of applying vibration-resistant pressure gauges in the oil and gas industry.*IOP Conference Series: Materials Science and Engineering* 012013

[9] Zemenkova M, Shalay V, Zemenkov Y, Kurushina E 2016 Improving the Efficiency of

Administrative Decision-Making when Monitoring Reliability and Safety of Oil and Gas Equipment *MATEC Web of Conferences* **73** DOI: 10.1051/matecconf/20167307001

[10] Kulikov A M 2016 On Choosing Multicomponent Multiphase System Separation Progress Optimization Criteria V. *IOP Conference Series: Materials Science and Engineering* **154** DOI: 10.1088/1757-899X/154/1/012039

[11] Bautin S P, Krutova I Y, Obukhov A G 2015 Twisting of a fire vortex subject to gravity and Coriolis forces. *High Temperature* **53-6** 928-930 DOI: 10.1134/S0018151X1505003X

[12] Bautin S P, Obukhov A G 2013 Mathematical simulation of the near-bottom section of an ascending twisting flow *High Temperature* **51-4** 509-512 DOI: 10.1134/S0018151X1302003X

IOP Conf. Series: Materials Science and Engineering **357** (2018) 012023 doi:10.1088/1757-899X/357/1/012023

Swirl-chamber experimental model creation

A M Kulikov

Industrial University of Tyumen, 38 Volodarskogo St., Tyumen, 625001, Russia

E-mail: lis-prepod@mail.ru

Abstract. The process of experimental studies organization of preparation and division of multiphase multicomponent systems by means of intensification of gas evolution processes is described.

1. Introduction

According to the "Energy Strategy of Russia for the period until 2030" and the order of the Government of the Russian Federation of November 13, 2009, No. 1715-r, the oil and gas industry is required to integrate machines and equipment into the production process, as well as methods and processes that allow minimizing energy costs and adverse environmental impact of technological processes [1, 2]. As a consequence, the development of new methods for designing, constructing and operating oil and gas facilities, as well as optimizing the operation of existing equipment is promising and relevant [7, 8].

2. Materials and methods

Hydrocarbon crude will remain the main energy carrier for a sufficiently long period of time; therefore, despite the possible reserve depletion, its extraction, processing and transportation will inevitably develop. The processes of preparation of liquid hydrocarbons for transportation under conditions of fields and special points have been thoroughly studied and have facilities support in the form of a number of processing units designed to prepare raw materials with a wide variety of physicochemical properties [3]. In the process of preparing oil and gas condensate for long-distance transport, it is necessary to isolate dissolved gases, to dehydrate and desalinate, to purge from mechanical impurities, i.e. to ensure the requirements for commercial oil and gas-condensate.

In modern conditions of market economy, high requirements are imposed on the quality of oil products, which must be ensured at the stage of hydrocarbon crude preparation. Among other things, high quality is expressed in the low content of alien impurities of water and dissolved gas [4]. Modern degassing systems use the principle of a multi-stage process, which prevents the loss of butanes, pentanes and gasoline fractions, but this significantly reduces the speed of the fluid and does not always provide the required degree of degassing [5].

It is difficult to expect that in the near future, the problem of separating the liquid phase of hydrocarbons in gas pipelines will be solved, since an insufficient part of the transported gas is adequately prepared for further transport in conditions of gas processing plants (GPP) or gas treatment plants (CGTP). Besides, even such relatively qualitative treatment of gas does not exclude the appearance of a liquid phase in the internal cavity of the gas pipeline. On older gas pipelines, where GPPs are not available, condensate losses reach tens of thousands of tons per year.

Consequently, it becomes necessary to develop more efficient degassing methods, which use the flow energy more rationally. Earlier in a number of papers, it was shown that the solution of this problem could be connected with the use of a vortex flow. The vortex motion of a gas-saturated liquid forms a field of centrifugal forces, in which components with different density are separated from each other. In addition, there is an intensive mass transfer of liquid volumes with different gas saturation, which inevitably leads to the formation of a gas-liquid division with a turbulized structure in which the process of gas evolution significantly increases. In some ways, processes in a swirl-chamber filled with a gas-saturated liquid can be compared to processes in a centrifuge. In both cases, there is a field of centrifugal forces, forcing components with different density to differentiate from each other [5,6].

The swirl-chamber is able to create flows with a high degree of twist, high velocities and accelerations, both in the lower part of the chamber filled with liquid, and in the upper one, where the gas phase is located. The absolute velocity of the particle motion can be represented in terms of its components. In the lower part of the chamber, the rotating liquid is subjected to intensive degassing, where the process is occurring throughout the volume of the liquid. Such massive degassing is caused by several reasons, the most important of which are the following:

- lowering the pressure in the chamber as compared with inlet pressure of the nozzle entry, causing the formation of gas bubbles in a liquid stream directed into the chamber;
- the existence of a fluid in the swirl-chamber, leading to intensive mass transfer of liquid volumes with different local gas saturation, which in turn contributes to the intensification of the degassing process.

The formation of a «gas-liquid» interface with a turbulized structure, which facilitates degassing from the surface of the liquid.

Thus, other things being equal, the presence of a swirl leads to a reduction in the degassing time.

The first of the known experimental studies of degassing a liquid in a swirl-chamber belongs to E.A. Abdullaev. In this paper, the authors investigated the crude condensate degassing processes of "Bahar" field (Azerbaijan). The experiments confirmed the principal possibility of using swirl-chambers as degassers for wet condensate and showed the advantages of using a swirl-chamber in front of ordinary traps [9].

The author drew attention to the fact that intensive rotation of the gas phase facilitates the separation of liquid droplets from the gas due to the swirling effect. In other words, the upper part of the swirl-chamber acts as a centrifugal gas separator, in which a sufficiently complete separation of the gas from the inclusions (droplets) of the liquid is achieved.

The degassing of a saturated liquid in a swirl-chamber was studied in an experimental unit, the scheme of which is shown in Fig. 1.

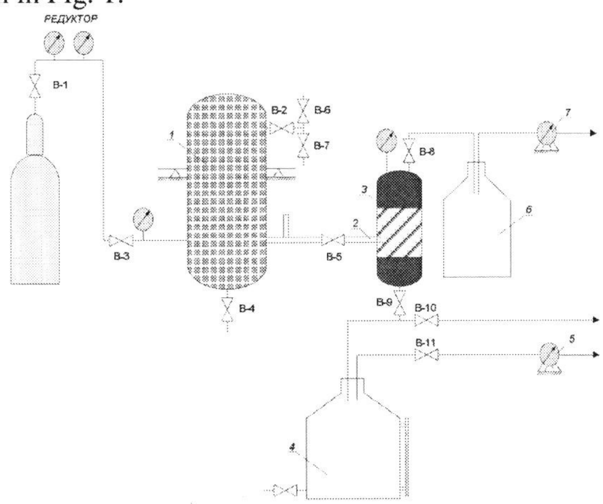

Figure 1. An experimental unit for studying the degassing process of liquid in a swirl-chamber.

In experiments carried out under laboratory conditions, the experimental unit was connected to a mixing tank, which was intended for preparation of the test sample.

In an experimental study of the degassing of any saturated liquids, the latter entered swirl-chamber 3 through valve B-5 and nozzle 2.

With the help of valves B-8 and B-9, set respectively on the output lines of gas and liquid, the necessary mode of operation of the camera was selected. By regulating the flow of gas and liquid through the swirl-chamber via these valves, it was possible to set in it a different height of the liquid column and various pressures. The pressure drop between the saturated liquid pressure in front of the chamber and in chamber 3 itself determined the mass flow velocity of the liquid from nozzle 2. The degassing gas from the chamber passed through valve B-8 and gas vents to control separator 6, which serves to determine the presence or absence of entrainment of the dropping liquid with degassing gases. After separator 6, gas flew into gas meter 7 for measurement.

The liquid degassed in the swirl-chamber, leaving the B-9 valve, could either enter the measurement through the B-11 valve (with the B-10 valve closed), or drain into the sewage system through the B-10 valve (with the B-11 valve closed) [10, 11].

By comparing it with the equilibrium gas saturation at the same pressure, the efficiency of liquid degassing in the swirl-chamber was estimated. As it was mentioned above, $\varepsilon = q_P/q_K$ was the quantitative value of the liquid degassing process efficiency.

According to an analytical study of the degassing of liquids in a swirl-chamber, it is necessary to establish the relation between the mass flow velocity of the liquid through the inlet nozzle of swirl-chamber u, the characteristic dimension of swirl-chamber D/h_l, and the coefficient of gas-yield intensity n. The mass flow rate of the gas-saturated liquid through the nozzle was determined by the results of the experiments. The value of D/h_l was determined during the experiments on degassing the liquid in the swirl-chamber by visual measuring the liquid level in the $h_{ж}$ chamber at the known diameter. The coefficient of gas emission intensity n was determined from the equation.

To ensure visual observation of the degassing of saturated liquid in swirl-chamber 1 and the possibility of regulating the liquid level in it, the chamber is made of polished organic glass. The chamber consists of the following main parts: a body frame, a cylindrical working cavity, a perforated washer, a bunker space, a lower cover, screws for securing the bottom cover to the body frame, nozzle inlets, and B-8 and B-9 valves. The body frame is designed for operating pressure of 7.0 MPa and a temperature not exceeding 50 °C. The cylindrical working cavity of chamber 2 has a diameter of 100 mm and a height of 180 mm [12].

The perforated washer can be installed at different heights from the bottom cover of the chamber. The perforated washer serves to untwist the flow of degassed liquid before exiting the chamber. Untwisting of the flow is necessary in order to reduce the crater depth at the gas-liquid interface. Under the washer, a bunker space is formed, in which the fluid no longer forms a vortex, but moves in the axial direction to the B-9 valve. The presence of bunker space allows lowering the hydraulic resistance at the outlet from the chamber and reducing the probability of gas sucking into the liquid outlet line through the crater at the gas-liquid interface.

The liquid column level in the chamber was set proceeding from the condition that prevents the gas from entering the liquid line.

This condition is the excess of the height of the liquid level in chamber h_l over the critical level h_{cr}, at which the crater gas core breaks into the drain hole. H_{cr} is calculated by the following formula:

$$H_{cr} = 0{,}5\, D\, (w_0 /\ gD)^{0{,}5} \quad (1)$$

As the pressure in the chamber increases, all other conditions being equal, the velocity must increase, and hence quantity $h_{кр}$. The height of the arrangement of the perforated washer was set, as a rule, equal to 0.25 h_{cr}, i.e., it was approximately 30% of the height of the column of the rotating liquid [13, 14].

The maximum height of the liquid column was limited by the required volume of the gas phase.

To reduce the depth of the crater, as mentioned above, perforated washer 3 was used. Moving the washer made it possible to change value $D/h_{ж}$ without disturbing the chamber's operation mode. Let us note that in some experiments value D/h_l varied due to the decrease or increase in the height of the liquid column in the chamber with washer 3 unchanged.

The nozzle inlets are tangential and have an oblique cut. The nozzles are located at different heights. The location of the nozzles at different heights makes it possible to estimate the effect of the relative height of nozzle $h_c/h_к$ on the degassing of the liquid in the chamber.

To choose optimal relative nozzle height $h_c/h_к$, a series of visual observations should be made of the depth of the crater on the surface of the liquid rotating in the chamber and the structure of the gas-liquid interface [15].

In a laboratory unit, the degassing of water saturated with carbon dioxide was studied at an inlet liquid temperature of 20 °C at the entrance to the chamber and at a gas saturation of 8.2 to 10.2 cm^3 / cm^3.

From the results of the experiments, it was concluded that the relative height of the nozzle location in investigated range $h_c/h_к = 0.3 \ldots 0.7$ does not affect the amount of fluid entrainment with degassing gases and the degree of degassing. Therefore, the height of the nozzle location should be chosen within these limits for other indicators not related to the amount of entrainment, for example, based on the organization of the best structure of the vortex, the gas-liquid interface or the minimum crater depth formed on the surface of the swirling flow.

Based on visual observations of the crater depth, it is recommended to take value $h_c/h_к$ equal to 0.5.

3. Conclusions

1. As a result of a comparative analysis of the existing methods for degassing gas-saturated hydrocarbon liquids, it has been established that the use of swirl-chambers allows the gas to be degassed with greater efficiency due to increased gas evolution, increased degassing, and higher gas evaporation rate.

2. When comparing the degassing devices at the MGL of GCS and the fields, positive aspects of the use of swirl-chambers were revealed, such as smaller dimensions and mass, simplicity of construction, less metal consumption, increased reliability due to the lack of moving parts, ease of installation and maintenance.

3. As a result of experimental studies, it was found that when the swirl-chamber is filled to 85% of the height and less, the entrainment of the liquid by the degassing gas flow is absent regardless of the relative height of the nozzle location. At the same time, only the column of the rotating liquid phase participates actively in the degassing process.

4. During the experiments, it was determined that with an increase in mass velocity u of fluid outflow through the swirl-chamber nozzle, the liquid degassing intensification' processes go on (the growth of the gas evolution coefficient).

5. An analytical dependence of the intensity of degassing (characterized by the coefficient of gas evolution n) on the residence time of the liquid in the chamber is obtained. As a result of the calculations it was found that with increasing a stay period of liquid in the chamber, the gas release coefficient sharply decreases, which indicates a higher process completion in the swirl-chamber.

References

[1] Trivus N A, Seidali-zade B M 1974 Optimum modes of condensate degassing step and its effectiveness, B Sat.: Development and exploitation of gas and gas condensate fields. *VNIIEGazprom* 6 17-22

[2] Zemenkov Y D, Shalay V V, Zemenkova M Y 2015 Expert systems of multivariable predictive control of oil and gas facilities reliability *Oil and Gas Engineering Procedia Engineering* **113** 312-315 DOI: 10.1016/j.proeng.2015.07.271

[3] Sinai E G, Lapin E A, Zaitsev Yu V 2002 *Separation of multiphase multicomponent systems.* (M .: Nedra)

[4] Bakhmat G V 1981 Use a vortex effect for degassing of crude condensate. *Oil and Gas Problems Tyumen* **49** 15-18

[5] Zemenkov Yu D, Markova L M, Prokhorov A D, Dudin S M 2009 *Collection and preparation of oil and gas.* (M.: Academy)
[6] Dudin S M, Zemenkov Y D, Shabarov A B, et al. 2016 Modeling the Phase Composition of Gas Condensate in Pipelines. *IOP Conference Series: Materials Science and Engineering* **154** DOI: 10.1088/1757-899x/154/1/012010
[7] Shalay V, Zemenkova M, Zemenkov Y et al. 2016 Modeling Parameters of Reliability of Technological Processes of Hydrocarbon Pipeline Transportation. *MATEC Web of Conferences* **73** 01029 DOI: 10.1051/matecconf/20167301029
[8] Zemenkov Y D, Shalay V V, Zemenkova M Y 2015 Expert systems of multivariable predictive control of oil and gas facilities reliability *Oil and Gas Engineering Procedia Engineering* **113** 312-315 DOI: 10.1016/j.proeng.2015.07.271
[9] Zemenkov Y D, Shalay V V, Zemenkova M Y 2015 Immediate analyses and calculation of saturated steam pressure of gas condensates for transportation conditions. *Source: Oil and Gas Engineering* **113** 254-258 DOI: 10.1016/j.proeng.2015.07.330
[10] Kulikov A M, Petryakov V A., Seroshtanov I V 2015 Optimization of separation processes of multicomponent multiphase systems. *Business magazine Neftegaz* **11-12** 36-40
[11] Kulikov A M 2010 The process of preparation of unstable gas condensate at the compressor station by swirl-chamber. New technologies - oil and gas region. *Proceedings of the scientific-practical conference of students, graduate students and young scientists* 102-105
[12] Kulikov A M 2016 On Choosing Multicomponent Multiphase System Separation Progress Optimization Criteria V *IOP Conference Series: Materials Science and Engineering* **154 (1)** DOI: 10.1088/1757-899X/154/1/012039
[13] Smulsky I I 1992 *Aerodynamics and processes in the swirl-chamber.* (Novosibirsk.: Nauka)
[14] Suhovich E P 1987 *Turbulence models for describing the anisotropy of turbulent transport in swirling flows. Methods of vortex flows with heterogeneous chemical. responds. Sredoy* (M .: Nauka)
[15] Mamadaliev R A, Kuskov V N, Zemenkov U D and Popova A A 2015 Influence of high-concentrated heat sources on alloying elements transition into the weld metal *Applied Mechanics and Materials* **770** 19-22
[16] Fedoseev S N, Valuev D V, Mamadaliev R A, Sokolov P N 2017 Application of integrated methods to improve the technological properties of steel *J. Key Engineering Materials* **736** 110-115
[17] Gusev A I, Kozyrev N A, Osetkovskiy I V, Kozyreva O A, Valuev D V 2017 Quality of metal deposited flux cored wire with the system Fe-C-Si-Mn-Cr-Mo-Ni-V-Co *J. Key Engineering Materials* **736** 23-28
[18] Mamadaliev R A, Kuskov V N, Popova A A, Valuev D V 2016 Alloying elements transition into the weld metal when using an inventor power source *J. IOP Conf. Series: Materials Science and Engineering* **127**

Thermodynamic properties of gas-condensate system with abnormally high content of heavy hydrocarbons

S A Zanochuev[1], A B Shabarov[2,3], S Yu Podorozhnikov[2] and A A Zakharov[3]

[1] OOO Tyumen Oil Research Center, 42 Maxim Gorky St., Tyumen, 625048, Russia.
[2] Industrial University of Tyumen, 38 Volodarskogo St., Tyumen, 625001, Russia
[3] Tyumen State University, 6 Volodarskogo St., Tyumen, 625003, Russia

E-mail: kaf_mms@utmn.ru, sergey_urevih@mail.ru

Abstract. Gas-condensate systems (GCS) with an abnormally high content of heavy hydrocarbons are characterized by a sharp change in both phase and component compositions with an insignificant decrease in pressure below the start pressure of the phase transitions (the beginning of condensation). Calculation methods for describing the phase behavior of such systems are very sensitive to the quality of the initial information. The uncertainty of the input data leads not only to significant errors in the forecast of phase compositions, but also to an incorrect phase state estimation of the whole system. The research presents the experimental thermodynamic parameters of the GCS of the BT reservoirs on the Beregovoye field, obtained at the phase equilibrium facility. The data contribute to the adaptation of the calculated models of the phase behavior of the GCS with a change in pressure.

1. Introduction

The extraction of crude hydrocarbons from gas condensate deposits is associated with constant phase transitions occurring in a wide range of pressures and temperatures during the fluid movement through the system "reservoir-well-gathering station". The change in the thermobaric parameters leads to a change in the phase and component composition of the gas-condensate mixture (GCM). Upon that, the higher the content of heavy hydrocarbons (HC C5 +) in the formation GCS is, the greater the influence of the change in thermobaric parameters on the phase addition of a reservoir hydrocarbon system [1] is. The uncertainty of the input data not only leads to significant errors in the forecast of phase compositions, but also often to an incorrect state estimation of the whole system [2].

2. Materials and methods

Under these conditions, a key role in studying the properties of a reservoir hydrocarbon system is played by experimental research methods that are aimed at determining the phase relation under different thermobaric conditions. To avoid significant errors, the design schemes should be oriented towards the use of information obtained from the results of actual measurements.

The gas-condensate system of the BT reservoir of the Beregovoye field under initial conditions is characterized by an abnormally high content of heavy hydrocarbons (over 400 g / m^3). The component composition (Table 1) and the properties (Table 2) of the gas-condensate mixture were used to calculate, according to the equation of state [3], the phase diagram (Figure 1), which shows that the temperature at which the system is located at the depth of the deposit is close to the critical temperature of the mixture. This allows us to classify the GCS as "near-critical" fluids.

A peculiarity of the phase behavior of such systems is a sharp change in properties with a slight decrease in pressure below the pressure of the beginning of condensation (phase transition) [4]. Moreover, in conditions of a significant excess of the critical saturation in the reservoir, problems associated with the combined filtration of the gaseous and liquid phases arise [5].

Experimental data on the change in the potential content of HC C5 + in the composition of the gaseous phase with a decrease in pressure (Fig. 2) show that when the pressure decreases by 6 MPa below the pressure at the start of condensation, the gaseous phase loses half of the initial "liquid" component. Such conditions arise in the operation of wells, even with a small production output [6].

Taking into account the relevance of studying the parameters of the GCS of the Beregovoye field, special experiments were performed within the framework of this research, and a set of experimental data was organized, which in future will allow specialists in hydrodynamic modeling to create an adequate PVT model adapted in the extended data range.

Table 1. The initial composition of the GCM

Component, group	Molar content, %
He	0.011
H_2	0.005
CO_2	0.002
N_2	0.908
CH_4	74.579
C_2H_6	6.426
C_3H_8	5.587
i-C_4H_{10}	1.284
n-C_4H_{10}	2.546
i-C_5H_{10}	0.824
n-C_5H_{10}	1.006
C_6	1.087
C_{7+}	5.737
Sum	100.000

Table 2. The initial properties of GCM

Parameters	Parameter value
Molar mass, g/mol	28.63
Density, kg/m^3	1.191
HC content C_{5+}, g/m^3	407.2
Degassed condensate density, kg/m^3	730.0

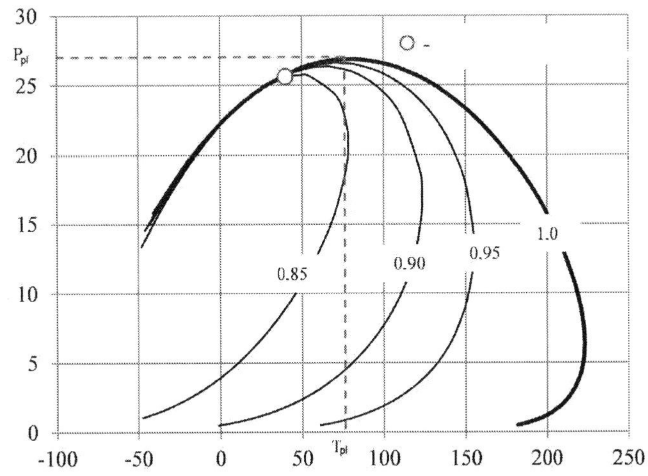

Figure 1. Calculated phase diagram of the GCS (Beregovoye OGCF, BT reservoirs)
Cipher is the volume fraction of the gas phase

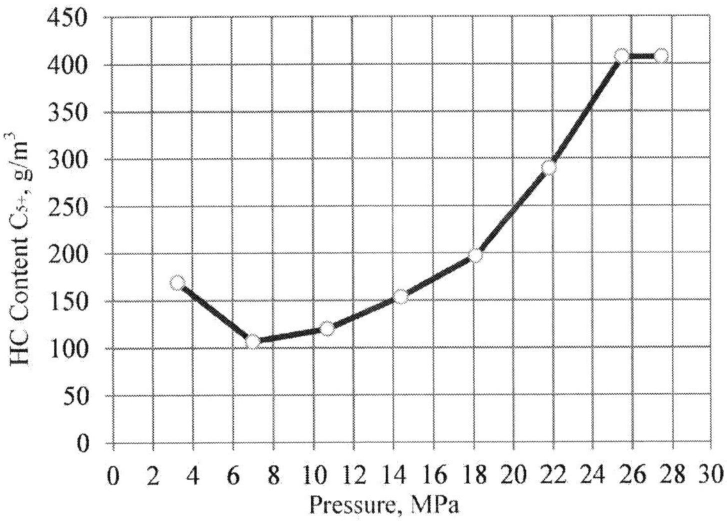

Figure 2. Change in the content of HC C5 + in the gaseous phase with a decrease in pressure.

To carry out the experimental studies, the measuring equipment included in the phase equilibria unit "PVT 3000GL System (Chandler Engineering)" with additional auxiliary equipment was used. The technical characteristics of the unit make it possible to carry out studies of hydrocarbon systems in a wide range of pressures and temperatures. The limits of the permissible pressure measurement errors are ±0.1%, temperature is ±0.5 ° C, density is ±0.75%. Accuracy of measuring the volume of the LF - hundredths of a cubic centimeter [8, 9, 10, 11].

The disadvantage of this type of units is the impossibility of turning over the measuring cell, which makes it impossible to provide reliable mixing of the system and establish thermodynamic equilibrium in the study of GCS [7]. In order to ensure the establishment of thermodynamic equilibrium of the phases, the experimental unit was modernized. An additional line was included in the circuit, which allowed the phases to be stirred by recycling ("bubbling") the gas through the precipitated liquid phase.

As an auxiliary equipment, a recombination cell is used to carry out the studies, which allows preparing the necessary hydrocarbon mixtures in a wide range of pressures (up to 100 MPa) and temperatures (up to 150 ° C). To measure the viscosity of the liquid and gaseous phases, the EV-1000 Vinci piston electromagnetic viscometer is used - a licensed analogue of the SPSL 440 Cambridge Electromagnetic Viscometer with an inaccuracy of 2.5% of the range of the value measured [12].

During the experimental studies, a clear dependence between the density and viscosity of the gaseous phase of the BT reservoir of the Beregovoye field on the pressure and content of HC C5 + in the composition (Fig. 3) was established. Studies were performed in the pressure ranges of 21 - 36 MPa, under conditions of single-phase formation gas at a reservoir temperature of 78 °C.

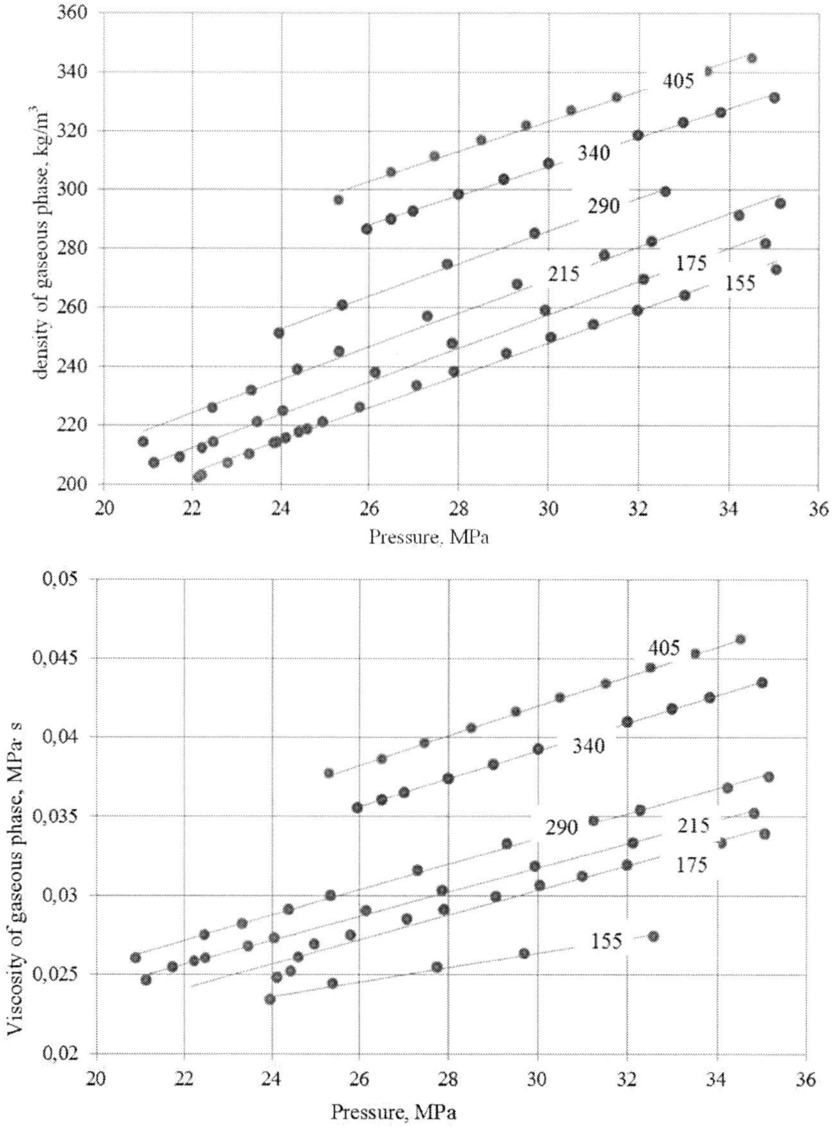

Cipher - content of HC C5 +, g/m^3

Figure 3. Dependence of the density and viscosity of the gaseous phase on the pressure and HC content of C5 + at a temperature of 78 °C.

In the thermobaric conditions of the coexistence of two phases, that is, at a pressure below the condensation start pressure, the properties of the gaseous phase depend not only on the pressure, but also on the amount of the released liquid phase. So, Fig. 4 shows the dependences of the relative volume of the liquid phase on the initial content of HC C5 + in the formation HC system at different pressures. The maximum saturation of the liquid phase under thermodynamic equilibrium conditions with a C5 + content of more than 400 g / m^3 was about 16% of the volume of the system at reservoir pressure.

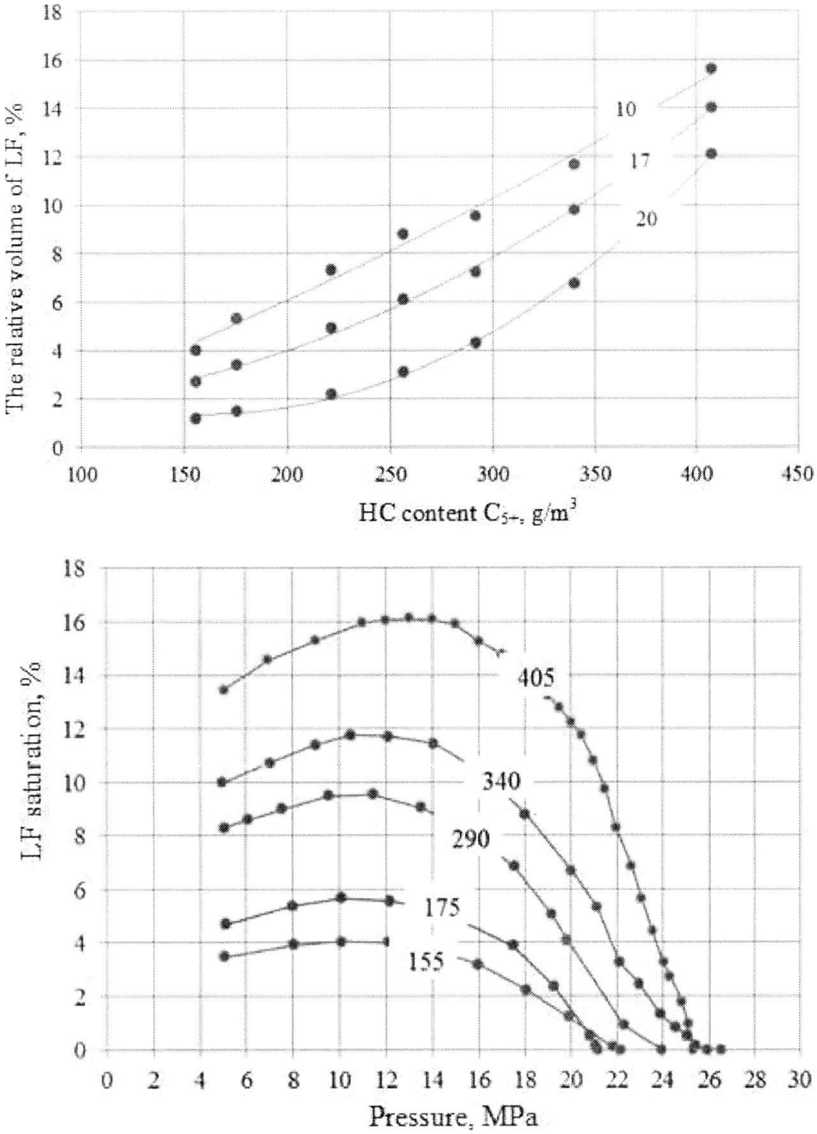

Figure 4. Dependence of the relative volume of the liquid phase (Beregovoye OGCF, BT reservoirs) on the content of hydrocarbon C5 + and pressure at a temperature of 78 °C

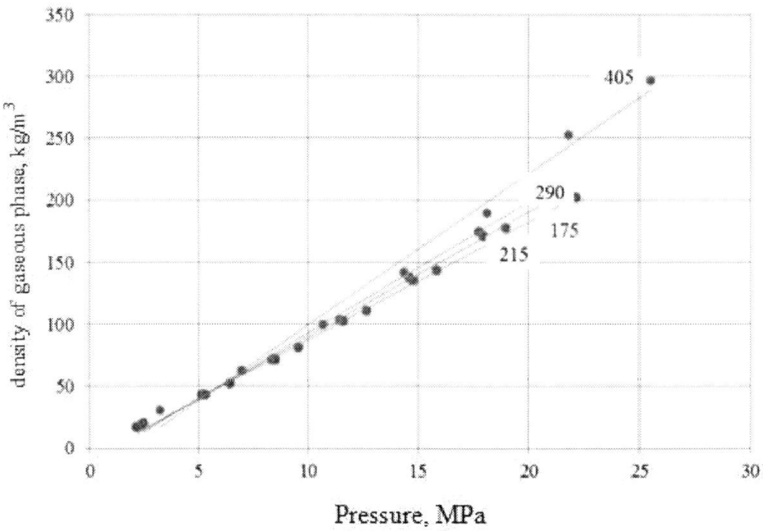

Figure 5. Dependence of the density of the gaseous phase in the two-phase area (Beregovoye OGCF, BT reservoirs) on pressure and content of HC C5 + at a temperature of 78 °C

3. Conclusion

Experimental studies have established that the density of the gaseous phase with a pressure drop below 14 MPa in the two-phase area is slightly dependent on the initial content of HC C5 +, but depends only on the change in pressure (Fig. 5).

As a result of the study, experimental data were obtained, which can form the basis of the adapted thermodynamic PVT model of formation fluid. The dependences created on the basis of the experimental data can also be used to characterize similar gas-condensate deposits with a C5 + hydrocarbon content close to 400 g / m^3 and a density of degassed condensate in the range of 720-750 g / m^3.

References

[1] Gritsenko A I, Gritsenko I A, Yushkin V V, Ostrovskaya T D 1995 *Scientific foundations of the forecast of the phase behavior of gas-condensate systems* (M: Nedra)

[2] Zanochuev S A 2011 *Forecasting reservoir losses and content UV C5 + in reservoir gas with pressure reduction based on the optimization problem. Actual issues of research of reservoir systems of hydrocarbon fields: a collection of scientific articles. Part 1. Ed. B.A. Grigoriev.* (M: Gazprom VNIIGAZ)

[3] Robinson D B 1979 Capability of the Peng-Robinson programs *Hydrocarbon Processing* **9** 269 – 273

[4] Yang T, Chen W-D, and Guo T 1997 Phase behavior of near-critical reservoir fluid mixture, *Fluid Phase Equilibria* **128** 183–197

[5] Shabarov A B, Zanochuev S A 2015 Method for calculating the changes in the component and phase composition of the gas-condensate mixture in the bottomhole formation zone. *Bulletin of Tyumen State University. Physical and mathematical modeling. Oil, gas, energy.* 1 (1)

[6] Ter-Sarkisov R M 1998 *Scientific basis for increasing the efficiency of gas condensate fields development: monograph* (M: Nedra)

[7] Lapshin V I, Nikolaev V A, Troitsky V M 2008 *Problems of technical support of reservoir oil and gas condensate systems thermodynamic studies in the fields of JSC «Gazprom».* (M .: VNIIGAZ)

[8] Shalay V, Zemenkova M, Zemenkov Y et al. 2016 Modeling Parameters of Reliability of Technological Processes of Hydrocarbon Pipeline Transportation. *MATEC Web of Conferences* **73** 01029 DOI: 10.1051/matecconf/20167301029

[9] Zemenkov Y D, Shalay V V, Zemenkova M Y 2015 Expert systems of multivariable predictive control of oil and gas facilities reliability *Oil and Gas Engineering Procedia Engineering* **113** 312-315 DOI: 10.1016/j.proeng.2015.07.271

[10] Zemenkov Y D, Shalay V V, Zemenkova M Y 2015 Immediate analyses and calculation of saturated steam pressure of gas condensates for transportation conditions. *Source: Oil and Gas Engineering* **113** 254-258 Published: 2015 DOI: 10.1016/j.proeng.2015.07.330

[11] Zemenkova M, Shalay V, Zemenkov Y, Kurushina E 2016 Improving the Efficiency of Administrative Decision-Making when Monitoring Reliability and Safety of Oil and Gas Equipment *MATEC Web of Conferences* **73** DOI: 10.1051/matecconf/20167307001

[12] Dudin S M, Zemenkov Y D, Maier A V, Shabarov A B 2016 Research and Design of Thermophysical Gas-Liquid Mixture Parameters in Product Pipelines. *IOP Conference Series: Materials Science and Engineering* **154** DOI: 10.1088/1757-899X/154/1/012021

Influence of surfactants on gas-hydrate formation' kinetics in water-oil emulsion

Yu D Zemenkov[1], A V Shirshova[2], E A Arinstein[2] and A N Shuvaev[1]

[1]Industrial University of Tyumen, 38 Volodarskogo Str., Tyumen, 625001, Russia
[2]Tyumen State University, 6 Volodarskogo Str., Tyumen, 625003, Russia

E-mail: muzemenkova@mail.ru, kaf_mms@utmn.ru

Abstract. The kinetics of gas hydrate formation of propane in a water-oil emulsion is experimentally studied when three types of surfactants (SAA (surface acting agent)) - anionic type emulsifiers - are added to the aqueous phase. It is shown that all three types of surfactants decelerate the growth of the gas-hydrate in the emulsion and can be considered as anti-agglutinating and kinetic low-dose inhibitors. The most effective inhibitor of hydrate formation in water-oil emulsion of SV-102 surfactant was revealed. For comparison, experimental studies of gas-hydrate formation under the same conditions for bulk water have been carried out. It is shown that in bulk water, all the surfactants investigated act as promoters (accelerators) of hydrate formation. A qualitative explanation of the action mechanisms of emulsifiers in the process of gas-hydrate formation in water-oil emulsion is given.

1. Introduction

The current problem of the oil and gas industry in the northern conditions is the formation of gas-hydrate plugs in oil production and transportation systems [1, 2]. To prevent gas-hydrate formation, various classical thermodynamic inhibitors (for example, methanol, glycol or their combinations) are usually introduced into the system, shifting the phase equilibrium curve to higher pressures and lower temperatures [3]. The mechanism of hydrating inhibitors of thermodynamic action consists in reducing the activity of water in an aqueous solution and, as a consequence, in changing the equilibrium conditions for the formation of hydrates. To achieve the required effect, conventional thermodynamic inhibitors are used in large quantities. This requires a large operating cost to prevent hydrate formation in field conditions.

At the moment, alternative, so-called anti-agglutinating and kinetic low-dose inhibitors are being developed and tested in the oil and gas industry [4].In view of their low concentration, they do not shift the phase equilibrium curves, but slow the formation of nuclei, the growth and adhesion of hydrate crystals, resulting in an easily transportable hydraulic fluid. In this case, there is no hydration on the walls of pipes and field communications. A great advantage of inhibitors of the kinetic type is the dosage, which is much lower than dosages of thermodynamic inhibitors. This makes it possible to reduce significantly operating costs. Kinetic inhibitors of hydrating also belong to the category of "environmentally friendly", which reduces the risks during transportation, storage and use. As a rule, a solution of organic polymers is used as kinetic inhibitors. It is assumed that the action principle of kinetic inhibitors is based on the prevention of the crystal hydrate nucleus formation by blocking the inhibitor of the crystallization center by a polymer molecule [5]. It is also known that one and the same chemical, depending on its concentration and depending on the combination with other substances, can act as an inhibitor and as a promoter [6], but the mechanism of such action has not been sufficiently studied.

The objective of this research is an experimental study of the kinetics of propane gas-hydrates growth in water-oil emulsion in the presence of surface-active substances - emulsifiers of anionic type.

2. Experimental part

In the experiments, a water-oil emulsion based on the oil of the Urnensky field of Tyumen region was used. With the help of a mixer, the following samples of emulsions with 50% volume content of distilled water were prepared: 1) without surfactant (control emulsion, sample No. 1); 2) adding sodium sulfanol (sample No. 2); 3) adding neonol AF 9-12 (sample No. 3); 4) adding CB-102 (sample No. 4). The volumetric dosage of the surfactant was 2% of the aqueous phase volume of the emulsion for all samples. The choice of surfactants was determined by the fact that they are all highly effective anionic emulsifiers, thermobaric stable in the investigated temperature and pressure range, are highly soluble in water, economically available and environmentally safe.

A mixer with a blade speed of 1300 rpm was used to stir the mixture of oil, water and surfactants. The resulting emulsion had a homogeneous, more viscous than oil, consistency and a lighter color. Fig. 1 shows a photomicrograph of an emulsion obtained by means of an optical microscope. As can be seen from the figure, the droplet size did not exceed 10 μm.

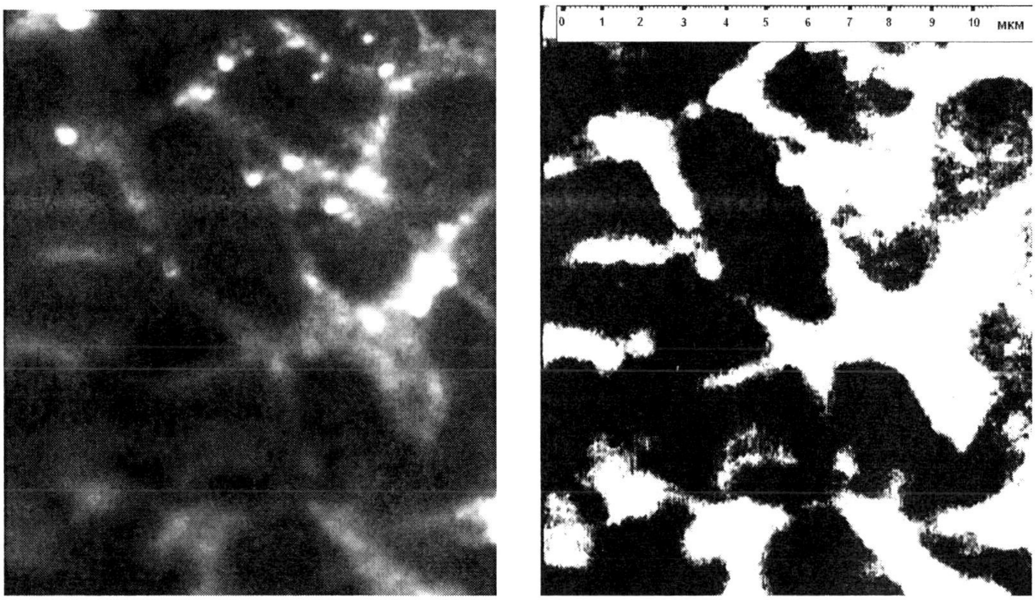

Figure 1. Microphotograph of water-oil emulsion 50/50%:
on the left - the original image; on the right - processed by the program PAINT.NET

The investigations were carried out at the automated installation "Gas hydrates" of Tyumen State University in accordance with the previously developed methods [7, 8, 9]. The results of the experiments were cyclic isothermal curves of gas pressure changes with time with the growth of gas hydrate from water-oil emulsion. One gas injection cycle lasted about 48 hours. For each sample of emulsions, 4 gas injection cycles were carried out under the same thermobaric conditions. It has previously been shown [2] that stable nucleation and growth of crystalline mass is observed by the beginning of the second cycle. That is why the comparison of the obtained experimental data of different emulsion samples started on the second cycle.

Approximating the P (t) dependence of the exponential, one can find relaxation time τ, which characterizes the rate of growth of the gas hydrate. The reciprocal of the relaxation time is kinetic growth coefficient β [2]. Fig. 2 shows the dependence of coefficient β on the cycle number. As can be seen from the figure, all three inhibitors slow down the growth rate of the gas hydrate in the emulsion, but they act differently at different stages of gas hydrate growth. Dependences of the gas hydrate growth coefficient on the cycle number for samples of emulsions No. 2 and No. 3 are similar - the maximum difference is observed in the 1st cycle and is no more than 8%. For the sample of emulsion No. 4 (with SA-102 surfactant), a significant growth retardation was observed in the 4th cycle (almost twice as compared to the control sample).

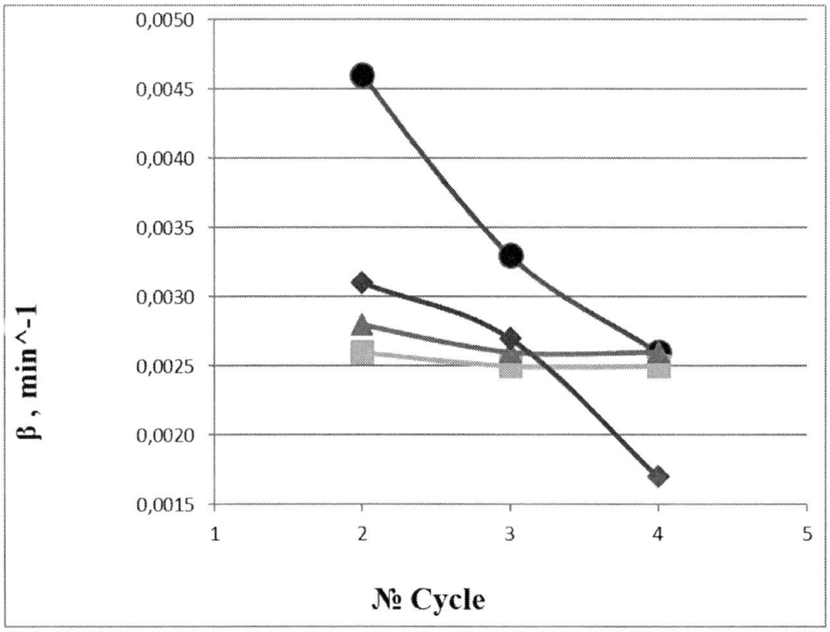

Figure 2. Dependence of the gas hydrate growth rate on the cycle number.

(sample No. 1; sample No. 2; sample No. 3; sample No. 4)

One of the important parameters characterizing the hydrate-containing environment is the coefficient of hydration, which determines the total fraction of moisture that has passed into the hydrate at a given moment in time.

Using the obtained experimental data of P-T parameters, it is possible to calculate the gas absorption by the equation of state of a real gas, taking into account the gas compressibility and solubility:

$$m_G = \frac{P_i \cdot V \cdot M}{R \cdot T_i \cdot z}, \quad (1)$$

where P_i— pressure at time t_i, ; V— reduced volume of low-pressure chamber; M— molar mass of propane; R— universal gas constant; T_i— temperature at the moment of time t_i; z— gas compressibility under given conditions.

From equation (1), it is possible to determine the mass of an ideal gas, converted into a hydrate per cycle:

$$\Delta m_{G-H} = \frac{\Delta PVM}{RTz}, \quad (2)$$

where ΔP— pressure change per cycle.

The gas mass difference at the beginning and at the end of the cycle when hydrate Δm_{G-H} forms allows one to calculate the mass of hydrate m_H and the mass of water m_{W-H} converted to the gas hydrate:

$$m_{w-H} = \Delta m_{G-H} \cdot 2{,}32, \quad (3)$$

where the coefficient of 2.32 is calculated based on the chemical formula of propane hydrate.

According to the data obtained, hydration factor K_H was calculated - the fraction of water converted into hydrate from the total amount of water in the sample.

$$K_H = \frac{m_{W-H}}{m_W} . 100\%, \quad (4)$$

where m_W – initial mass of water in the sample.

The dependence of the hydration coefficient on the cycle number allows one to monitor the relative total mass increment of the gas hydrate by the end of each cycle and, thus, to compare the kinetics of different emulsion samples' growth. Figure 3a shows the results of calculations. As can be seen from the figure, all three surfactants under study reduce the interaction of gas and emulsion in comparison with the control sample, i.e. they are inhibitors of gas hydrate formation in water-oil emulsions. The greatest effectiveness of inhibition was found on the sample of emulsion № 4 (with the addition of surfactants CB-102).

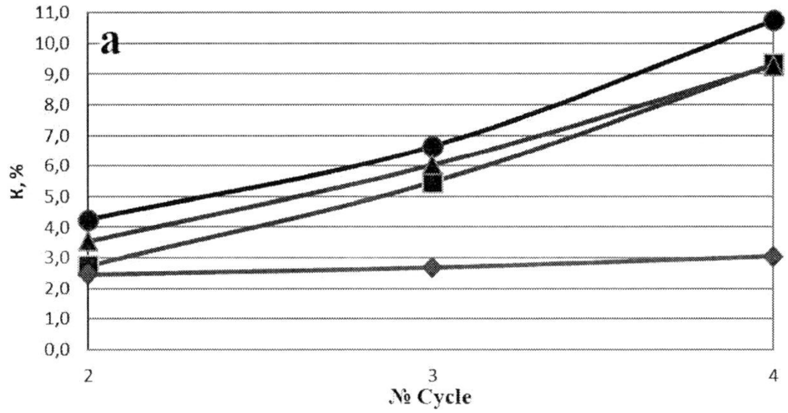

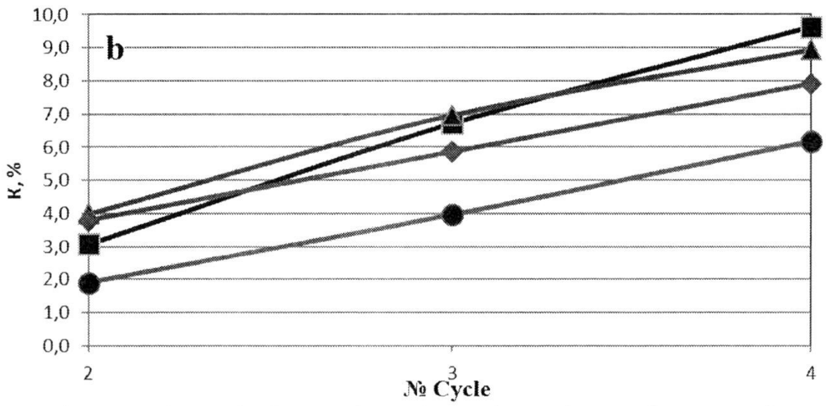

Figure 3. Dependence of the gas hydrate ratio on the cycle number: a - for water-oil emulsion, b - for water and aqueous solutions of surfactants

(sample No. 1; sample No. 2; sample No. 3; sample No. 4)

For comparison, similar experiments were carried out for bulk water (sample № 1) and for three corresponding surfactant solutions in distilled water at the same concentration. Fig. 3b presents the results of these studies. As can be seen from the figure, for bulk water, as expected, all three surfactants increase the rate of growth of the gas hydrate, i.e. they are promoters of gas hydrate formation. The offered mechanism for accelerating the growth rate of gas hydrate formation in an aqueous solution of a low-concentration surfactant was studied in detail earlier in research [10, 11]. The opposite effect discovered - the inhibitory effect of the surfactants studied in a dispersed medium - the water-oil emulsion - can be easily explained by the properties of emulsifiers. When the emulsifier is added to the water-oil emulsion at the water-oil interface, a shell is formed from the emulsifying substance, which prevents the gas from penetrating into the aqueous phase of the emulsion, and prevents the water droplets from converging (sticking together). Consequently, the emulsifiers in the initial stage of gas hydrate formation prevent the formation of nucleus of the gas hydrate in the emulsion and then inhibit the formation and growth of agglomerates from the nucleation centers of crystal hydrates. The higher efficiency of inhibition of CB-102 surfactants in comparison with other surfactants under study can be explained by the higher interfacial tension of its aqueous solution and oil [12, 13]. Thus, according to modern classification, the surfactants under study can be considered as anti-agglomerating and kinetic low-dose inhibitors of gas hydrate formation in water-oil emulsion.

To explain the high efficiency of inhibition of CB-102 surfactants in comparison with other surfactants, additional measurements of the surface and interfacial tension between aqueous solutions of surfactants and oil were carried out using a drop-test method. The results of these studies are presented in Table 1. As can be seen from the table, the interfacial tension between an aqueous solution of a surfactant based on CB-102 and oil is several times higher than that of other surfactant solutions.

Table 1. Results of surface and interfacial tension measurements

NAME		σ, mN/m
Surface tension	Distilled water	(71±2)
	oil of Urnenskoye field	(34±2)
	Distilled water + surfactant (neonol AF 9-12)	(33±5)
	Distilled water + surfactant (sodium sulfanol)	(31±2)
	Distilled water + surfactant (CB-102)	(26±2)
Interfacial tension between 2% aqueous surfactant solution and oil	Water-Oil	(8.2±0,2)
	A solution of surfactants (neonol AF 9-12) in distilled water - oil	(3,2±0,1)
	A solution of surfactant (sodium sulfanol) in distilled water - oil	(2,1±0,3)
	A solution of surfactant (CB-102) in distilled water - oil	(16±1)

3. Conclusions

An experimental study of the kinetics of gas hydrate formation in water-oil emulsion in the presence of three highly effective emulsifiers of anionic type: sodium sulfanol, neonol AF 9-12, and CB-102, has been carried out.

When processing the obtained experimental data by the method of relaxation curves, it was found that all inhibitors reduce the kinetic coefficient of growth of the gas hydrate in the emulsion, but act differently at different stages of gas hydrate growth.

The calculation of the gas hydrate ratio confirms the inhibitory effect of all three surfactant types, which can be considered as anti-agglomerating and kinetic low-dose inhibitors. The most effective inhibitor of hydrate formation in the water-oil emulsion is CB-102.

To explain the high efficiency of inhibition of CB-102 surfactants in comparison with other surfactants, surface and interfacial tension measurements between aqueous solutions of surfactants and oil were carried out. It was found that the interfacial tension between an aqueous solution of a surfactant based on CB-102 and oil is several times higher than that of other surfactant solutions.

It is shown that in bulk water, unlike the emulsion, all three surfactants investigated act as promoters of hydrate formation, which is coherent with the already known studies.

A qualitative explanation of the surfactants action mechanisms in the process of gas hydrate formation in water-oil emulsion is given.

References

[1] Molchanov D A, Shirshova A V, Politova S S 2015 Determination of the depth of gas hydrate formation in oil wells, in view of the mode of its operation and the composition of associated gas. *Bulletin of Tyumen State University. Physico-mathematical modeling. Oil, gas, energetics.* **2 (2)** 22-28

[2] Shabarov A B, Shirshova A V, Gasheva S S 2014 Influence of physical and chemical properties of oil on gas hydrate formation in water-oil emulsions. *Proceedings of universities. Oil and gas.* **2** 88-93

[3] Istomin V A, Kvon V G 2004 *Prevention and dissolution of gas hydrates in gas production systems.* (M .: Gazprom)

[4] Carroll J 2007 *Hydrates of natural gas.* (M.: Tehnopress)

[5] Faresov A V, Ponomarev A I 2014 Study of the technological characteristics of hydrate-forming inhibitors of kinetic type during application and utilization at oil and gas production facilities. *Oil and gas business: electronic scientific journal* **1** 137-147

[6] Nesterov A N, Reshetnikov A M, Manakov A Yu, Rodionova T V, Paukshtis E A, Asanov I P, Bardakhanov S P, Bulavchenko A I 2015 Promotion and inhibition of gas hydrate formation by oxide powders. *Journal of Molecular Liquids.* **204** 118–125

[7] Dudin S M, Zemenkov Y D, Shabarov A B, et al. 2016 Modeling the Phase Composition of Gas Condensate in Pipelines. International Scientific-Practical Conference of Students, Graduate Students and Young Scientists Transport and Storage of Hydrocarbons *IOP Conference Series: Materials Science and Engineering* **154** DOI: 10.1088/1757-899x/154/1/012010

[8] Shirshova A V, Danko M Yu 2016 Growth and dissociation of gas hydrates in water-oil emulsions. *Proceedings of universities. Oil and gas* **5** 95-101

[9] Shabarov A B, Danko M Yu, Shirshova A V 2011 Designing a machine for the accelerated growth of gas hydrates from micron-sized ice particles in a gas stream. *Bulletin of the Tyumen State University. Physico-mathematical modeling. Oil, gas, energetics.* **7** 46-51

[10] Amelkin S V, Melnikov V P, Nesterov A N 2000 Growth kinetics of gas hydrates in dilute solutions of non-electrolyte inhibitors. *Colloidal Journal* **62-4** 450-455

[11] Akulov K, Golik V, Chekardovsky M, Serebrennikov A 2016 Research of Spined Heat-Exchanging Pipes. *MATEC Web of Conferences* **73** DOI: 10.1051/matecconf/20167301027

[12] Akulov K A, Petryakov V A, Mostovaya N A 2017 The Use of Foamed Concrete for Oil Spill Treatment and Carbon Capture. *IOP Conference Series: Earth and Environmental Science* **50** DOI: 10.1088/1755-1315/50/1/012042
[13] Pimnev A, Aginey R, Yakubovskaya S, Alexandrov M 2016 Tank Structural Fatigue Fracture in Hydrocarbon Transportation *MATEC Web of Conferences* **73**
[14] Brand A E, Chekardovskiy S M, Akulov K A 2017 The Analysis of the Development Dynamics and Structural Balance of Solar Energy in the World. *IOP Conference Series: Earth and Environmental Science* **50** DOI: 10.1088/1755-1315/50/1/012034

IOP Conf. Series: Materials Science and Engineering **357** (2018) 012026 doi:10.1088/1757-899X/357/1/012026

Power law of distribution of emergency situations on main gas pipeline

K S Voronin and K A Akulov

Industrial University of Tyumen, 38 Volodarskogo St., Tyumen, 625000, Russia

E-mail:voronin_tsogu@mail.ru

Abstract: The article presents the results of the analysis of emergency situations on a main gas pipeline. A power law of distribution of emergency situations is revealed. The possibility of conducting further scientific research to ensure the predictability of emergency situations on pipelines is justified.

1. Introduction

Experience in the operation of main pipelines shows that most of the damage occurs suddenly without external apparent causes. It is believed that these disruptions have a fatigue nature [1, 2]. The main method for assessing the strength reliability is to determine the safety margin. The number of cycles of variable load is usually 10^5-10^6 per year [2]. The absence of a general theory of destruction complicates the theoretical analysis of the mechanisms determining this process. The complexity of the object and the multistage nature of the process of crack formation in the pipe, taking into account the stochastic nature of the operating loads, are the justification that the reliability problem should be solved on the basis of probabilistic methods.

2. Materials and methods

Indeed, the reliability of the pipeline element $H_i(t)$ is defined as the probability of the element's quality parameter remaining in the specified range of permissible values for time t. The point is that with time, the chosen quality parameter will necessarily go beyond the range of admissible values due to the stochastic dynamics of the studied system. In this connection, it becomes necessary to determine the probabilistic characteristics, namely the probability density of the parameters of the system behavior and external influences.

When solving the problem of statistical dynamics, it is generally assumed that the external influence and the parameters of the system behavior can be characterized by determinate or random variables. Random variables are determined by the corresponding densities of probability distribution, and stationary random functions are characterized by mathematical expectation and correlation functions. Most often in papers on failure prediction as a result of long-acting cyclic loads, the ones that most significantly influence the strain-stress state of the structure are chosen as input parameters. It is very often assumed that the distribution of these parameters obeys the normal law [3, 4, 5].

At the initial stage of analyzing the reliability of the pipeline system, homogeneous groups of elements are distinguished. In this paper, such group of elements is the linear part of the pipeline. Consequently, the investigated system can be represented in the form of series coupling of elements.

Let us consider the definition of the system reliability, where the reliability of each element is taken $H_к$. A pipeline section of length *L* is taken as an element for operation period *t*. In (Fig. 1), a system flow-chart is shown.

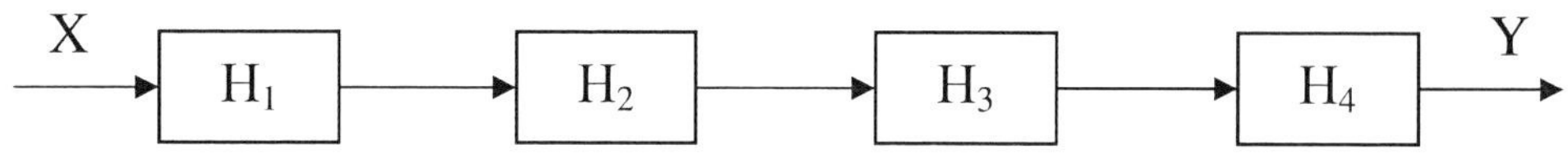

Figure 1. System flow-chart of the pipeline

The probability of failure-free operation of each element is taken P(t, L) = H(t, L). Elements interact in such a way that their failures ($R_к = 1 - P_к$) are independent. In this case, the system remains operational if all its elements work reliably. The probability of a joint event consisting of several independent events will be equal to the product of the probabilities of the realization of each event [8, 9].

Such distribution is most often found in nature, and it reflects the linear response of the system to a stationary external disturbance. At the same time, the reliability of the linear part of a pipeline of length *L* consisting of *N* sections during operation period *t* is determined by the formula:

$$P(t, L) = H(t, L) = \prod_{i=1}^{N} H_i(t, \Delta L_i) \qquad (1)$$

where *L* – pipeline length;
N – number of pipeline sections;
t – operational period of the pipeline;
ΔL_i – length of the ith pipeline section.

This approach is valid if the failures at each allocated section are independent (due to the linearity of the system). However, in main gas pipelines, the gas flow regime is unstable. A significant pressure change along the pipe length through time is observed. The pressure gradient when the turning off the compressor of the intermediate station makes 0.3-0.4 MPa [2]. The generated high and low pressure waves cause coherent stresses at high distances. This can lead to a violation of linearity in the pipeline. Consequently, the distribution of the parameters responsible for the state of the system may deviate from the normal distribution law.

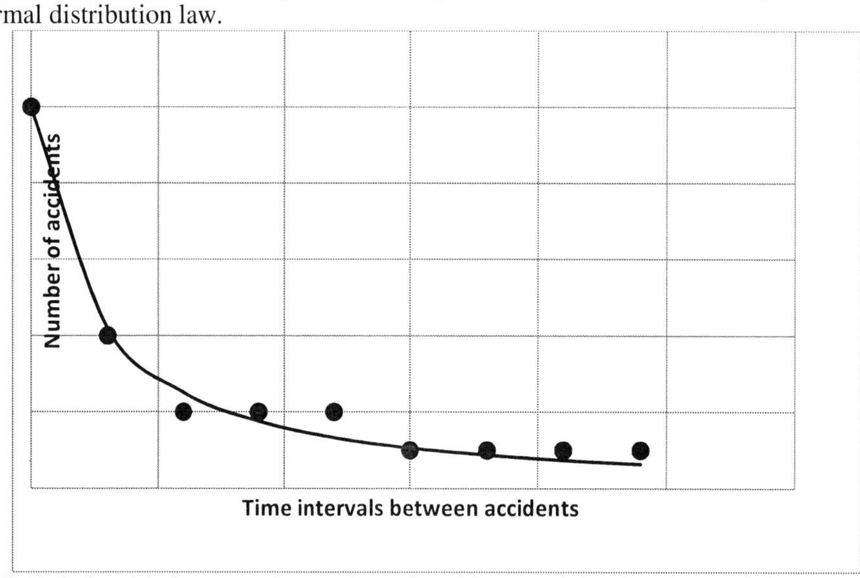

Figure 2. Relationship between the number of accidents and the time intervals between them

By the example of one of the gas pipelines in the north of Tyumen region, the laws of distribution of emergency and catastrophic situations during 6 years of operation are considered.

When analyzing emergency situations, probability density ρ(x) can be replaced by the dependence of the size of the event on its rank. The economic damage caused by the accident can be chosen as the rank of events [10, 11].

From the analysis of dependence U(x) (Fig. 2), an approximating dependence can be obtained. The curve of emergency behavior demonstrates the power law dependence:

$$U(x) \sim x^{-(1+\alpha)} \tag{2}$$

This distribution differs significantly from the normal one. A normal distribution is characterized by a rapid decrease in power dependence. This means that large accidents are very rare events and the probability of their implementation falls rapidly to zero.

It follows from the graph (Fig. 2) that the probability of the event of a large accident is not equal to zero in the pipeline system. Indeed, the catastrophic event occurred on the considered gas pipeline [12, 13]. Its scale exceeds far the average statistical emergency situations. In Fig. 3 and 4, a comparison of the exponential, normal, and power distributions is shown.

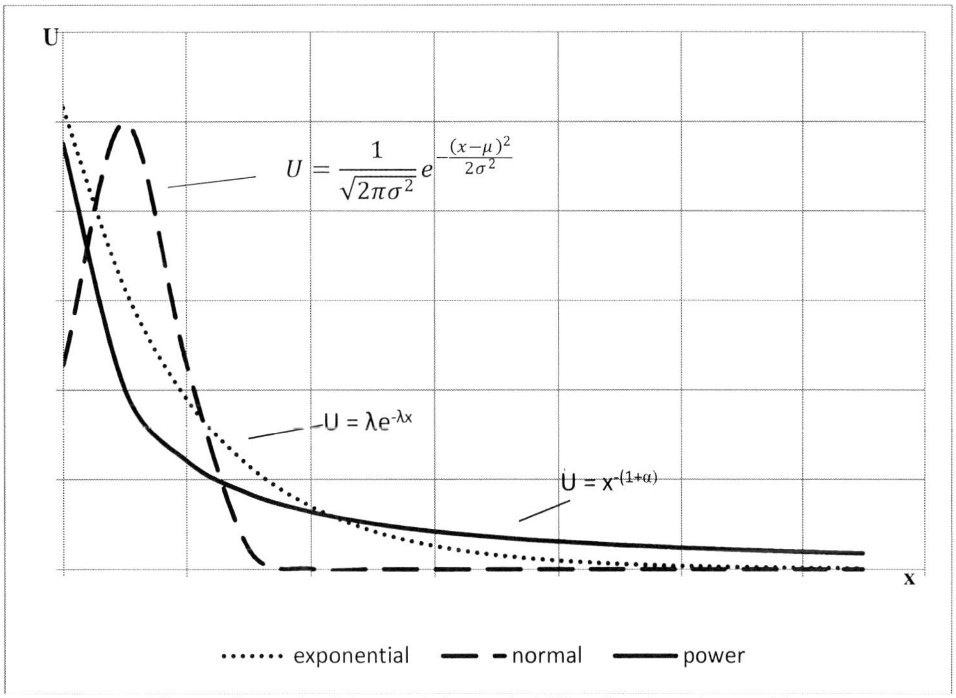

Figure 3. Typical form of the probability density of quantities

In the papers [1, 6, 7], the description of emergency and catastrophic phenomena from the point of view of nonlinear dynamics is given. It is known that the power law of the distribution of random events is observed in systems with feedback. This means that the disturbance can spread through the pipeline as far as possible. Pressure jumps for an unstable gas flow in a pipe can be considered as a disturbance object.

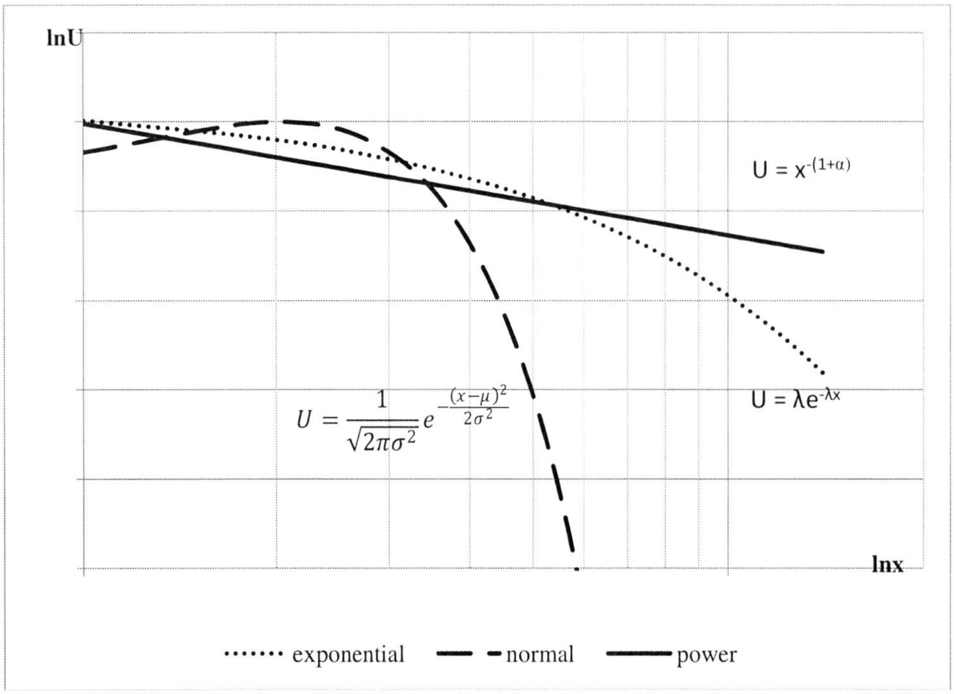

Figure 4. Typical form of the probability density of quantities in logarithmic coordinates.

3. Conclusion

This approach opens a program for developing an algorithm for monitoring emergency and catastrophic events. At the first stage, it is necessary to develop methods that allow us to minimize the information about the processes occurring in metal caused by a number of disturbances and aggressive environmental influences in the functional dependence. At the second stage, methods should be developed for analyzing the probability of transmission of disturbances in a hierarchical environment. The fact is that the causes of the appearance of cracks lie at lower structural levels (defects, dislocations, micro cracks). Thus, further research is required to ensure the predictability of accidents on main gas pipelines.

References:

[1] Venttsel 'E S, Ovcharov V A 2000 *The theory of probability and its engineering applications.* (Moscow: Vysshaya skola)

[2] Harinovsky V V 2000 *Reliability and resource of gas pipeline constructions.* (Moscow: Nedra)

[3] Box D, Jenkins G 1974 *Time series analysis, forecast and management.*

[4] Ostreikovskii V A 2003 *The reliability theory.* (Moscow: Vysshaya shkola)

[5] Kashian R, Rao A 1983 *Constructing the dynamic stochastic models on experimental data.* (Moscow: Nauka)

[6] Malinetsky G G, Podlazov A V, Kuznetsov I V 2006 National system of scientific monitoring. *Novoe v sinergetike* **1** 28-48

[7] Agrafenin S I, Perov S N 2006 Methodology of ensuring the reliability of pipeline systems in their design. *Neftyanoe hozyaistvo* **11** 102-106

[8] Voronin K S, Ogudova E V 2016 The Effect of Dynamic Processes in the System "Pipe-Soil" on the Pipeline Deviation from Design Position. Transport and Storage of Hydrocarbons. *IOP Conf. Series: Materials Science and Engineering* **154** 012019 doi:10.1088/1757-899X/154/1/012019

[9] Voronin K S 2016 Forecasting and Evaluation of Gas Pipelines Geometric Forms Breach Hazard. Transport and Storage of Hydrocarbons. *IOP Conf. Series: Materials Science and Engineering* **154** 012020 doi:10.1088/1757-899X/154/1/012020

[10] Dudin S, Voronin K, Yakubovskaya S, Mutavaliev S 2016 Modeling Hydrodynamic State of Oil and Gas Condensate Mixture in a Pipeline. *MATEC Web of Conferences.* DOI: 10.1051/matecconf/20167302021

[11] Pirogov S P, Cherentsov D A, Gulyaev B A 2016 Prospects of applying vibration-resistant pressure gauges in the oil and gas industry. *IOP Conference Series: Materials Science and Engineering* 012013

[12] Pirogov S P, Cherentsov D A, Chuba A Y 2016 Study of elastic sensing elements for vibration-resistant pressure gauges. *IOP Conference Series: Materials Science and Engineering* 012015

[13] Pirogov S P, Cherentsov D A 2016 Theoretical foundations of the design of vibration-resistant manometers. *Measurement Techniques* 1-5

[14] Kozyrev N A, Galevsky G V, Valuev D V, Shurupov V M, Kozyreva O E 2015 Surfacing with Tungsten-containing Ores *J. IOP Conference Series: Materials Science and Engineering* **91(1)**

[15] Mamadaliev R A, Kuskov V N, Popova A A, Valuev D V 2016Alloying Elements Transition into the Weld Metal When Using an Inventor Power Source *J. IOP Conference Series: Materials Science and Engineering* **127**

[16] Fedoseev S N, Valuev D V, Mamadaliev R A, Sokolov P N 2017 Application of integrated methods to improve the technological properties of steel *J. Key Engineering Materials* **736** 110-115

[17] Saprykina N A, Saprykin A A, Matrunchik M S 2014 Formation of Surface Layer of Cobalt Chrome Molybdenum Powder Products with Differentiation of Laser Sintering Modes *J. Applied Mechanics and Materials.* **682** 294-297

[18] Gizatulin R A, Kozyrev N A, Valuev D V, Valueva A V, Serikbol A 2016 Research on operational characteristics of imported rails under conditions of east siberian railway *J IOP Conf. Series: Materials Science and Engineering* **127**

[19] Valuev D V, Serikbol A 2014 Stress state and its definition in hardfacing heat-resistant steel cold rolling mill rollers *J. Advanced Materials Research* **1040** 703-707

[20] Malushin N N, Valuev D V, Valueva A V, Serikbol A 2014 Testing of metallurgical equipment on contact strength and wear resistance *J. Applied Mechanics and Materials* **682** 53-57

Complex solution of problem of all-season construction of roads and pipelines on universal composite pontoon units

A V Ryabkov[1], N A Stafeeva[1], V A Ivanov [1] and A F Zakuraev[2]

[1] Industrial University of Tyumen, 38 Volodarskogo St., Tyumen, 625000, Russia
[2] Center for Road Strategic Studies, lane Minaevsky 2, Moscow, 117601, Russia

E-mail: an10091980@gmail.com

Abstract. A complex construction consisting of a universal floating pontoon road for laying pipelines in automatic mode on its body all year round and in any weather for Siberia and the Far North has been designed. A new method is proposed for the construction of pipelines on pontoon modules, which are made of composite materials. Pontoons made of composite materials for bedding pipelines with track-forming guides for automated wheeled transport, pipelayer, are designed. The proposed system eliminates the construction of a road along the route, ensures the buoyancy and smoothness of the self-propelled automated stacker in the form of a "centipede", which has a number of significant advantages in the construction and operation of the entire complex in the swamp and watered areas without overburden.

1. Introduction

Currently, the subsurface method of laying pipelines with low lifting properties of bogs of the third type is predominantly used (practically complete absence of mineral soil). A considerable amount of soil expansion is observed due to unsteady degradation and metamorphic processes in flooded areas, permafrost soils and in soils with deep seasonal freezing. This leads to the movement of pipelines and the loss of their longitudinal and transverse stability with subsequent escape from the underground position to the day surface with the formation of domes of different configurations. This can lead to a critical value of the strain-stress state of the pipeline wall and its subsequent destruction [1, 2] (Fig. 1).

Figure 1. Domes in the form of vertical spatial curves

In this connection, the task is to create all-season complexes for a full cycle of pipeline construction, described in detail in Podgorbunsky's paper.

2. Materials and methods

The base of the complex is two main sledge platforms of catamaran type. The distance between the pontoon bodies of the main sledge platforms is such that it is possible to place two auxiliary roller platforms between them, on which all technological operations are performed.

On the main sledge platforms (including on the collector platform), there are portal frames on which the following equipment is located. There is a crane with the removed running base on the main sledge platform. On the additional main sledge platform there is an upper structure with crossing bridges, a closed area for the working personnel of the complex, and a load-lifting mechanism providing vertical and horizontal movements of the machine attached to it for cleaning and insulation of welded joints (Fig. 2).

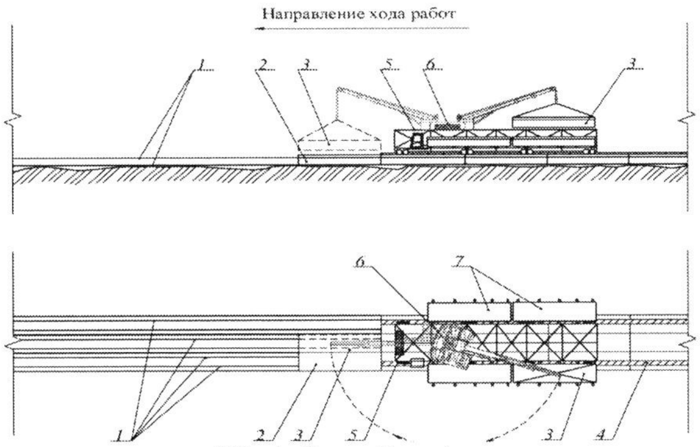

Work flow direction

Figure 2. A general view of the assembling and laying complex for year-round construction of main pipelines under conditions of Siberia

1 - tensioned carrier wires;
2 - pontoon unit laid on the wires;
3 - laid pontoon unit to be laid;
4 - tracking route;
5 - automated wheeled traffic;
6 - crane mounted on automated wheeled traffic;
7 - pontoon units laid on automated wheeled traffic.

Two main stages can be identified when creating new alternative technological solutions for laying pipelines through bogs:

— the development of technology for the construction of the main pipelines, which combines methods used for laying pipelines and excludes the technology of building pipelines using traditional methods;

— creation of an automated full-cycle assembling and laying complex.

For this, at both stages the authors solved the following main tasks:

— development of technology for pipelines construction in bog areas and with water-saturated soils;

— development of specialized machines, mechanisms and devices that provide a complete construction process;

— development of integrated mechanization and automation schemes.

A rational organization consists in the application of a combined summer and winter organization of construction and installation jobs in order to rationally distribute the resources of production units for the pipelines construction. In this case, it is necessary to significantly reduce the volume of winter construction due to the selection of sections of the route where it is possible to lay the main roads in the summer [3, 4].

In this regard, an important role will be played by increasing the capacity of territorial construction and installation organizations in areas of construction concentration, as well as the development of a network of mobile organizations and easily built inventory mobile bases in the form of unit structures.

All newly created versions of full-cycle complexes for year-round construction in bogs and watered areas should meet the following basic requirements: provide for work in the summer period on bogs of all types and watered areas; have a design that provides for work with existing traction means; to have the necessary stability in performing the technological operations and a floating margin at a nominal load (for complexes based on sledge platforms and platforms with air off-loading); ensure the work performance on the full technological cycle.

The technology of year-round construction in wetlands with the use of semi-stationary mechanized complexes is in many respects similar to the considered technology.

The use of the mentioned technology is considered economically efficient in conditions when the bogginess of the routes exceeds 20% and assumes the maximum use of water-flooded routings in the spring, summer and autumn periods. The technology of semi-stationary mechanized complexes is also based a combination of traditional technologies. Sites are selected on a dry section of the route. A construction site is built for installation and insulation of pipe joints, storage of basic structures, construction the launching way for pipeline paying. The interval between the sites of technology of semi-stationary mechanized complexes is 10 - 20 km.

The experience of building pipelines in bogs and water-flooded areas on similar technological schemes exists also in foreign practice. However, the technologies developed for constructing pipeline crossings through bogs using combined methods in the first stage, contain a number of significant shortcomings along with certain advantages.

In particular, there is a slight increase in the efficiency and rate of construction; complexity and high relocation frequency of machines and mechanisms; short work duration between relocations; labor intensity and high cost of additional engineering training (arrangement of the site, entrances, etc.); absence of a united front and route of work performance; dependence of the technology of works performance on the specific route conditions, which, in the final analysis, causes the absence of unified technological schemes; impossibility of laying pipelines through bogs and water-flooded areas under particularly difficult conditions, with a long extension of bogs and lacustrine-boggy areas and the absence of dry ones; absence of original progressive technical and technological solutions.

Due to the imperfection of such combined technologies, in the second stage of development, the main attention was paid to the creation of all-season full-cycle complexes operating on a unit technology, regardless of the season and specific hydrogeological conditions of construction.

In further developments, in accordance with the adopted technology, the complexes for the year-round construction of pipelines in bogs and water-flooded areas were finally divided into transport complexes and assembling-excavating-laying ones [5, 6].

With this kind of division, a different degree of complexity of the construction development was revealed. The simplest one is the creation of transport complexes using already existing air cushion support platforms. Non-self-propelled platforms with load-carrying capacity of 20, 40 and 60 tons were invented by the institution Sibkomplektmontazh, which can be used as one assembly with caterpillar tractors. The use of self-propelled transport platforms on an air cushion SAVR-5GD and SAVR -40 is even more promising. Caterpillar trucks with a hydraulic drive on special long-stroke controlled mountings or self-propelled transport-technological platforms are used there similar to a platform with a rotary-propeller engine PSTT-I.

Technological schemes were developed which provided for the synchronous promotion of all sledge platforms and schemes in which the platforms could move independently with the help of a special wire gripper device. The complex forward movement was carried out by a pulling winch and anchors, arranged on pre-selected dry sections.

A version was also developed, in which the movement of the complex was provided by a special universal traction device with heavy lifting capacity. In addition, it was initially assumed that the pipeline would be rigidly connected to sledge platforms, that is, all platforms would be equipped with stationary fixed supports.

However, all these developments had a number of common shortcomings.

A non-self-propelled version and impossibility to perform work in the absence of hard soil to fix the anchors; the platforms equipment by any kind of propeller (rotary-screw, caterpillar, etc.) was considered inexpedient, due to the complexity and unreliability of the structure and the large mass of the platforms; the inoperability of the complex in difficult relief conditions; a large number of pontoon platforms and their extended placement along the length of the installed pipeline caused platform overloads from the weight of the raised section of the pipeline. This, in turn, required an increase in the geometric dimensions and load-carrying capacity of the platform and led to a significant increase in tractive effort necessary for their promotion.

Thus, the analysis showed that in the case of load equalization, the height of the pipeline end above the soil surface at the site of the centering and welding of the root (first) fillet can reach 5-6 meters for a pipeline with a diameter of 1420 mm. Such lifting heights values are for the pipeline from separate 12-meter pipes. These disadvantages are, to some extent, removable, but the performance rate of welding, installation and insulation-packing works is reduced so much that the proposed technological schemes become economically inefficient.

In parallel with the work on the creation of assembly and packing systems based on sledge platforms and platforms with the air cushion, a number of original developments were made. Despite the individual technical and technological solutions that deserve attention, the application of any variant of the assembling and packing systems considered above is not a radical solution of the problem.

Thus, at present, an important task is the development of technology for year-round construction and the appropriate design of the installation and packing complex, combining the advantages of the alternative considered technological solutions and free from their shortcomings.

According to our proposal, for the first time when designing pipelines and along path routings of a new type, the above-mentioned shortcomings can be reduced or completely eliminated [7].

The essence of the method is as follows:

Profile of a complex structure, constructed of composite materials (pontoons for laying) with track-forming guides will provide for the increased reliability along the road, floating and ease of movement for the self-propelled lifting and loading device, designed in the form of a "centipede", which has a number of significant advantages in the construction and operation of the entire complex.

This complex is realized using high-strength composite pontoons made of basalt-plastic composite and mounted under the base of the main pipeline laid on all types of bogs and permafrost soils.

The composite pontoon assembly for the pipeline route base includes a volumetric body, the component parts for the pipeline base made of multi-link units consisting of central and side pontoons, composite longitudinal and transverse wires (ropes) [8].

Each internal spatial power frame of the unit is equipped with solid thin-walled I-beams, connecting elements (locks) of units of the type "male-female" [9, 10]. The deck part of the pontoons is made of sheet material, supported by beams of the I-section. The traffic way is made of beam channels with the caps upwards.

This system during construction completely excludes the use of excavators, pipe layers and pipe carriers for digging trenches. All this creates significant saving of fuel and resources in the construction of such pipeline route and almost excludes the burden removing. The profile of the earth's surface due to the peculiarities of the formation of the relief by natural influences is largely closer to curves with a variable radius, which gives a better negotiation of such route into the terrain of the area [7].

Pontoons in the longitudinal and transverse direction are fastened together by composite wires. Before pulling the composite wires, it is necessary: to check their integrity and compliance with the requirements; to carry out the installation of equipment for unhindered unwinding of wires. The technology of the wires pulling is (see Fig. 3), that between the two closest block-technological wells the point of supply and the point of reception of wires are selected. Longitudinal wires are fixed at certain distances in the land area by "anchors", installed in wells or small buildings, with hydraulic lifts to maintain a constant level of the pipeline in the pontoon system so that they can withstand greater pulling forces and change the geometric position of the track.

Composite wires are suggested to be stretched using a low-power ground-effect machine developed by the authors. In this case, it performs the work on the tension of the composite cables (Fig. 3).

Figure 3. Tensioning the wires with ground-effect machine over the bog:
1 – ground-effect machine for stretching composite cables above the bog; 2 – composite wire; 3 – automated pipe layer of the "centipede" type; 4 – bog

Also on the pontoons, an oil and gas pipeline is installed and fixed from lateral displacements. The mentioned three constituent components are fixed in a single bundle and in this state the system has a rigid structure with a high degree of reliability, durability and floating under any changes in the surface water level in the swamp and in the pipeline, which has the appropriate design of the strain-stress state compensators. In addition, such pontoon road can be operated year-round, under all weather conditions for repairing routings.

For example, in winter, with large snowdrifts and the presence of ice, a vehicle can be launched in front of the column, which will have a splitting and raking effect, cleansing track for unhindered moving of any kind of equipment.

In connection with the mentioned above, the goal is to build a fundamentally new combined pontoon routing made of composite materials of standard sizes assembled into a single "path" with pontoon properties with the establishment of special metal ruts along the entire length of the route connected along and across by the wires to lay the pipeline system.

An example of a completed project for the construction of a pipeline bed and a routing for all-season use made of high-strength pontoon units using an automated "centipede" complex is shown in Fig. 4 – *a, b, c, d*.

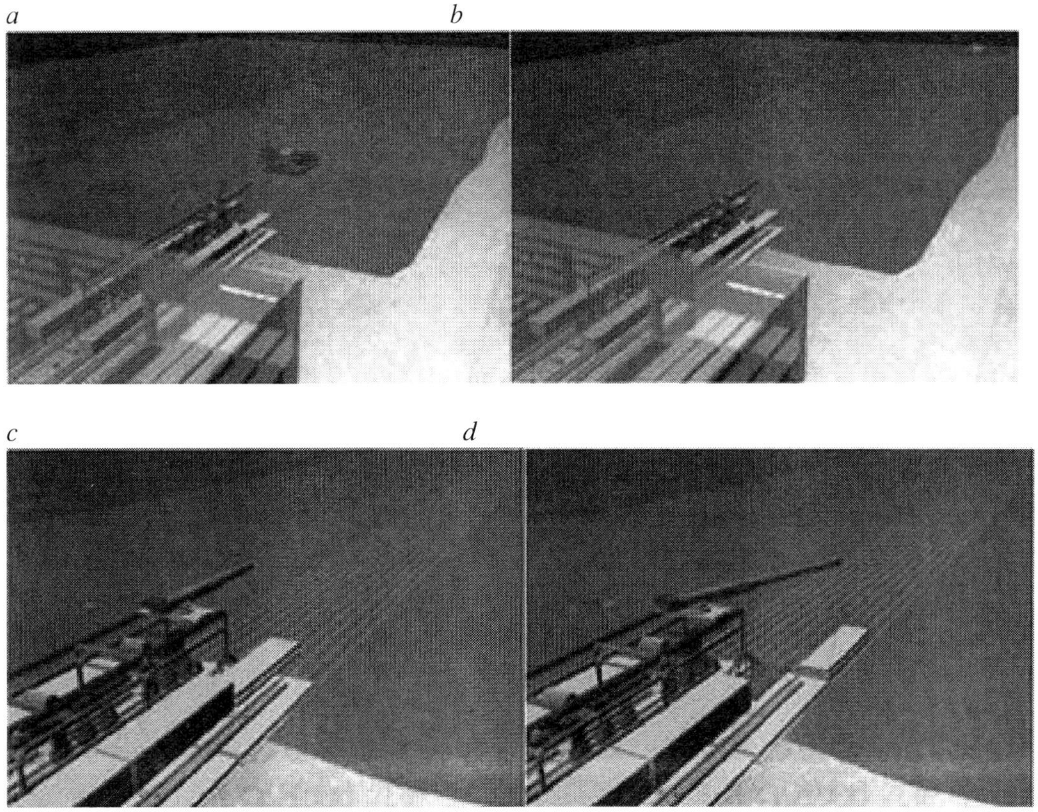

Figure 4. Construction of a pipeline bed made of composite pontoon units: a – wire routing through the obstacle with the help of ground-effect machine, b – fastening the wire in the well, c – pontoon unit laying on the wires, d – fixing the pontoon unit on the wires

The result of the new achievements in the field of technological and technical solutions for laying the large diameter pipelines through bogs and water-flooded areas resulted in the creation of a fundamentally new technological scheme for the production of both welding and installation works and the development of a general design solution for the installation and laying complex in the automatic mode for year-round construction of pipelines under conditions of Western Siberia and the Far North.

3. Conclusion

As a result of the conducted scientific researches and design work, it is possible to state the following advantages of the proposed technology for the construction of bases for the main pipeline and along the routing:

1. Straightforward configuration of the pipeline routing instead of bypassing the bogs, which reduces the total length of the system by 44%.
2. All-season construction of the pipeline system.
3. The rational principle of tension of longitudinal high-strength composite wires (ropes) by a surface-effect airborne ship
4. Exclusion from the construction of parallel along the routing paths.
5. Reducing the volume by removing burden by 90%.
6. Full automation of the welding production system.

7. Construction of modular warehouses for storage of emergency stocks of pipes.
8. The proposed new materials and construction technology eliminate the thawing of permafrost soils.
9. The connection of pontoon units is made rigid due to the use of the lock type "male-female".
10. Due to the adjustment in the tension of the longitudinal wires, let us avoid the influence of the cycling of the raising and lowering (breathing) of the surface of the bog, which ensures that the appearance of the pipeline domes.
11. Available temperature control of pumped liquid gas and oil.
12. Free access to the pipeline for carrying out the diagnostic work.
13. The covering of the pipeline excludes the impact of solar radiation on waterproofing compound.
14. Protection of the pipeline by a casing excludes hooligan actions of poachers.
15. The constriction of the roadway running parallel to the laid pipeline provides for a year-round access to the system by means of transport units.
16. Pontoons made of basalt-plastic composite have a high strength. They are not subjected to corrosion, have an unlimited lifetime, high rigidity for bending and high stability. They have no embrittlement, are stable to wave oscillations and environmentally friendly.

The implementation of the developed technology and construction methods will ensure a long-term, reliable, environmentally friendly and economically viable option for the nature management of pipeline systems, both in Russia and abroad.

References:

[1] SNiP 2.06.04-82* *Loads and impacts on hydraulic structures (wave, ice and ships).*
[2] SNiP 3.04.03-85 *Protection of building structures and buildings against corrosion.*
[3] Dudin S, Bahmat G, Zemenkov Y, Voronin K, Shipovalov A 2017 Strategy for monitoring and ensuring safe operation of Russian gas transportation systems *MATEC Web of Conferences* **106** DOI: 10.1051/matecconf/201710606004
[4] Dudin S, Voronin K, Yakubovskaya S, Mutavaliev S 2016 Modeling Hydrodynamic State of Oil and Gas Condensate Mixture in a Pipeline. *MATEC Web of Conferences.* DOI: 10.1051/matecconf/20167302021
[5] Shantarin V D, Zemenkova M Y, Zemenkov Y D, et al. 2016 Development of Thermophysical Hydrocarbon Wastes Pyrolysis Model (in the Case of Wood). *IOP Conference Series: Materials Science and Engineering* **154** DOI: 10.1088/1757-899x/154/1/012012
[6] Zemenkova M Y, Zemenkov Y D, Shantarin V D, et al. 2016 Developing Mathematical Provisions for Assessment of Liquid Hydrocarbon Emissions in Emergency Situations. *IOP Conference Series: Materials Science and Engineering* **154** DOI: 10.1088/1757-899x/154/1/012011
[7] Ryabkov A V, Shipovalov A N, Seroshtanov I V 2013 All-season construction of pipelines on pontoon units in the regions of Siberia and the Far North
[8] Ryabkov A V, Ivanov V A, Zakuraev A F 2014 *Development of a new technology for laying the pipelines on composite pontoon units under the conditions of Siberia and the Far North.* (Tyumen, Tyumen Press House)
[9] Vlasov V Z 1959 *Thin-walled elastic rods.* (Moscow: Fizmatgiz)
[10] Volmir A S 1956 *Flexible plates and shells.* (Moscow: Gosizdat)

Technologies of polytechnic education in global benchmark higher education institutions

V A Kurushina, E V Kurushina and M Y Zemenkova

Industrial University of Tyumen, 38 Volodarskogo St., Tyumen, 625000, Russia

E-mail: kurushina.tsogu@yandex.ru, muzemenkova@mail.ru

Abstract. The Russian polytechnic education is going through the sequence of transformations started with introduction of bachelor and master degrees in the higher education instead of the previous "specialists". The next stage of reformation in the Russian polytechnic education should imply the growth in quality of teaching and learning experience that is possible to achieve by accumulating the best education practices of the world-class universities using the benchmarking method. This paper gives an overview of some major distinctive features of the foreign benchmark higher education institution and the Russian university of polytechnic profile. The parameters that allowed the authors to select the foreign institution for comparison include the scope of educational profile, industrial specialization, connections with the leading regional corporations, size of the city and number of students. When considering the possibilities of using relevant higher education practices of the world level, the authors emphasize the importance of formation of a new mentality of an engineer, the role of computer technologies in engineering education, the provision of licensed software for the educational process which exceeds the level of a regional Russian university, and successful staff technologies (e.g., inviting "guest" lecturers or having 2-3 lecturers per course).

1. Introduction

Reforms in the Russian education led to the formation of a pool of federal and research universities from among the best Russian universities. Prospects for the development of higher education in the regions are connected with the possibility of forming benchmark universities. The Expert Council of the Ministry of Education and Science of the Russian Federation identified the list of regional universities, on the basis of which benchmark universities will be established. Growth of the management efficiency of the preparation of engineers is one of the basic conditions of modernization of the Russian economy [1]. The Russian university selected for this study is a regional university of engineering profile in a medium-size city of Russia and the students' center of the region. The main task of this Russian university is reforming the existing system of engineering education in order to provide the region with engineering personnel with the skills to conduct the next industrial revolution. The best foreign practices in the field of higher education should be studied in order to form the development program for the Russian university. In the present research, the differences in education technologies are investigated as a comparison between the current characteristics of engineering education in the Russian university and the technical university of the world level, chosen according to the criteria listed in Section 2.

The aim of this paper is to identify the main differences between the engineering higher education in the world-class university and in one of the engineering universities in Russia in order to determine what kind of qualitative changes in the education process the Russian university can benefit from in the possible future.

The paper is structured as follows. General introduction to this research is given in Section 1. Section 2 provides the characteristics of the research method and the details of choice of the foreign benchmark university. In Section 3, four major differences between the analyzed Russian and foreign engineering education are considered: course contents and software, information resources, staff solutions, and quality control means. The main outcomes of this paper are given in Section 4.

2. Research method

Benchmarking is selected for this study as the method of objective systematic comparison. As noted by O.V. Lenkova [2], introduction of benchmarking as an independent management tool is related to the Strategic Planning Institute in Cambridge (USA). In this institute, it was found in 1972 that the knowledge of the best experience of other organizations, which are successful in the similar conditions, allows one to take effective decisions in the field of competition. G. Watson, the past leader of the American Society for Quality (ASQ), considers benchmarking as a systematic and continuous measurement process: assessment of processes of the enterprise and comparison with the processes of the world-leading enterprises to collect information which will help the organization to take action to improve its' performance [3]. According to Y. Ohinata [4], the process of adopting the method should be preceded by the procedures of comparison and identification of weaknesses of the organization.

Within the framework of this study, when selecting benchmark universities, the industrial profile of higher education institutions and its' connection with solving the regional tasks was in priority. The selection was performed taking into account that the university should be located in the medium-size city and constitute a major student center, but not the main city of the state. At the university acting as a benchmark, schools of geology and engineering should exist; the number of university students has to be comparable to the number of students in the Russian university (about 15 thousand people). The university should train the future staff for large corporations in the region, including oil and gas industry.

Considering these criteria and the list of universities with diplomas recognized in the Russian Federation [5], the authors selected one of the universities of the world level as a benchmark (which will be referred as the benchmark university or reference university further). Some attractive characteristics of this university for the Russian university are:

- position in the international rankings of THE and QS (should be in the top 200 universities);
- high proportion of masters and post-graduate students (over 25% of the number of students);
- high percentage of foreign students (over 30%);
- compliance of university research activities with international quality standards;
- university employees awarded with Nobel Prize.

Two universities are compared with each other in this paper in the following key aspects: content of courses which are not present in the Russian university; dominating mindset of lecturers and students and related use of resources; personnel for course teaching; and ways to control the education result. The authors attempt to define the ways for improvement for the Russian university in each considered aspect.

3. Comparison results

3.1. Structure of educational programs in the benchmark university

Comparison of educational technologies in the selected reference university and the Russian university allowed identifying advantages of the reference university related to the structure of programs in engineering and to the degree of informatization of the subjects. Specifically, the course of engineering mathematics is followed by the course of design with computational and programming skills. An example of such course may include 6 weeks of programming in Matlab, 3 weeks of tasks performed in Solidworks and 1 week of exercises in SprutCam. A design assignment with an open end to complete in 2-3 weeks is also normally included in such course. This subject may be given to the second year students of different specialties, for example, mechanical and civil engineering students.

As for example given above, the Solidworks software allows designing two- and three-dimensional solid bodies separately and in assembly, and also liquids in a closed volume, with the further modeling of loads and deformations. Students' designs from Solidworks are loaded into the SprutCam software where the user selects a machine for the production of parts from the workpiece, sets the operations and location of the tool to obtain the object of the desired size and shape. Students have the opportunity to see the work of the machine in the laboratory in accordance with the previously created operations in SprutCam. The programs like Solidworks, Abaqus and Ansys can be involved for design exercises at different levels of education, from the first attempts made by undergraduate students to the research by PhD students.

The Matlab software is applied in many general and specialized disciplines in the benchmark university. For example, one of the specialized courses of the third year at the reference university first involves solving problems without information technology, by hand and a calculator, and then solving problems in the Matlab environment using the functions of integrating differential equations and all the studied lectures' material.

Thus, an undergraduate student of the university, chosen as a benchmark, finishes studies familiar with the various software, basics of programming and three-dimensional modeling. In comparison with the Russian university, the students have stronger adaptive qualities for the world of computerized technology. The connection of the mathematics course through the calculation and programming course to the specialized courses is often missing in the Russian education for the majority of engineers. The Russian specialties imply either a pure engineering, practice-related education, or a mathematical and programming education, but not a combination of them. Since the advanced engineering idea now speaks English and the coding language, the task of transformation for the Russian university is to break this linguistic barrier before the resources of a new knowledge for engineering students.

3.2. Formation of a new mentality of an engineer in the era of information technologies

The modern engineer exists in the time of the fifth technological order, lives in the conditions of strengthening globalization processes in the economy [6, 7] and migration of the population, availability of the Internet with access to the continuously updated information and the best practices from all over the world. In these conditions, former priorities of mentality of engineers should be left in the past along with the aging generation of technology. The mentality of an engineer of the new generation assumes: 1) flexibility of mind; 2) high susceptibility and desire for a new knowledge; 3) ability to abandon previous ideas and experiences, which do not work anymore; 4) knowledge of foreign languages and programming; 5) communication skills and ability to work in a team to achieve a common goal.

Reforms of polytechnic education should begin, firstly, with an increase in the level of information culture, since Russia is behind the developed countries in the transition to the fifth technological order. It is largely due to the inadequate formation of its core, including computer equipment, software and information services [8]. The university chosen as a benchmark annually spends more than 3.5 million dollars for the work of the library, of which around 400 thousand dollars is spent on access to databases of the latest scientific papers and books. The university owns software licenses sufficient to cover the students working in the classroom at the same time, for example, 100 licenses per program, or it is available for all the students without limitation.

In the Russian university, the information resources appear more limited, but there is also a related problem. The students are supposed to follow the course instructions, including the list of resources, which students should use to obtain the correct answers. Thus, the education becomes largely the knowledge transfer from the lecturer to students, where the primary goal is to avoid the information loss. The initiative of students to suggest the alternative solutions often is not appreciated. This problem is not just something typical for one lecturer, one university, or one country, it is a general phenomenon known as the fixed mindset in education [9].

The benchmark university selected for the current comparison, on the contrary, puts the main focus on the students' self-study, encourages students to look for their own answers, create their own solutions, and discussions with the lecturer about the material are appreciated. Real case studies are often given as design assignments, where it is possible to discover many answers and to progress from a simple to a complex solution. This way of teaching and learning requires a continuous effort from all the participants of the process and leads to the shift from thinking about the result of teaching as a well-known, clear outcome to the continuously improving outcome.

One of the related advantages of the studied benchmark university is that each year the course and its assignments are reviewed and updated by the lecturers. This is partially the result of the updated requirements of the accreditation procedure, but also the result of responding to the students' feedback. The feedback in the reference university is collected in the middle of the term to reveal the problems during teaching the course, and also at the end of term to allow the lecturer make changes for the next year. At the same time, the feedback of students in the Russian university is rarely applied as an instrument of improvement for the ongoing teaching.

3.3. Strategy of formation of the teaching staff and education technologies in the benchmark university

The next important component of reforming the polytechnic education is a successful staff strategy that ensures the achievement of the goals stipulated by the Bologna process: transparency, quality, growth, efficiency and skill [10]. To provide the high-quality education, the benchmark university uses a modular system and technology of inviting "guests" – specialists from companies, competent in certain topics. Companies send their employees to demonstrate the professional level of knowing and applying the material required for work in the form of 1-2 hour classes at the university. For example, every second or third lecture of the course of offshore extracting facilities for master's students of the second year of study is held by "guests" from companies.

In addition, the course lectures and practical classes are not given all by the same teacher. There is a coordinator at the head of the course – usually the main lecturer, who selects 1-2 more lecturers and the necessary number of demonstrators for the course and links the theoretical blocks and practical assignments with each other. Demonstrators are invited personnel from post-graduate students or even undergraduate students of the fourth year of studies competent in demonstrating certain software or in solving problems for practical classes. Recruiting the additional junior staff to conduct classes at the bachelor's level is the distinctive feature of the benchmark university.

The supporting staff in the practical class has the following functions: answers students' questions on solving problems; solves problems that arise during the class; by all means creates a favorable, safe environment for effective teaching. Demonstrators are "good" when they can explain and show the solution of the problem in different ways (so that students understand it), and also provide the necessary connection with the lecture material. Requirements for demonstrators include: 1) a complete bachelor's degree (or a candidate needs to be a fourth-year student) in the study area; 2) an expert level in the subject; 3) an understanding of what knowledge and skills a student should have at the end of the course; 4) experience in demonstrations (appreciated).

Favorable environment in the classroom includes both physical (lighting, ventilation, room temperature), and psychological and social factors. Particular attention in the benchmark university is paid to creating equal opportunities for students. The course lecturer is obliged to provide students with the material in various forms: books, articles, audio recordings, presentations, etc. For health reasons, students are provided with separate rooms for each examination. At the level of demonstrator, during practical classes, it is necessary not to prioritize when communicating with students from different countries, of different age, gender, religion, orientation, etc.

The qualification of lecturers in the field of higher education in the country of the benchmark university is regulated using system of five spheres of competence. The basic principles of this system are the accessibility of education for all and the depth of knowledge of the subject. One of the most important areas is the design of the learning environment, which includes all the teaching means (simulators, computers, programs, lectures, etc.), as well as the limitless support from demonstrators and lecturers.

The goal of the benchmark university is to provide not just "good" education, but "excellent" education. A prerequisite for the high-quality education is the effective feedback so that the student can independently regulate the learning process. In the reference university, the lecturers aim to follow the feedback principles, as formulated in [11]. According to them, the lecturer "helps to clarify what good performance is (goals, criteria, expected standards); facilitates the development of self-assessment (reflection) in learning; provides high-quality information to students about their training; encourages teacher and peer dialogue around learning; encourages positive motivational beliefs and self-esteem; provides opportunities to close the gap between the current and desired performance; provides information to teachers that can be used to help shape teaching" [11].

If students do not share (at least partially) the concepts of their lecturer with respect to evaluation objectives (as well as criteria and standards), then the feedback information they receive is unlikely to lead to the desired result, as noted in [12]. In this case, it will be difficult for students to estimate the discrepancy between the required and the actual level of performance. It is also important to note that feedback not only leads the students towards academic goals, but, over time, it will also play a role in clarifying the goals themselves [13].

3.4. Quality control of teaching

All the educational programs in the reference university are accredited through the bodies, specialized for different kinds of engineering: chemical engineering, mechanical engineering, civil engineering, etc. These institutions obtain the licenses to accredit the educational programs and update their requirements for universities regularly. At the same time, the Russian educational programs for bachelors and masters are currently regulated by one body – the Ministry of Education and Science.

Quality control of education at the benchmark university is provided by: 1) the duration of tests (the exam lasts 2-4 hours); 2) recruiting staff not associated with the course to control the examination process (invigilators); 3) the anonymity of works submitted for evaluation; 4) showing test results in the electronic system.

Examination of students in the reference university is performed in the written form only. The examination assignments pass the double check after they are created by the lecturer: the internal check with the other lecturer in the same university, and the external check with the lecturer from a different university. The examinations are held in the fall and spring session time and provide the main contribution to the students' marks.

In the Russian university, there is the system of three attestations for students per term, including the final test, and the students' marks are given as results of tests in the electronic system, or written tests, or oral examination. Additional examinations are held after the end of term for students who did not pass the regular tests during the term successfully. The examinations are normally conducted by the same lecturer who taught the subject, and the contents of tests are not checked with other lecturers. However, there is the system of external and independent control of students' knowledge in Russia – FEPO, which implies that selected groups of students in university can be tested for the knowledge of particular subjects during the term.

The authors of this study suggest that the practice of recruited independent personnel (invigilators) to conduct written examinations in the reference university is the most perspective to improve the examination process in Russia.

4. Conclusions

The differences examined in this paper between the Russian and the foreign practices of engineering education are given in Table 1. It is worth noting that the actual differences are not limited to this list, and if other universities are selected for comparison, the results may differ. The described distinctions can serve as the guidelines for qualitative changes in the Russian education in general and in certain higher education institutions in particular.

Table 1. Comparative analysis of characteristics of the process of engineering education in two universities

Characteristics	Reference university	Russian university
1. Provision of licensed software for students	Software is provided for all the students to meet all their needs expected while they study towards the degree.	Very basic resources like MS Office are provided. Students are encouraged to take software courses, which can be beneficial for final thesis and future career, on their own outside of university.
2. Access to scientific papers	Access is provided for all the students to databases like sciencedirect.com, onepetro.org, etc., where it is possible to find the latest papers published in the leading journals of the field throughout the world.	Students can access the papers from the free system of the Russian citation index elibrary.ru.
3. Courses of computations and programming	Necessary part of engineering education providing the link between engineering mathematics and specialized disciplines, and also providing the possible basis for the final year thesis.	There is a course of general informatics, including some programming tasks, which is not related to mathematics course or specialized courses.
4. Tutorials and practical hours of specialized courses	Exercises are solved by hand on tutorials, and then similar tasks can be solved in code or in specialized software.	Most of exercises are solved by hand with scientific calculator, or in Excel.
5. Course staff	Course coordinator, 1-3 lecturers on the course, guest lecturers, demonstrators. The course is created or modified according to feedback from the previous year by a team of teachers.	In majority of cases, 1 lecturer and, possibly, but not necessarily, 1-3 demonstrators are assigned to the course. All decisions on tasks and forms of teaching are responsibility of the single lecturer.
6. Feedback from students	Feedback is collected regularly in multiple ways, including the written feedback in the middle and in the end of course, but not limited to it.	Feedback is collected occasionally, depending on the course lecturer.
7. Examination	Conducted by the personnel which was not involved in teaching the subject.	Conducted by the staff teaching the course.
8. Growth, initiative, creativity in students learning	Exercises with multiple solutions or open end, and design exercises based on real cases are the necessary part of learning. Students are encouraged to use any information resources they can find, develop new skills and their own ways to solve the tasks. Discussions with lecturers, demonstrators and among the students on the subject are strongly encouraged.	Students are supposed to follow the course instructions and instructions for the particular task in order to obtain the defined answer. Students may apply information from not listed sources if it leads to the same final answer. Discussions can be a part of the education process, but not widely encouraged.

As the result of this study, adoption of the best teaching practices to educate engineers in Russia can take the form of: changes in the structure of educational programs; increased spending for information technologies; increase in the number of tasks with multiple solutions and open end; encouraging discussions in the classroom; increase in the number of lecturers per subject; improving teaching based on the students' feedback; shifting the focus from the knowledge transfer to growing students' skills of self-education. It can become one of the steps for reducing the delay in science and technology observed in Russia. A modern engineer has to change along with evolutionary changes in technology and learn continuously. The task of a benchmark university is to provide its graduates with these advantages.

References

[1] Zemenkova M, Shalay V, Zemenkov Y, Kurushina E 2016 Improving the Efficiency of Administrative Decision-Making when Monitoring Reliability and Safety of Oil and Gas Equipment. *MATEC Web of Conferences* **73** 07001 DOI: 10.1051/matecconf/20167307001
[2] Lenkova, O.V. (2014). Approaches to understanding of the Benchmarking. *Fundamental research*, 3-4, pp. 790-794.
[3] Watson G H 1993 How process benchmarking supports corporate strategy. *Planning Review* **21(1)** 12-15
[4] Ohinata Y 1994 Benchmarking: the Japanese experience. *Long Range Planning* **27(4)** 48-53
[5] Decree of the Government of the Russian Federation of May 21, 2012 N 811-r Moscow "On approval of the list of foreign educational organizations that issue documents of foreign states on the level of education and (or) qualifications recognized in the territory of the Russian Federation." URL: http://rg.ru/2012/05/25/obrazovanie-dok.html
[6] Neganova V P, Smirnova E A, Petrov V P 2016 Improving the system of marketing communications at the consumer market. *Russian Journal of the Economic Theory* **2** 178-181
[7] Kurushina E V, Kurushina V A 2014 Evolution of economic development aims. Assessment of the smart growth. *Life Science Journal* **11(11)** 517-521
[8] Glazyev S Yu 2010 *The strategy of advanced development of Russia in the conditions of the global crisis* (Moscow: Economics)
[9] Ehrlinger J, Mitchum A L, Dweck C S 2016 Understanding overconfidence: Theories of intelligence, preferential attention, and distorted self-assessment. *Journal of Experimental Social Psychology* **63** 94-100
[10] Enders J, de Boer H F, Westerheijden D F 2011 *Reform of Higher Education in Europe.* (Sense Publishers) URL: https://www.sensepublishers.com/media/871-reform-of-higher-education-in-europe.pdf
[11] David J N, Macfarlane-Dick D 2006 Formative assessment and self-regulated learning: a model and seven principles of good feedback practice. *Studies in Higher Education* **31(2)** 199-218
[12] Hounsell D 1997 Contrasting conceptions of essay writing. *The experience of learning* **2** 106–125
[13] Sadler D R 1989) Formative assessment and the design of instructional systems. *Instructional Science* **18(2)** 119–144

IOP Conf. Series: Materials Science and Engineering **357** (2018) 012029 doi:10.1088/1757-899X/357/1/012029

Selection of optimal conditions for preparation of emulsified fuel fluids

V A Ivanov, V I Berg and M D Frolov

Industrial University of Tyumen, 38 Volodarskogo St, Tyumen, 625000, Russia

E-mail: ivanov_v_a@list.ru, berg_vi@mail.ru

Abstract. The aim of the article is to derive the optimal concept of physical and chemical effects, and its application to the production of water-fuel emulsions. The authors set a research task to attempt to estimate the influence of the surfactant concentration on such indicator as the time before the beginning of emulsion breaking. The analysis, based on experimental data, showed that an increase in the concentration of sodium lauryl sulfate is expedient to a certain point, corresponding to 0.05% of the total mass fraction. The main advantage of the model is a rational combination of methods of physical and chemical treatment used in the production of emulsions.

1. Introduction

At present, the problem of disposal of industrial wastes, as well as oil residues, including hydrocarbon and water-containing components, is acute. One can note the regularity of depressurization of pipelines and tanks at facilities associated with the extraction, transport, storage and distribution of hydrocarbon raw materials. These substances pollute the environment, including water. Elimination of the consequences of such accidents and incidents is extremely resource-costly [7]. In addition, the oil product into which the water has been trapped is no longer suitable for burning or for laboratory analysis. In the work, it is proposed, in case of breach of the integrity of the equipment for transportation and storage of petroleum products, to create water-fuel emulsions from the spilled substances for their subsequent combustion in boiler rooms. This method is also proposed to be used to reduce the level of harmful emissions (for example, nitrogen oxide) resulting from fuel combustion [10].

In this paper, it is proposed to create an emulsion of a hydrocarbon represented by diesel fuel with water by means of a complex physical and chemical effect using thermal and cavitation treatment with preliminary dissolution of the surfactant in the dispersed phase [5]. In order to find the most optimal method of creating emulsions that increases the time of breaking, it is proposed to dissolve the solubilizers in the dispersion medium, with the addition of salt represented by sodium chloride as the most water-soluble one.

A study on the complex effect of physical and chemical treatment was carried out on the basis of an industrial prototype of a cavitation homogenizer (RF patent No. 2620606). The component parts of the structure include an electric motor that provides a mechanical effect for the production of restructured water, which in turn is connected via a pipeline to a storage tank. The prepared water storage chamber is also equipped with a pipeline and a system of uniform water supply to the mixing chamber, the process being carried out by the operation of the process valve. A similar solution has been applied to the supply of a liquid hydrocarbon into the homogenizer chamber. In the process of restructuring the

IOP Conf. Series: Materials Science and Engineering **357** (2018) 012029 doi:10.1088/1757-899X/357/1/012029

constituent components of the final water-fuel emulsion, they are uniformly heated. This is done to equalize the viscosity of the components of the prepared mixture. It is possible to install a separate tank providing direct supply of surfactants improving the quality of the final product. The rotor shaft of the mixing grates homogenizer providing rotation is driven by an electric motor. Further, the liquid is discharged into a storage tank, from which samples of the finished water-fuel emulsion are directly taken.

An analysis of the existing literature has shown that the greatest stabilization of water-fuel emulsions can be achieved by using anionic-type surfactants with a hydrophilic-lipophilic balance (HLB) index of 10.5, which corresponds to the optimal index for creating emulsions with oil and petroleum products of low viscosity [8]. It should be noted that surfactants, creating the necessary hydrophilic properties, are not readily available in the existing market. Due to this, it is proposed to use a mixture of emulsifiers to obtain the required parameters. The possibility of using the most common emulsifier in the industry with the maximum possible number of HLB, 40, of sodium lauryl sulfate (SLS) is also considered. When the equilibrium concentration in the aqueous phase is reached with the given surfactant, the breaking rate will vary inversely proportional to the logarithm of the total concentration of sodium ions that are included in the emulsifier [8], as well as in the previously added sodium chloride, and the equation takes the form:

$$\vartheta_{omc.} = \frac{1}{\ln C_{Na+}} \tag{1}$$

There is an opinion regarding the application of ultrasonic treatment for the purpose of carrying out a repeated process of cavitation [2] as a necessary condition for the creation of a water-fuel emulsion that is not dissolved over a long period of time. In the installation used by the authors, it is envisaged to use the equipment for this treatment as an intermediate stage between the primary and secondary purging of the components of the mixture.

2. Materials and methods

The effectiveness of the use of a surfactant, sodium lauryl sulfate, is shown in the graph (Fig.1). The dependence of the emulsion breaking time on the concentration of the emulsifier taken on the total mass fraction in percent is shown.

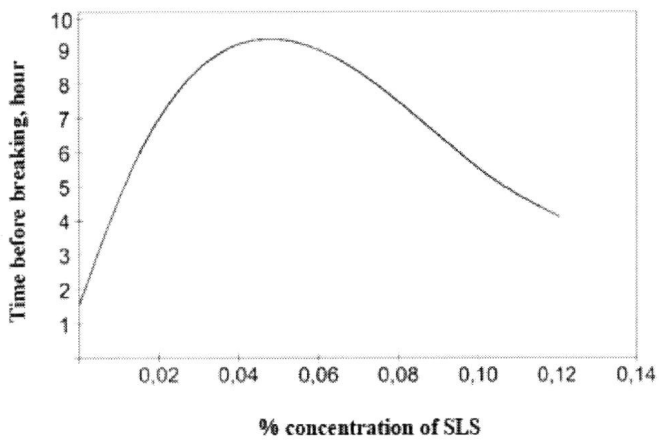

Figure 1. Dependence of time before breaking on emulsifier concentration

One should note the advantages and design features of the device designed for improved dispersion of liquid hydrocarbon raw materials with water [1]. In the pilot installation, it was possible to eliminate the appearance of a transit flow, while increasing the efficiency of the rotor apparatus and improving the homogeneity of the medium being processed by pulsating the flow velocity. The creation of these conditions naturally increases the degree of homogenization. The high quality of the homogeneity of the liquids to be mixed is achieved by first equating the viscosities of the dispersed phase and the dispersion medium by heating them to different temperatures, as well as restructuring the original components that passed the preliminary filtration [4].

When the rotor rotates, the rotor grates in close proximity to each other create a pressure corresponding to the boiling point of the liquids at a given temperature. As a result of induced cavitation, there are water and hydrocarbon balls, which subsequently collapse on a conical plate, the coalescence rate being determined by the formula:

$$\vartheta = K_1\left(\frac{-0{,}24\psi^2 - \Theta_f \sum H}{RT}\right) \tag{2}$$

where K_1 – hydrodynamic collision factor, equal to $4Fkr/3\eta_c$; η_c – aqueous phase viscosity; B – a constant equal to 0.24, which is derived empirically from the theory of coalescence kinetics; ψ – surface potential; Θ_f – proportion of the surface of particles coated with emulsifiers (if present in liquids); $\sum H$ – total energy barrier.

The power required to break a total energy barrier is minimized by a reduced number of revolutions per minute, and is in the standard range of 0.1-10 hp for this type of installation [6].

The application of cavitation emulsification is due to the need to reduce the surface tension of particles to simplify their intermolecular interaction. In addition, this increases the contact area between the dispersed phase and the dispersion medium [3]. The optimum position of the rotor relative to the stator and the possibility of varying the speed of rotation eliminates the need to create overpressure, which in turn leads to optimization of the energy inputs required for the operation of this experimental installation. Depending on what type of hydrocarbon fuel will be used to produce the emulsion, and therefore what common energy barrier will have to be destroyed, the number of revolutions per minute varies from 1000 to 3000 steps per 500 rpm.

3. Conclusion

Based on the results of pilot tests, it was possible to establish that during the preparation of the fuel fluid, a finite equalization of viscosities, creation of commensurate polymer chains, stabilization of the physical and chemical properties of the sample occurred on this device.

Further, this method is planned to adapt to various types of hydrocarbons, including other fuel fluids represented by fuel oil, furnace fuel, oil sludge, etc. Also, the resulting water-hydrocarbon emulsion is proposed to be subjected to ultrasonic cavitation and repeated homogenization to reduce the size of micellar formations, and an increase in the breaking time.

The indicators achieved through physical processing, expressed by cavitation, correspond to laboratory studies on the production of water-fuel emulsions using surfactants. It should be noted that the approximate equality in the breaking time indicates that the technological process proved to be no less effective, in contrast to the chemical process. Taking into account the difference in the indirect purpose of cavitation in this area, and the direct role of emulsifiers in the formation of emulsions, it can be concluded that the time before breaking will further increase when these methods are applied in order to achieve the best result.

References

[1] Toropov S Y, Berg V I, Petryakov V A, Mostovaya N A 2016 Calculation of the thermal parameters and change in the share of emissions of harmful substances during combustion of water-fuel emulsion *IOP Conference Series: Materials Science and Engineering, "Transport and Storage of Hydrocarbons"*

[2] Abramzon A A, Abramova I V 1972 Some features of stabilization of an emulsion by high-molecular surfactants *Colloid magazine* **34** 444-446

[3] Likhanov V A, Lopatin O P 2013 Improving the performance of the diesel by using an ethanol-fuel emulsion *News of international academy of agrarian education* **4** 170-173

[4] Levina E Yu 2014 A review of the properties of modern water-fuel emulsions *Potential of modern science* **6** 37-39

[5] Levina E Yu 2014 A review of methods for obtaining water-fuel emulsions *Potential of modern science* **5** 36-39

[6] Pimenov Yu A, Efimova N L, Novikov B A, Pokrovsky A V, Sakhin V S 2005 Emulsion fuels for diesel engines *Innovations* **2**. 103-105

[7] Berg V I, Mostovaya N A, Brand A E 2017 *Application of water-fuel emulsion as a way to increase the efficiency of implementation of the energy saving program of PJSC "AK" Transneft "* (Tyumen) 20-23

[8] Kumar R, Kuloor N R 1970 The formation of bubbles and drops *Adv. Chem. eng.* **8** 255-375

[9] Kulagina T A 2004 *The use of Thermomechanical Effects of Cavitation in Various Processes* (Cheboksary)

[10] Zemenkov Y D, Zemenkova M Y, Berg V I, Gordievskaya E F 2016 Technology of ultrasonic treatment of high-viscosity oil from yarega oilfield to improve the rheological properties of oil *IOP Conference Series: Materials Science and Eng ineering. "Transport and Storage of Hydrocarbons"*

IOP Conf. Series: Materials Science and Engineering **357** (2018) 012030 doi:10.1088/1757-899X/357/1/012030

Oscillations of manometric tubular springs with rigid end

D A Cherentsov[1], S P Pirogov[1,3], S M Dorofeev[2] and Y S Ryabova[2]

[1] Industrial University of Tyumen, 38 Volodarskogo St., Tyumen, 625000, Russia
[2] Tyumen Higher Military Engineering Command School, 1 Lev Tolstoi St., Tyumen, 625001, Russia
[2,3]State Agrarian University of the Northern Trans-Urals, 5 Repablic St., Tyumen, 625003, Russia

E-mail: cherentsovda@bk.ru

Abstract. The paper presents a mathematical model of attenuating oscillations of manometric tubular springs (MTS) taking into account the rigid tip. The dynamic MTS model is presented in the form of a thin-walled curved rod oscillating in the plane of curvature of the central axis. Equations for MTS oscillations are obtained in accordance with the d'Alembert principle in projections onto the normal and tangential. The Bubnov-Galerkin method is used to solve the equations obtained.

1. Introduction

The use of mechanical pressure gauges in some cases remains uncontested and is regulated by regulatory documentation. The need to carry out activities aimed at increasing the vibration protection of MTS serving as sensitive elements of mechanical pressure gauges is caused by an increase in the requirements for the reliability and precision of instruments for measuring pressure. The main methods of protection are detuning from resonance frequencies and vibro-damping with fluid. Studies in this area are performed in [1-4]; however, in these works the effect of the MTS tip, which significantly affects frequency characteristics, is neglected.

2. Materials and methods

The tips can be used to increase the stroke of elastic elements [5-6].

Typically, for a number of springs, the same tips are used at different pressures, so depending on the thickness of the tube, the mass ratio of the tip and the mass of the tube varies over a wide range (line 2 of Table 1).

A comparison of the experimental and calculated values of the frequencies of free oscillations [1] showed that deviations may be from 10 to 80% (see line 5 of Table 1) and increase with increasing tip mass.

In [2], the mass of the tip is considered by introducing the decreasing coefficient of the frequency of free oscillations determined by the experimental method. The influence of the mass of the tip on the parameters of the oscillations attenuation and, as a result, on the attenuation rate of oscillations has not been theoretically studied to date.

The mathematical model used allows us to take into account the influence of the viscosity of the damping fluid and the mass of the MTS tip and is represented as a curved rod that oscillates in the plane of curvature of the central axis (Fig. 1).

Table 1. Value of frequencies of natural oscillations of steel springs

№	Rated pressure, MPa	0.1	0.4	1.0	2.5	4.0	6.0	10.0
1	Wall thickness, mm	0.2	0.3	0.4	0.6	0.7	0.8	1
2	m_{tip}/M_{tubes}	0.965	0.523	0.197	0.132	0.116	0.101	0.08
3	Frequency (experiment), Hz	49.3	95.5	157	197	234	251	277
4	Frequency (calculation without regard to the tip), Hz	89.3	160	198	235	274	275	301
5	Deviation, %	81.1	67.5	26.1	19.3	16.6	9.6	8.7
6	Frequency (calculation with regard to the tip), Hz	52.8	101.1	161.1	213.5	237	264.3	290.3
7	Deviation, %	7.1	5.9	2.6	8.4	1.3	5.3	4.8

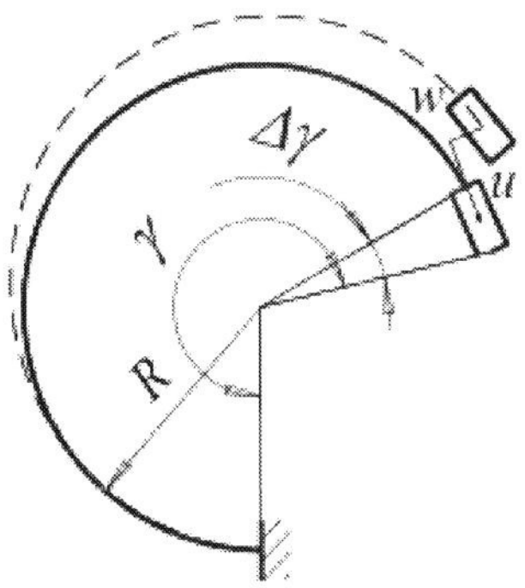

Figure 1. Curved rod.

Figure 2 shows the tangential - τ and the normal - n displacement of an infinitesimal element.

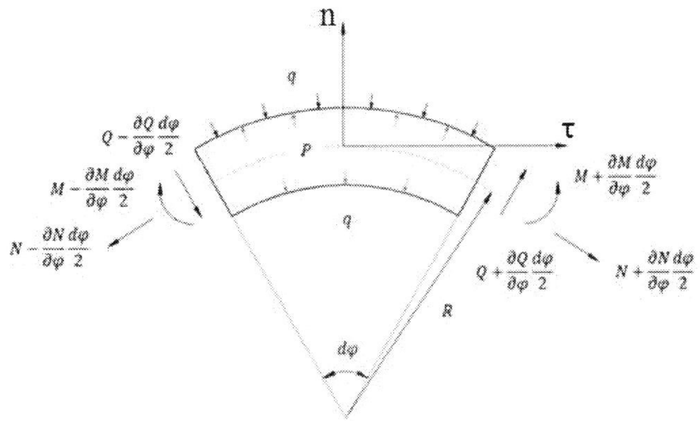

Figure 2. Infinitesimal rod element.

where q – the force of resistance to movement in a fluid (only the normal component is taken into account): $q_n = \beta v_n$; $q_\tau = 0$); β – the coefficient of damping fluid resistance; Q – the transverse force; N – the longitudinal force; M – the bending moment; R – the radius of curvature of the central axis; $d\varphi$ – the angle of an infinitesimal element cut from a curved rod.

Projecting all the forces, including the forces of inertia, on axes τ and n, let us obtain the system of equations of the MTS motion:

$$m(\varphi)\frac{\partial^2 w}{\partial t^2} + \left(1 + \frac{b}{R}\right)\beta\frac{\partial w}{\partial t} - \frac{\partial Q}{R\partial\varphi} + \frac{N}{R} = 0,$$
$$m(\varphi)\frac{\partial^2 u}{\partial t^2} - \frac{Q}{R} - \frac{\partial N}{R\partial\varphi} = 0, \tag{1}$$

where $m(\varphi)$ – the apparent density (mass).

The following boundary conditions are imposed on this system of differential equations. At the base of the manometric spring, at $\varphi = 0$, the longitudinal and transverse displacements, as well as the angle of rotation of the tube cross-section, are zero (the main boundary conditions).

At the free end ($\varphi = \gamma$), near the concentrated mass, the MTS distortion can be neglected. This assumption leads to the following boundary conditions at the free end: $M(\gamma) = 0$; $N(\gamma) = -m_n\frac{\partial^2 u}{\partial t^2}$; $Q(\gamma) = -m_n\frac{\partial^2 w}{\partial t^2}$ (natural boundary conditions).

The initial conditions do not matter, since only the frequency characteristics of MTS are determined.

The problem is solved by the Bubnov-Galerkin method. Displacements *u* and *w* are given as:

$$u = \psi_1(\varphi)a_1(t) + \psi_2(\varphi)a_2(t) + \cdots \psi_n(\varphi)a_n(t) = \sum_{i=1}^{n}\psi_n(\varphi)a_n(t)$$
$$w = \zeta_1(\varphi)b_1(t) + \zeta_2(\varphi)b_2(t) + \cdots \zeta_n(\varphi)b_n(t) = \sum_{i=1}^{n}\zeta_n(\varphi)b_n(t) \tag{2}$$

where $a_1, a_2 \dots a_n, b_1, b_2 \dots b_n$ – unknown functions of variable t;

$\psi_1, \psi_2 \dots \psi_v, \zeta_1, \zeta_2 \dots \zeta_v$ – basic functions of variable φ.

Power functions are used as basis functions:

$$\psi_i(\varphi) = \varphi^i; \quad i = 1, \dots, n.$$
$$\zeta_j(\varphi) = \varphi^{j+1}; \quad j = 1, \dots, n. \tag{3}$$

Such system of functions is complete; the functions are linearly independent and satisfy the main boundary conditions.

To estimate the convergence of the solution, the authors studied the change in the values of oscillations attenuation parameters for a different number of functions included in the displacement approximation.

To obtain satisfactory results, it was sufficient to preserve five basic functions.

Calculations of natural oscillations frequencies taking into account the rigid tip are given in lines 7 and 8 of Table 1. As can be seen from the table, deviation from the experimental data is from 4 to 8%, which indicates a good accuracy of the results obtained.

The results of studies of the influence of the tip mass on the frequency of attenuating oscillations and the oscillation attenuation coefficient are presented as surfaces depending on the viscosity of the damping fluid and the mass of the tip for a sample with parameters R=30mm, γ, =30 degrees, h=0.2mm, a=5 mm, 2 mm (Fig.3).

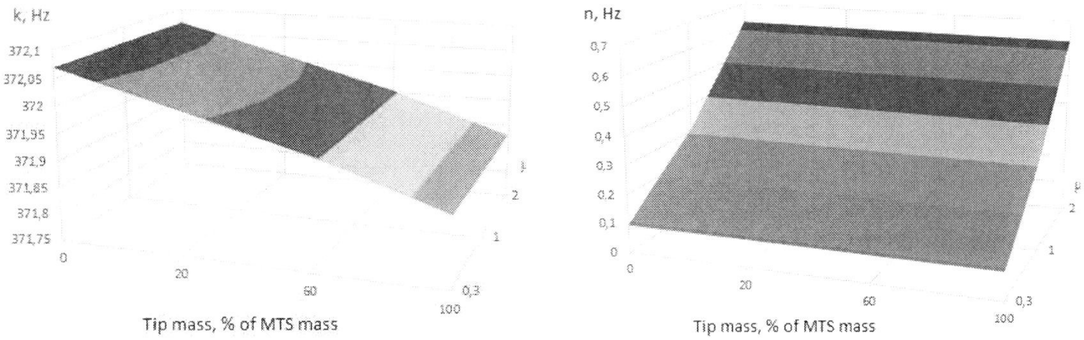

a) Frequency of oscillations b) Coefficient of attenuation

Figure 3. Changes in the oscillations attenuation parameters

Analysis of the presented dependences shows that an increase in the mass of the tip leads to a decrease in the frequency of attenuating oscillations and a slight decrease in the oscillation attenuation coefficient.

A decrease in the attenuation coefficient leads to an increase in the attenuation rate; the results of the investigation of springs with a tip and without it are shown in Fig.4

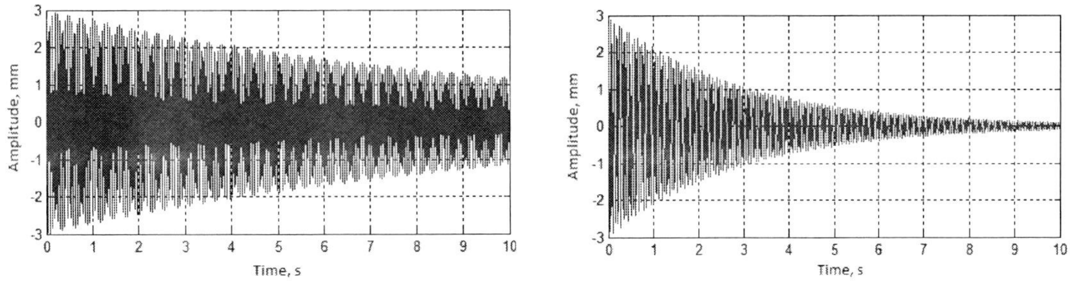

a) Without a tip b) With a tip

Figure 4. Amplitudes of oscillation attenuation

Since the change in the mass of the tip affects the oscillation process, the limiting value of damping fluid viscosity must undergo a change, at which aperiodic motion begins to take place. Below, Table 2 provides viscosity values at which aperiodic movement begins with and without a tip.

Table 2. Dynamic viscosity value

Dynamic viscosity value, Pas	Sample number	
	1	2
Without a tip	1000	1200
With a tip	1200	1400
Deviation, %	16.7	14.3

The obtained results show that as the mass of the tip increases, the limiting value of dynamic viscosity decreases, but it is possible to increase the mass only to a certain limiting value. Table 3 shows mass values as a percentage of the MTS mass at which the system will cease to oscillate.

Table 3. Dynamic viscosity value

Sample number	1	2
MTS mass, g	2.84	13.23
Tip mass, g	13.3	66.7
Tip mass, %	368.3	404.2

3. Conclusion

Numerical experiments confirm a significant influence of the tip mass on the MTS oscillation process. An increase in the mass of the tip leads to an increase in the total mass of the system, and this in essence is one of the methods of vibration protection - vibro-damping. Thus, an increase in the mass of the tip can be used as an independent method of vibration protection, and in combination with the method of detuning from resonance frequencies and vibro-damping of oscillations with a fluid, thereby increasing the effectiveness of protection.

References

[1] Pirogov S P 2009 *Manometric tubular springs.* (St.Petersburg.: Nedra)
[2] Chuba A Y 2007 *Calculation of the natural oscillation frequencies of manometric tubular springs: thesis for Candidate of Engineering degree.* (Tyumen)
[3] Pirogov S P, Cherentsov D A, Chuba A Y 2015 *Oscillations of manometric tubular springs: monograph.* (Tyumen, TyumGNGU)
[4] Pirogov S P, Cherentsov D A 2015 Theoretical foundations of the design of vibration-resistant manometers. *Measurement Techniques* **8** 38-41
[5] Tyunin N I 1969 To the calculation of the movement of the extension tip of a manometric spring. *Devices and control systems* **6** 47-48
[6] Voronin K S, Ogudova E V 2016 The Effect of Dynamic Processes in the System "Pipe-Soil" on the Pipeline Deviation from Design Position. Transport and Storage of Hydrocarbons. *IOP Conf. Series: Materials Science and Engineering* **154** 012019 doi:10.1088/1757-899X/154/1/012019
[7] Mikhlin S G 1970 *Variational methods in mathematical physics.* (Moscow. Nedra)
[8] Voronin K S, Ogudova E V 2016 The Effect of Dynamic Processes in the System "Pipe-Soil" on the Pipeline Deviation from Design Position. Transport and Storage of Hydrocarbons. *IOP Conf. Series: Materials Science and Engineering* **154** 012019 doi:10.1088/1757-899X/154/1/012019
[9] Voronin K S 2016 Forecasting and Evaluation of Gas Pipelines Geometric Forms Breach Hazard. Transport and Storage of Hydrocarbons. *IOP Conf. Series: Materials Science and Engineering* **154** 012020 doi:10.1088/1757-899X/154/1/012020
[10] Dudin S, Voronin K, Yakubovskaya S, Mutavaliev S 2016 Modeling Hydrodynamic State of Oil and Gas Condensate Mixture in a Pipeline. *MATEC Web of Conferences.* DOI: 10.1051/matecconf/20167302021
[11] Bautin S P, Krutova I Y, Obukhov A G 2015 Twisting of a fire vortex subject to gravity and Coriolis forces. *High Temperature* **53-6** 928-930 DOI: 10.1134/S0018151X1505003X
[12] Bautin S P, Obukhov A G 2016 Mathematical simulation of the near-bottom section of an ascending twisting flow *High Temperature* **51 – 4** 509-512 DOI: 10.1134/S0018151X1302003X
[13] Kulikov A M 2016 On Choosing Multicomponent Multiphase System Separation Progress Optimization Criteria V *IOP Conference Series: Materials Science and Engineering* **154 (1)** DOI: 10.1088/1757-899X/154/1/012039

Methodology for reducing energy and resource costs in construction of trenchless crossover of pipelines

V S Toropov

Industrial University of Tyumen, 38 Volodarskogo Str., Tyumen, 625000, Russia

E-mail: vladimir.s.toropov@gmail.com

Abstract. The paper suggests a set of measures to select the equipment and its components in order to reduce energy costs in the process of pulling the pipeline into the well in the constructing the trenchless pipeline crossings of various materials using horizontal directional drilling technology. A methodology for reducing energy costs has been developed by regulating the operation modes of equipment during the process of pulling the working pipeline into a drilled and pre-expanded well. Since the power of the drilling rig is the most important criterion in the selection of equipment for the construction of a trenchless crossover, an algorithm is proposed for calculating the required capacity of the rig when operating in different modes in the process of pulling the pipeline into the well.

1. Introduction

Today the method of horizontal directional drilling for the construction of pipelines is one of the most common as it has many advantages and extensive positive experience in the application of laying the utilities [1]. Horizontal directional drilling units have a high cost, and their operation modes are such that machines and equipment involved in construction can quickly fail [2].

A cost-efficient approach to the operation of this equipment can significantly reduce the cost of construction, which is quite high, compared with other trenchless methods of laying pipelines.

2. Materials and methods

The first step in reducing the costs of constructing the communication transitions using horizontal directional drilling technology is selection of the most appropriate equipment and components for the work performance.

The right choice of equipment and components can save energy due to:

1. The use of a less powerful drilling rig that can develop the required pulling force when laying a pipeline on this technology [3].

2. The use of drill pipes with a smaller diameter and, as a consequence, with a less mass and a more flexibility for drilling and pulling, reduces the bending radius of the well as well as the weight of the drilling string and the required effort to move it, which ultimately results in less earthwork [4].

Hardness coefficient of the rock is the main criterion for choosing the drilling equipment, in particular, the rock cutting tool.

The amount of rock extracted from the well during pilot drilling and expansion, which directly depends on the coefficient of soil softening, determines not only the consumption of drilling mud for removal of cuttings, but also the costs and volumes of earthwork for equipping the pits at the working site for the waste drilling mud.

Thus, this indicator influences the cost of constructing the pipeline crossover through several items of expenditure.

The amount of soil to be extracted from the well will be equal to:

$$V_{gr} = \frac{k_d \cdot L_{bh} \cdot \pi \cdot D_{bh}^2}{4} \tag{1}$$

where L_{bh} – well length;

D_{bh} – well diameter;

k_d – coefficient of soil softening where drilling is performed

Dependence of the soil volume that must be removed from the drilled well on the well diameter with a well length of 200 m for different soil types is shown in Figure 1

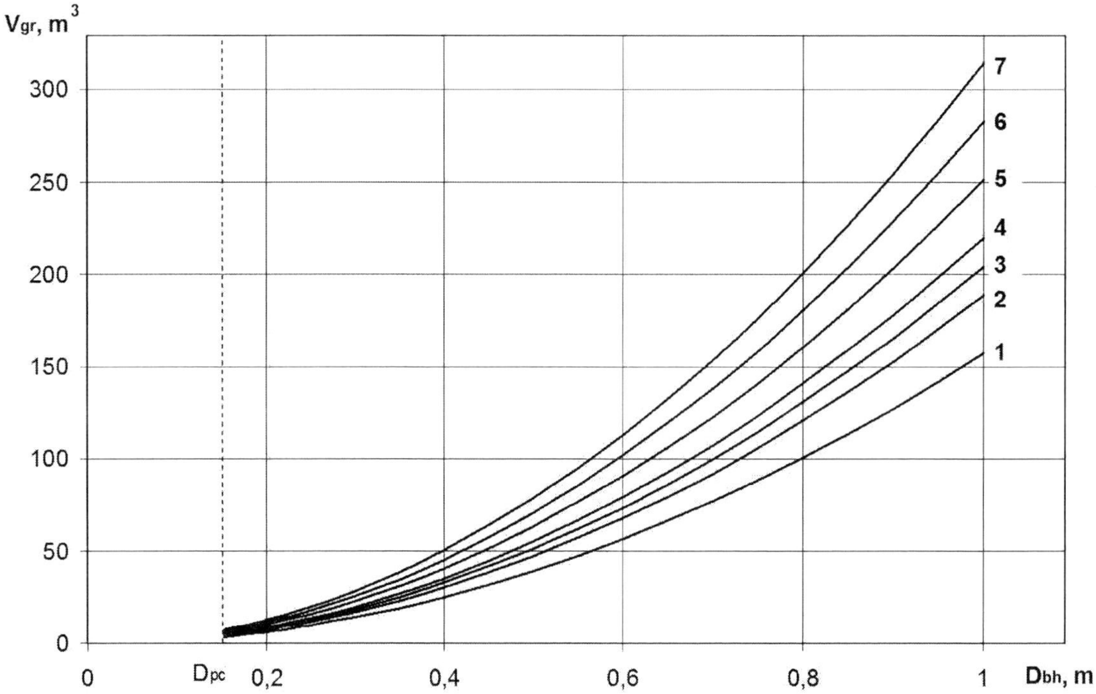

Figure 1. Dependence of the soil volume, which must be removed from the drilled well, on the well diameter with a well length of 200 m for different soil types: 1 – the initial well volume; 2 – sand, sandclay; 3 – vegetative soil, clay, loam, gravel; 4 – half-rocks; 5 – rocks of moderate strength; 6 – hard rocks; 7 – very hard rocks; D_{bh} – the diameter of the pilot well.

The dependence shows that, all other things being equal, the soil volume can differ significantly from the initial volume of the well for different soil types. In addition, stronger soils tend to have a higher soil softening coefficient, which is an additional argument in favor of a thorough study of the ground conditions in the crossover section, and, if possible, the choice of a ground crossing path with a lower soil softening coefficient.

Measures for the selection of equipment and its components can be carried out after calculating the geometric parameters of the profile of the projected well, starting from the parameters of the pipeline being laid, as well as calculating the operating modes of the equipment during the pipeline pulling operation, in order to determine the necessary power characteristics of the drilling rig [5, 6].

To calculate the basic parameters of the pulling process, it is necessary to determine the geometric parameters of the profile of the projected pulling.

IOP Conf. Series: Materials Science and Engineering **357** (2018) 012031 doi:10.1088/1757-899X/357/1/012031

The main parameters of the well profile are:
1. The well length in the plan (over the day surface).
2. The burial depths at the lowest point of the profile.
3. Well length.
4. Well diameter.
5. Spudding angle.

The calculation of the profile can be made in several variants, depending on the initial data.

The well profile, the length of the horizontal crossover, the maximum burial depths and the diameter of the working pipeline are determined as the initial data for construction. Then the bend radius of the transition profile and the pipeline depth at each point of the profile are calculated.

The actual well length is calculated by equation:

$$L = R_c \alpha_x = R_c \left(\arcsin\left(\frac{L_h}{2R_c}\right) - \arcsin\left(\frac{\frac{L_h}{2} - x}{R_c}\right) \right) \tag{2}$$

where R_c – bend radius of the well profile;
x – horizontal coordinate;
L_h – well length in the plan;
a_x – central angle of the well profile section.

In equation (2), variable x is assumed to be equal to the total well length in the plan.

The bend radius must be compared with the minimum allowable bending radius of the pipeline being laid for steel pipelines and with the minimum allowable bend radius of the drilling string for polyethylene pipelines. And, if necessary, the radius must be increased by changing the crossing length or the maximum burial depths in the plan.

It is expedient to select the spudding angle from the working range of the drilling rig angles, when calculating the bending radius of the profile.

Next, the maximum operating rate of pulling the pipeline must be set. Its value for a specific drilling rig will be limited by the maximum rate of carriage movement.

Due to the fact that the process of pulling the pipeline has a cyclical pattern, the calculation of the process power characteristics must be carried out separately for each cycle, choosing a cycle of pipeline movement, for which initializing the rig will require the greatest power [7].

Therefore in future, having received the value of the average power needed to disperse the drilling string within one cycle at different profile points while pulling the pipeline of the developed diameter, it is necessary to compare the result with the maximum power of the drilling rig. If the required power is too high, the maximum operating rate must be reduced while the pulling process.

Calculation of the resistance force at the beginning of the braking section is made using equation [8]:

$$\sum F_{res} = \lambda \cdot \rho_f \left(L_{bh} - L - x\right) \cdot \frac{\pi\left(D_{bh}^2 - D_{dr}^2\right)}{8D_{bh}} \cdot \left(\frac{dx}{dt}\right)^2 + \frac{\left(L_p - L - x\right)}{L_p} \cdot m_p \cdot g \cdot k_{rf} + \\ + \frac{\pi}{4}\left(D_{bh}^2 - D_{dr}^2\right) \cdot \left(\sqrt{R_c^2 - x^2} + H_{\max} - R_c\right) \cdot \rho_f \cdot g, \tag{3}$$

where λ - coefficient of the hydraulic resistance;
ρ_f – mud density;
D_{bh} – well diameter;
D_{dr} – drill pipe diameter;
L_{bh} – well length;
R_c – well profile curvature radius;
H_{max} – maximum burial depths of the well profile relative to the earth's surface.

Thus, the values of the sum of the resistance forces are obtained, which must be overcome when the pipeline moves at a given maximum operating rate at each particular point in the well profile as the pipeline and the drilling string enter it [9].

After this, the acceleration in the braking section and the length of the braking section are calculated.

Next, it is necessary to determine the version of the cycle, that is, the presence or absence of a uniform motion section.

In the case when a uniform motion section is present, it is necessary to specify its length and calculate the amount of the drilling rig operation related to moving the pipeline on the uniform motion section.

The length of the acceleration section is found as follows:

$$l_I = l_{dr} - l_{III} - l_{II}. \tag{4}$$

Further, it is necessary to calculate the amount of operation of the drilling rig for moving the pipeline in the acceleration section.

In the case when the uniform motion section is absent, the length of the acceleration section will be equal to:

$$l_I = l_{dr} - l_{III}. \tag{5}$$

And after that, the work of the drilling rig in the acceleration section is determined.

Then, in the case of the presence of a uniform motion section, the propagation time for the speeding up and uniform motion sections will be:

$$t_I + t_{II} = \sqrt{\frac{2l_I}{a_I}} + \frac{l_{II}}{v_{\max}}. \tag{6}$$

And in the case of absence of a uniform motion section, the propagation time for the acceleration section will be equal to:

$$t_I = \sqrt{\frac{2l_I}{a_I}}. \tag{7}$$

After that, the required capacity of the installation is calculated to move the pipeline to a length equal to that of the drill pipe, according to the calculation methodology presented in [10].

The pulling mode is determined by a combination of such parameters as maximum working speed, speeding up in the acceleration section, the presence or absence of a uniform motion section and its length [11]. The length of the used drill pipes is also significant.

Using the above described calculation sequence, it is possible to select the pulling mode, most profitable in terms of power, the maximum back pull force of the rig or the time of work performance.

Selecting the equipment according to peak values of tractive power calculated for the entire well profile, it is possible to select the pulling mode with the least energy costs that occur when pulling the pipeline and the drilling string during the pulling process [12].

3. Results and Discussion

Calculation of the parameters of the pulling process at various operating modes of the drilling rig requires a fairly large amount of calculations, in connection with the need to take into account a variety of factors that affect the amount of tractive power and the required drilling rig power [13].

Therefore, for calculations using the methodology described in this paper, an algorithm was developed that allows selecting the most cost-effective mode of pulling the pipeline.

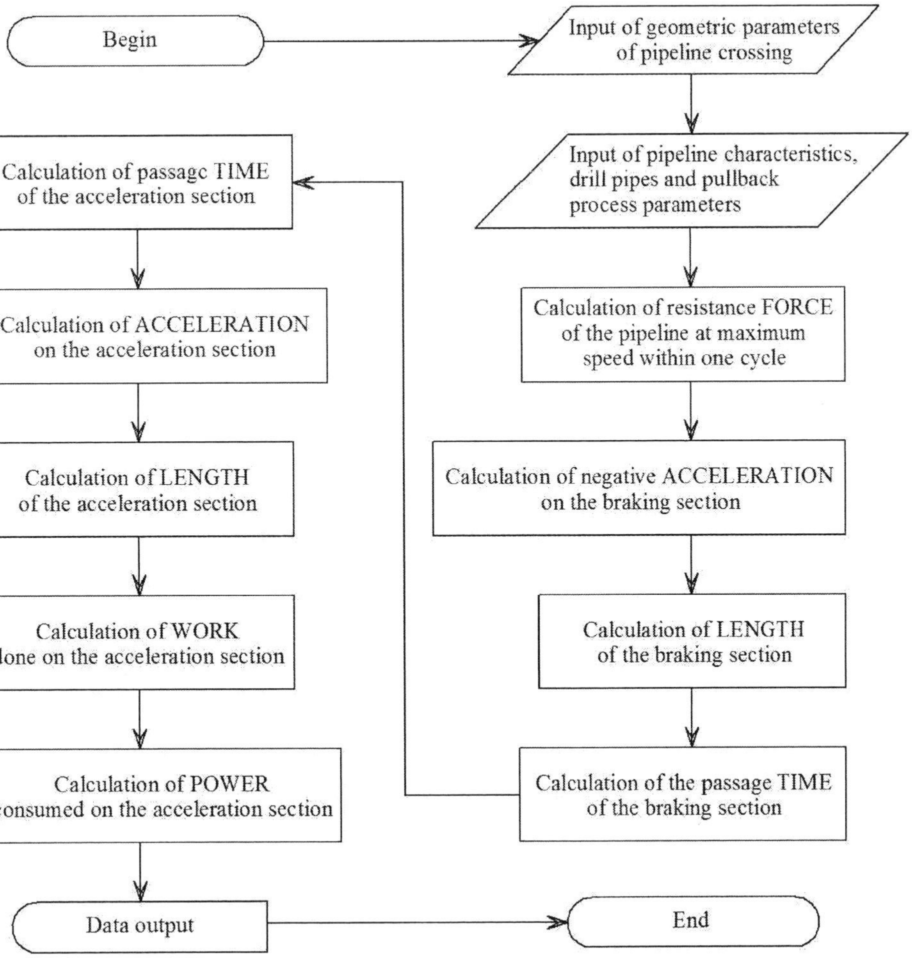

Figure 2. A flow-chart of the algorithm for calculating the required power of the rig when operating in different modes

The algorithm allows solving the following problems:

1. Determination of the geometric parameters of the well profile in accordance with the tasks facing the designer and the capabilities of the drilling rig.

2. Calculation of the instantaneous value of the traction power for the rig, which is necessary for pulling, at all points of the well profile, at different operating modes of the rig.

3. Calculation of the value of the rig power, which is necessary to perform the work of pulling the pipeline into the well, in this or that mode of operation.

The flow-chart of the algorithm is shown in Figure 2.

Calculation using this algorithm requires the following data:

1. Geometrical parameters of the transition profile:

1.1. The transition length over the day surface.

1.2. Maximum burial depths.

2. Parameters of the laid pipeline:

2.1. Pipeline diameter.

2.2. Material of the pipeline.

2.3. Weight of lineal meter of pipeline.

3. Drilling pipes parameters:

3.1. Drilling pipe weight.

3.2. Length of the drilling pipe.
3.3. Diameter of the drilling pipe.
4. Drilling mud parameters:
4.1. Drilling mud density.
4.2. Tangential shear stress of the drilling mud.
4.3. Coefficient of plastic viscosity of the drilling mud.
5. The parameters of the pulling process:
5.1. Maximum working speed of the pipeline during the pulling process.
5.2. The current coordinate of the well profile.
6. Characteristics of the used auxiliary equipment:
6.1. Coefficient of roller friction against the pipeline wall for a section of the pipeline located outside the well.

The calculation results are:

1. The instantaneous value of the thrust force of the rig, which is necessary for pulling, at all points of the well profile, at different operating modes of the rig.
2. The value of the rig power, which is necessary to perform the work on pulling the pipeline into the well, in this or that mode of operation.

4. Conclusion

The calculation results for the algorithm given in the work are the obtained values of the parameters of the pulling process at various operating modes of the drilling rig. They can be the basis for a reasoned choice of a drilling rig for work performance, and can also serve as indicators when taking decisions on the choice of operating modes of the used rig for horizontal directional drilling [14].

Thus, the developed set of measures for selection of the equipment and its components for pulling the pipeline is very relevant in a situation where the bulk of the work performers make crossings mainly on low-power rigs [15].

The rig possibilities can be expanded also by applying the proposed methodology of reducing the energy costs in terms of regulating the operating modes of the rig while pulling the working pipeline.

For the practical application of the proposed methodology, the algorithm for calculating the required power, proposed in the paper, can be used.

References

[1] Willoughby D 2005 *Horizontal Directional Drilling Utility and Pipeline Applications* (New York; The McGraw-Hill Companies)

[2] Antipiev V N, Zemenkov Yu D and Shabarov A B 2002 *Technical and parametrical diagnostics in pipeline systems* (Tyumen; Vector Book)

[3] Zemenkova M Yu, Shipovalov A N, Dudin S M and Zemenkov Yu D 2007 System analysis in oil and gas objects control processes *News of higher educational institutions. Oil and gas* **5** 116-9

[4] Zemenkov Yu D 2008 *Technique and technologic processes at transport of energy resources*: two volume instructional aid. Vol 1 (Tyumen; Vector Book)

[5] Shalay V, Zemenkova M, Zemenkov Y, et al. 2016 Modeling Parameters of Reliability of Technological Processes of Hydrocarbon Pipeline Transportation. *MATEC Web of Conferences* **73**, 01029 DOI: 10.1051/matecconf/20167301029

[6] Toropov V S, Ponomareva T G, Toropov S Yu, Toropov E S 2016 Quantitative assessment of changes in the pipeline - well system in the case of an emergency situation in horizontal directional drilling. *Izvestiya vysshikh uchebnykh zavedeniy. Oil and gas* **5** 71-77

[7] Smirnov V V and Zemenkov Yu D 2013 Using the finite element method for calculation of intense deformed condition of above ground pipelines constructed on permafrost soil *Pipeline transport: theory and practice* **4** 18-23

[8] Toropov V S, Tamer O S, Toropov S Yu, Nikiforov N A 2016 *Dynamics of pipeline pulling*

process by horizontal directional drilling. International Scientific-Practical Conference of Students, Graduate Students and Young Scientists *"Transport and Storage of Hydrocarbons"* **154**

[9] Toropov E S, Vorobieva T I, Toropov S Yu, Toropov V S 2015 *Finding the Displacement of Elastically Bent Pipeline Disposed on Roller Supports Under the Action of the Axial Shear Force.* International Scientific-Practical Conference of Students, Graduate Students and Young Scientists *"Transport and Storage of Hydrocarbons"*, **154**

[10] Toropov V S, Temirbaev R M, Toropov E S, Toropov S Yu 2016 *Interaction Between the Pipeline and Additional Equipment for Trenchless Technologies.* International Scientific-Practical Conference of Students, Graduate Students and Young Scientists *"Transport and Storage of Hydrocarbons"* **154**

[11] Zemenkov Y D, Shalay V V, Zemenkova M Y 2015 Expert systems of multivariable predictive control of oil and gas facilities reliability *Oil and Gas Engineering Procedia Engineering* ***113*** 312-315 DOI: 10.1016/j.proeng.2015.07.271

[12] Zemenkova M, Shalay V, Zemenkov Y, Kurushina E 2016 Improving the Efficiency of Administrative Decision-Making when Monitoring Reliability and Safety of Oil and Gas Equipment *MATEC Web of Conferences* **73** DOI: 10.1051/matecconf/20167307001

[13] Dudin S, Bahmat G, Zemenkov Y, Voronin K, Shipovalov A 2017 Strategy for monitoring and ensuring safe operation of Russian gas transportation systems *MATEC Web of Conferences* **106** DOI: 10.1051/matecconf/201710606004

[14] Voronin K S, Ogudova E V 2016 The Effect of Dynamic Processes in the System "Pipe-Soil" on the Pipeline Deviation from Design Position. Transport and Storage of Hydrocarbons. *IOP Conf. Series: Materials Science and Engineering* **154** 012019 doi:10.1088/1757-899X/154/1/012019

[15] Voronin K S 2016 Forecasting and Evaluation of Gas Pipelines Geometric Forms Breach Hazard. Transport and Storage of Hydrocarbons. *IOP Conf. Series: Materials Science and Engineering* **154** 012020 doi:10.1088/1757-899X/154/1/012020

IOP Conf. Series: Materials Science and Engineering **357** (2018) 012032 doi:10.1088/1757-899X/357/1/012032

Effect of section shape on frequencies of natural oscillations of tubular springs

S P Pirogov [1,2], A Yu Chuba[2] and D A Cherentsov [1]

[1]Industrial University of Tyumen, 38 Volodarskogo Str., Tyumen, 625000, Russia
[2]State Agrarian University of the Northern Trans-Urals, 5 Repablic Str., Tyumen, 625003, Russia

E-mail: cherentsovda@bk.ru, voronin_tsogu@mail.ru

Abstract. The necessity of determining the frequencies of natural oscillations of manometric tubular springs is substantiated. Based on the mathematical model and computer program, numerical experiments were performed that allowed us to reveal the effect of geometric parameters on the frequencies of free oscillations of manometric tubular springs.

1. Introduction

Tubular springs are the main elastic sensing element of manometric devices. Pressure gauges often have to work in the conditions of vibrations and pulsations of the working environment, which is typical for the oil and gas industry, so manometric tubular springs in these devices oscillate, which makes it difficult to accurately record instrument readings and adversely affects the precision of measurements. It is necessary to prevent the manifestation of these shortcomings at the design stage. Great attention should be given to the dynamic calculations of tubular springs, in particular, the definition of vibration resistance.

Vibration frequency is the main characteristic of vibration resistance; therefore, an important task is to study the influence of factors on this characteristic, in particular, the influence of geometric parameters of a tubular spring. Mathematical models of oscillatory motion of tubular springs are given in [1-3.]

2. Materials and methods

The manometric tubular spring is a tube bent with radius R of the central axis and having a cross-sectional shape different from the circumference (Fig. 1). The cross-section has one or two mutually perpendicular axes and is arranged so that one of its axes is an extension of the radius of curvature of the central axis.

One end of tube 2 is hermetically sealed by tip 4 and the other end is fixed in holder 1. Through the opening in the holder, pressure in the inner cavity of the tubular spring changes, leading to a deformation of the cross-section and changes in the curvature of central axis 3, and the free end of tube performs displacement λ. The free end of the tube is connected to the mechanism of the manometer, by means of which the movement is transmitted to the indicator of the device.

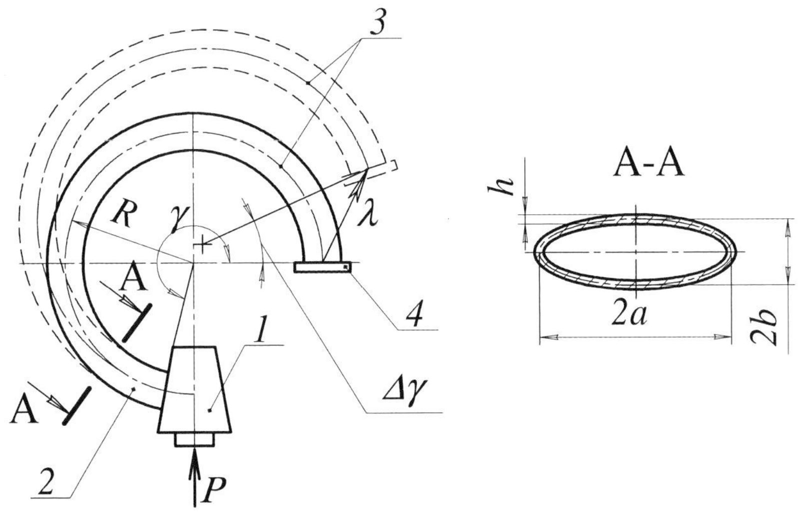

Figure 1. Manometric tubular spring
1 – holder; 2 – tube; 3 – central axis; 4 – tip

The geometric parameters of the tubular spring determining its dimensions are the radius of curvature of the longitudinal axis – R, the central angle of the opening – γ, the wall thickness – h, the major semiaxis – a, and the minor semiaxis – b.

At present, manometric springs of constant cross-section along the length of the tube are the most common; the main cross-sectional shapes are elliptical, flat-oval and 8-shaped. To improve the characteristics of manometric springs, a number of spring designs with a cross-section and wall thickness varying along the central axis [6-11] are proposed, which may have better performance characteristics than springs of constant cross-section. There are tubular springs with a variable cross-section of the tube, both along the perimeter of the cross section, and along the longitudinal axis of the tube. In this case, the variables can be either the dimensions or the shape of the section.

3. Study

To compare the properties of tubes of a constant cross section, their shape was represented by a universal section (Fig. 2).

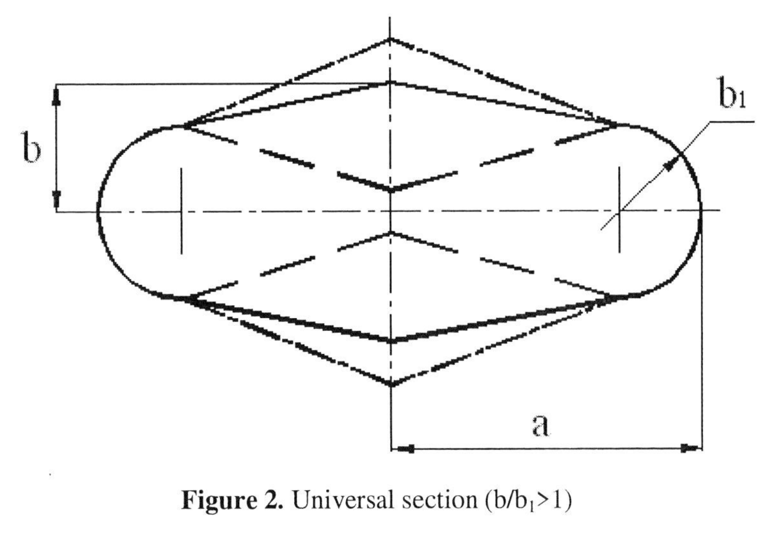

Figure 2. Universal section ($b/b_1>1$)

By changing ratio b/b_1, one can move from elliptical ($b/b_1>1$) to 8-shaped ($b/b_1<1$). For a flat-oval cross-section, $b/b_1=1$.

For all compared springs, the same geometric parameters will be: R=40mm h=0.4mm. The cross-sectional height of all springs is the same.

The graph of the dependence of the natural oscillation frequency on the shape of the cross-section is shown in Fig.3

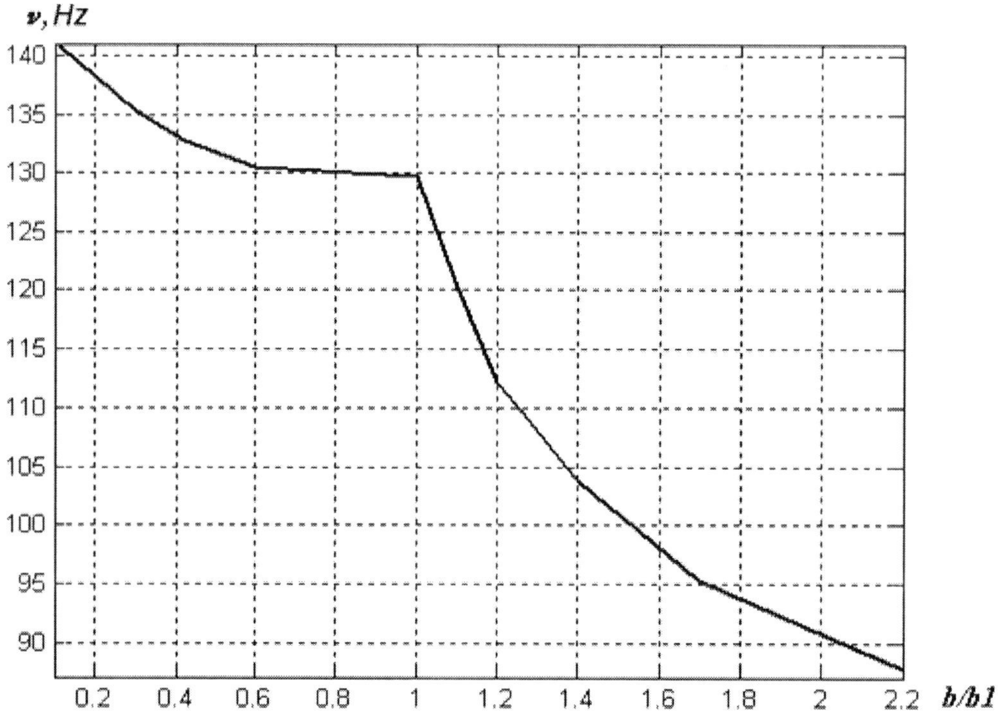

Figure 3. Dependence of the natural oscillation frequency on the shape of the cross-section

The graph shows that the most rational is the flat-oval shape; for the elliptical one there is a significant reduction in frequency, and for the 8-shaped - a slight increase.

Let us investigate the effect of changing the cross-sectional shape of springs of a variable cross-section on the example of nine brass springs with the geometric parameters: the radius of the workpiece tube is r=6mm, the radius of curvature of the central axis is R=40mm, the wall thickness is h=0.4mm, the value of the central angle of springs is $\gamma=230^{\circ}$. The cross-section of the springs varies from the rigidly fixed to the free end as follows:

1) from 8-shaped to flat-oval;
2) from 8-shaped to elliptical;
3) from flat-oval to 8-shaped;
4) from elliptical to 8-shaped;
5) from elliptical to flat-oval;
6) from flat-oval to elliptical;
7) from flat-oval to more flattened flat-oval;
8) from elliptical to more flattened elliptical;
9) from 8-shaped to more flattened 8-shaped.

Flat-oval sections have the following characteristics: $a/r=1.34$, $b/b_1=1$; elliptical: $a/r=1.4$, $b/b_1=1.42298$; 8-shaped: $a/r=1.26$, $b/b_1=0.1965$. Fig. 4 shows the results of the calculations.

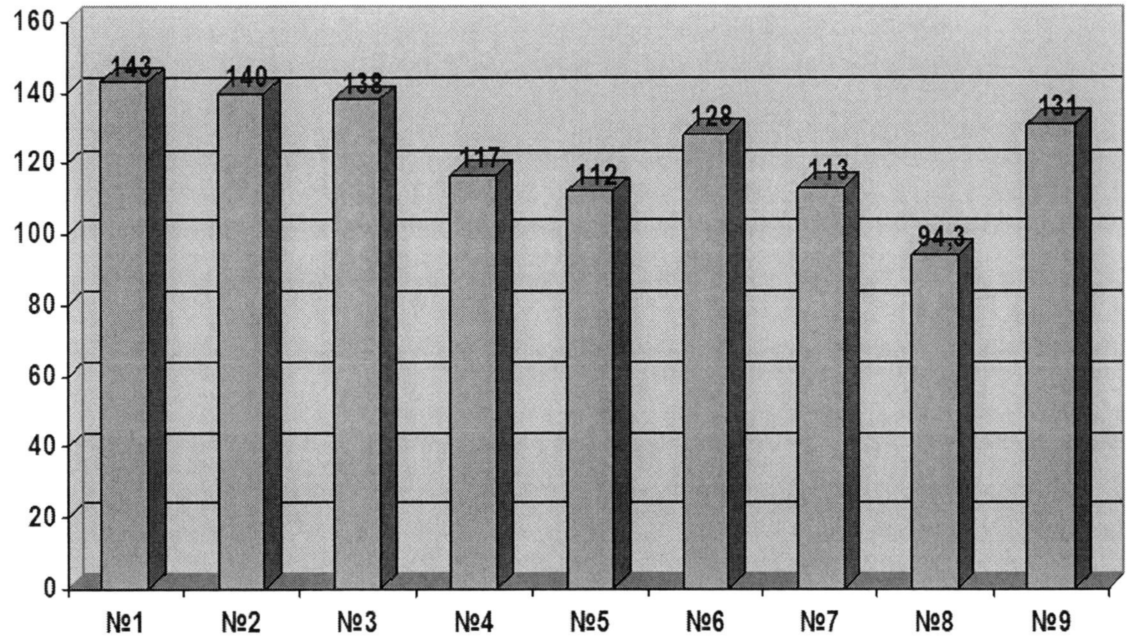

Figure 4. Effect of changing the shape of the cross-section of the spring on its natural oscillation frequency

4. Conclusion

Analysis of the results of numerical experiments showed that for tubes with a constant wall thickness, the most advantageous is the flat-oval cross-section.

It has been confirmed that variable-section springs have higher frequencies of natural oscillations than springs of a constant cross-section. In particular, it has been found that springs with a variable cross-section ranging from 8-shaped to flat-oval, as well as springs made from conical tubes, have frequencies of natural oscillations 20-40% higher than similar ones of a constant cross-sections.

References

[1] Chuba A Y, Pirogov S P, Dorofeev S M 2007 Determination of the frequencies of natural oscillations of a tubular spring of variable cross-section length. *Oil and gas* **2** 70-77

[2] Chuba A Y 2016 Algorithm for calculating the frequencies of natural oscillations of manometric tubular springs. *International Scientific Journal "Innovative Science"* **6-2** 164-166

[3] Chuba A Y, Samakalev S S, Pirogov S P 2007 *Certificate of official registration of the computer program 2007612005 of the Russian Federation. Software complex "PKRMTP" for calculation of manometric tubular springs. № 2007611194*

[4] Shalay V, Zemenkova M, Zemenkov Y, et al. 2016 Modeling Parameters of Reliability of Technological Processes of Hydrocarbon Pipeline Transportation. *MATEC Web of Conferences* **73** 01029 DOI: 10.1051/matecconf/20167301029

[5] Zemenkov Y D, Shalay V V, Zemenkova M Y 2015 Expert systems of multivariable predictive control of oil and gas facilities reliability *Oil and Gas Engineering Procedia Engineering* **113** 312-315 DOI: 10.1016/j.proeng.2015.07.271

[6] Pirogov S P, Ponomareva T G 1997 *Patent 2093805 RF, IPC 7, G 01 L 7/04. Manometric tubular spring № 96105858/28*

[7] Zemenkov Y D, Shalay V V, Zemenkova M Y 2015 Immediate analyses and calculation of saturated steam pressure of gas condensates for transportation conditions. *Oil and Gas Engineering* **113** 254-258 DOI: 10.1016/j.proeng.2015.07.330

[8] Zemenkova M, Zemenkov Y, Gladenko A, Podorozhnikov S 2016 Estimation of Emissions during Monitoring of Pipelines in the Dynamic Mode of Operation. *5th International Scientific Conference Integration, Partnership and Innovation in Construction Science and Education* 86 DOI: 10.1051/matecconf/20168604053

[9] Stepanov S, Cheremisin N, Sokolov S, Altunin A and Shabarov A 2012 Computational method for relative permeability curves based on the bernoulli generalized porous media equations ECMOR 2012 13th European Conference on the Mathematics of Oil Recovery

[10] Dudin S M, Zemenkov Y D, Maier A V, Shabarov A B 2016 Research and Design of Thermophysical Gas-Liquid Mixture Parameters in Product Pipelines. *IOP Conference Series: Materials Science and Engineering* **154** DOI: 10.1088/1757-899X/154/1/012021

[11] Kurushina V, Kurushina E, Shabarov A 2016 Technology Audit: Assessment of Innovative Portfolio *MATEC Web of Conferences* **73** DOI: 10.1051/matecconf/20167307028

[12] Voronin K S, Ogudova E V 2016 The Effect of Dynamic Processes in the System "Pipe-Soil" on the Pipeline Deviation from Design Position. Transport and Storage of Hydrocarbons. *IOP Conf. Series: Materials Science and Engineering* **154** 012019 doi:10.1088/1757-899X/154/1/012019

Analysis and optimization of indicators of energy and resource consumption of gas turbine and electric drives for transportation of hydrocarbons

V V Golik[1], M Yu Zemenkova[1], I V Seroshtanov[2] and Z V Begalko[1]

[1] Industrial University of Tyumen, 38 Volodarskogo St., Tyumen, 625000, Russia
[2] Gazprom transgaz Yugorsk, 15 Mira ыltr., Yugorsk, 628260, Russia

E-mail: darkpaint@mail.ru, muzemenkova@mail.ru, seroshtanoviv@mail.ru

Abstract. The paper presents the results of the analysis of statistical indicators of energy and resource consumption in oil and gas transportation by the example of one of the regions of Russia. The article analyzes engineering characteristics of compressor station drives. Official statistical bulletins on the fuel and energy resources of the region in the pipeline oil and gas transportation system were used as the initial data.

1. Introduction

Main oil and gas pipelines are important components of the energy supply system of the Russian economy. Therefore, the problem of reliability and efficiency of their functioning is considered as one of the priority. Sufficiently significant facilities in solving this problem are gas turbine and electric drive systems for pumping and compressor stations of main pipelines.

In the current practice of operating domestic main oil pipelines (MOP), the drive of all pumps of oil pumping stations (OPS) is electric, and in the system of main gas pipelines (MGP), the drive of gas compressor units of compressor stations (CS) is mostly ф gas turbine.

2. The role of energy consumption and resource consumption in the transport of oil and gas

Russia has a large under-utilized energy-saving potential, which, as far as its ability to solve the problem of ensuring economic growth of the country is comparable to the growth in production of all primary energy resources.

Lack of energy can be some factor in restraining the economic growth of individual countries. According to analysts, until 2020, the rate of reduction in energy intensity in the absence of a coordinated state policy on energy efficiency can slow down sharply. This could lead to an even more dynamic growth in the demand for energy resources within the country. There are enough oil and gas reserves in Russia; however, an increase in hydrocarbon production and the development of transport infrastructure require significant investments.

In the paper, the analysis and estimation of statistical indicators of energy and resource consumption in oil and gas transportation in the Tyumen region is presented as an example. To achieve this task, statistical bulletins were used as the initial data: cost of production and sales of products in Tyumen region for 2015-2016; use of fuel and energy resources in Tyumen region in the pipeline oil and gas transportation system for 2015-2016.

The total energy consumption of Russia in recent years has been about 990 million tons of fuel equivalent. When bringing the introduction of energy-saving and energy-efficient equipment to the level of the EU member countries, energy consumption would decrease to 650 million tons of fuel equivalent. In other words, about 35% of energy is lost.

Barriers to the development of energy conservation and energy efficiency in any country can be divided into four main groups:

- lack of motivation;
- lack of information;
- lack of experience in project financing;
- lack of organization and coordination.

There are two ways to solve the emerging problem of increasing energy and resource efficiency:

- the first is an extremely capital-intensive way of increasing oil and gas production and constructing new power generation facilities;

- the second is much less expensive, associated with ensuring economic growth in the country through increased efficiency in the use of fuel and energy resources.

It should be noted that in practice a complex of the first and second variants is necessary with an undoubted priority of energy efficiency.

Systematic work in the field of energy conservation and energy efficiency in various sectors and spheres of the Russian economy began after the adoption of the Federal Law of the Russian Federation of 23.11.2009 No. 261-FZ "On energy conservation and on improving energy efficiency and on introducing amendments to certain legislative acts of the Russian Federation".

In 2010, the Ministry of Energy of Russia jointly with CJSC "APBE", LLC "CENEF" and FGA "REA" developed the State Program of the Russian Federation "Energy saving and energy efficiency improvement for the period until 2020" ("SPEE-2020"), which was approved by the Government of the Russian Federation on 21.10.2010 and by the decree of the Government of the Russian Federation No. 2446-r dated 27.12.2010.

The program is designed to be a tool to address the massive task of reducing the energy intensity of GDP by 13.5% by 2020 through energy-saving measures.

To realize the measures prescribed in the Program, it is necessary to attract funding from various sources, since program activities are inconceivable without analysis and cost estimation [1].

It is also worth mentioning that the project "Energy strategy of the Russian Federation for the period until 2035" sets the task of developing new highly efficient gas pumping units with gas turbine and electric drives as well as more sophisticated domestic construction materials for the creation of these units. [2]

3. Assessment of the efficiency of gas turbine and electric drives

A significant proportion of accidents on main oil (MOP) and gas pipelines (MGP) is accounted for by oil pumping stations (OPS) and compressor stations (CS), respectively.

With existing backups on MOP and MGP in the absence of cascading failures, failure of one element does not lead to failure of the entire system, but only worsens the efficiency of its operation. Reliability problems can be solved by both rational use and distribution of costs for maintenance and development of the system, as well as costs for backups, and by increasing these costs. Therefore, the problem of reliability here is a technical and economic problem.

In recent decades, Russia has pursued a policy of establishing "parity" of domestic prices for electricity, liquid fuel and fuel gas, which to some extent affects the cost-effectiveness of energy resources for their consumers, in particular, for the OPS and CS drive systems of the main oil and gas pipelines.

The choice of a rational drive system of MGP CS is the subject of numerous studies and justifications, which do not allow one to make unambiguous decisions so far. However, a number of known causes and circumstances predetermined by now a very clear relationship between the types and expediency of using MGP CS gas turbine (GGPU) and electric (EGPU) drives.

The competitiveness of gas turbine and electric drives depends primarily on CS power supply conditions. Usually CS with electric drive GPU are built on sections of gas pipelines passing through areas with a developed power system that has reserves of electrical power. Selection of the parameters and type of the compressor GPU drive is carried out on the basis of technical and economic calculations. To calculate the reduced costs for the selected options, enlarged technical and economic indicators are used obtained as a result of experience in the construction and operation of main gas pipelines.

An important argument in favor of EGPM is high efficiency factor of electric motors (about 0.95) compared to low efficiency factor of gas turbine motors (0.25-0.29). However, it is more appropriate to consider the energy utilization efficiency factor. In this case, when electric drive stations are supplied with power from TPP with an efficiency of 0.3-0.35 through the electrical distribution systems, the energy efficiency factor of the electric drive and gas turbine CS are practically equal [3].

Many well-known technical, energy, environmental, reliability, regulatory, economic, price and other characteristics and indicators, as well as advantages and disadvantages of both systems of GPU drives, are so different or converging that in the absence of unity of approaches and opinions on the significance, priority and ranking of individual factors or groups of factors, the solution of a problem of an optimum choice of a drive system is strongly complicated. A systematic approach to the solution of the problem is needed on the basis of a scientific assessment of the significance and priority of the factors characterizing the systems of drives in question or their combinations at CS under different conditions and types of security of the energy base of their location areas [4].

The question of comparing the advantages of gas turbine and electric drive is of great importance. Large current costs for electricity due to the high tariff for it encourage replacement of the existing electric drive units for gas turbines.

Nevertheless, in a number of cases, for example, in areas with low costs for electricity, and also for a number of other reasons, it is impossible to eliminate the use of electric drive units with a number of undeniable advantages over gas turbine drives. An electric motor produced by "Electrotyazhmash-Drive" SDG2-12500-2R UHL3.1 and a gas turbine unit MS5002E produced by "Nevsky Zavod" in cooperation with GeneralElectricOil & Gas (NuovoPignoneS.pA) were compared. (Fig. 1, 2) (Table 1).

Figure 1. GTU MS5002E [5]

Figure 2. Electric motor SDG2-12500-2R UHL3.1 [6]

In this regard, it should be noted that in the domestic practice of pipeline transport of oil and petroleum products at pumping stations, the use of purely electric pumps was largely determined by the ideological significance of domestic electrification. While in the foreign practice of transport of oil and petroleum products, the gas turbine drive of OPS pumps with various power supply systems is mainly used [7].

Table 1. Comparison of the basic parameters of electric and gas turbine engines

Parameter	Electric motor SDG2-12500-2R UHL3.1	Gas turbine unit MS5002E
Power MW	12.5	32
RPM speed	3000	5714
Efficiency %	97.8	36
Weight kg	36000	93000

In this regard, it should be noted that in the domestic practice of pipeline transport of oil and petroleum products at pumping stations, the use of purely electric pumps was largely determined by the ideological significance of domestic electrification. While in the foreign practice of transport of oil and petroleum products the gas turbine drive of OPS pumps with various power supply systems is mainly used [7].

In the energy aspect, the electric drive from thermal power plants and the gas turbine drive of OPS are practically equally effective.

From the economic point of view, calculation of the effectiveness of the compared OPS systems is quite routine, requiring generalization and evaluation of a large number of initial data under current conditions, previous design decisions, conditions for the future operation of the oil pipeline and foreign experience.

At the same time, it can be noted that with an unregulated electric drive of pumps, lower costs for buildings and equipment are significantly increased by the cost of constructing a power supply system. With a gas turbine drive of pumps, the composition and cost of associated equipment (installations for the selection, preparation and use of fuel) are significantly increased.

GTE in the drive of main oil pipeline pumps can be used in areas with different climatic conditions. It is preferable to use them in areas with a relatively low average annual ambient temperature, i.e. in the regions of the Far North and Siberia.

In the world practice, there are enough examples of the extensive use of gas turbine engines on main oil pipelines, including Trans-Alaska, with a length of about 1300 km in northern conditions close to Western Siberia and the Far North of Russia [12, 13].

A special place in the operation of GTE on oil pipelines takes the use of pumped fuel. Such fuels as kerosene, gasoline, diesel, associated gas and natural gas directly correspond to the requirements for fuel for gas turbines. Natural gas is efficiently used when a gas pipeline runs parallel to an oil pipeline. Associated gas can be used if it is pumped with oil. Pumped crude oil can be used after preparation in special block distillation units. After rectification of the oil, 25-30% of the wide fraction of light hydrocarbons (NGL) is obtained, and the residue (fuel oil) is pumped into the oil pipeline. The use of such system is economically advantageous in areas remote from powerful sources of electricity. Utilization of combustion products solves the problem of OPS heat consumption.

The gas turbine drive is the most economical to regulate the delivery of a pumping unit, and a controlled electric drive system is the least economical.

In the domestic practice of pumping unit drives of main oil pipelines, there are also some examples of the evaluation and use of a controlled electric and gas turbine drive. Special attention should be paid to the development of Aviadvigatel for the Sakhalin-2 project.

However, in order to fully assess the effectiveness of a particular type of the OPS drive, further technical and economic assessment is necessary, taking into account the reliability for the specific operating conditions. In areas with low cost for electricity supplied from hydroelectric or atomic power plants, electric OPS drives can be more economical than gas turbine drives. At the same time, in the conditions of frequent changes in the operation mode of the oil pipeline, the OPS gas turbine drive may be preferable [14].

A useful foreign experience in the design and construction of OPS and the linear part of the main oil pipelines, including the experience of aboveground laying in permafrost conditions, should be taken into account in the construction of new main oil pipelines.

Total efficiency of the value-added use of energy resources in the economy (taking into account losses during extraction, transportation, etc.) by 1975, even in industrially developed countries, like the United States, was estimated at 10-15%.

In the economy of the USSR, the utilization efficiency of energy resources was roughly estimated (according to accounting data for 1985) to be 10.6%. However, even this single and very rough estimate allows us to conclude that the level of energy use in the economy of the country is much lower than it was previously assumed [8].

It has always been assumed that there are large reserves of increasing the efficiency of energy resources (energy and fuel), i.e. a component of energy saving. However, in practice the weight of the energy resources used is not estimated by their indicator (utilization efficiency), but by their share of value in the total costs of the economy.

At present, the share of energy costs in the total costs of the economy in Russia averages about 13%, and for certain regions and areas of activity (industries) it deviates up to 50%, more or less than the average. So in Tyumen region, it is about 7%.

As for the share of costs for raw materials and materials in Russia, on average, it is 2-3 times higher than the share of energy costs, and for individual industries, this difference reaches significantly higher multiplicities. This was also about 50 years ago in the USSR.

Relatively low shares of energy costs in the country's economy, as well as in the oil and gas sector, with relatively high shares of costs for stock and raw materials, do not cause much enthusiasm for energy saving and energy efficiency among the managers of these industries. But broad propaganda of energy saving often determines the need for their participation in campaigns to develop their energy-saving programs.

4. Ratio of the shares of costs for oil and gas transportation and costs for using fuel and energy resources in cost parameters

In the practice of enterprise management, continuous monitoring of energy and resource efficiency indicators is used. Comparative analysis makes it possible to assess the influence of various factors on the energy and resource consumption of production. For this research, the following parameters are of interest when analyzing the cost of oil and gas transportation: expenses for stock and raw materials; expenses for fuel and energy (based on data from the statistical bulletins of the industry in Tyumen region for 2015-2016).

A large share in the cost for production and sales of products is taken up by costs associated with the cost for "fuel and energy"; the figure reaches 1/5 of the total costs.

In the below mentioned statistical data, the dynamics of costs in 2015-2016 for the indicator "stock and raw materials" for the transportation of oil and oil products is observed, which amounted to 287.68 c.u. (2015) and 306.88 c.u. (2016), which is respectively 3% and 2% of the total costs (Table 2) [9].

Over 2015-2016, there has been a stable share of the costs for "fuel and energy" (mainly electricity) - 18%, which in monetary terms is 1,739.86 c.u. in 2015, 2,318 c.u. - in 2016.

To bring the data of Tables 4 and 5 to the state of liquidity, the following transfer scale of units of energy types in fuel equivalent (f.e.) was used:

- Electricity - 1 kWh = 0.32 kg of f.e.;
- Heat power -1 Gcal = 170 kg of f.e.;
- Fuel (gas) - 1m^3 of gas = 1.13 kg of f.e.

Table 2. Pipeline transportation of oil and oil products

	2015	2016
Costs for the production and sale of products (goods and services) c.u., (%)	9,539.27 (100%)	12,816.7 (100%)
Expenses for the purchase of raw materials, stock materials, purchased semi-finished products, components purchased for production, but sold on the side without processing. c.u., (%)	287.68 (3%)	306.88 (2%)
Fuel consumption expenses. c.u., (%)	153.88 (1.6%)	125.64 (0.98%)
Energy expenses. c.u., (%)	1,739.86 (18%)	2,318 (18%)

In gas transport, a comparative cost analysis showed that for "stock and raw materials" costs are from 6.4% (6,203.65 USD in 2015) to 4.4% (4,449.38 USD in 2016). The total costs for "fuel and energy" (mainly gas) fluctuate within 11-13% (Table 3) [10].

Table 3. Pipeline transportation of gas and gas products

	2015	2016
Costs for the production and sale of products (goods and services) c.u., (%)	96,469.82 (100%)	99,370.58 (100%)
Expenses for the purchase of raw materials, stock materials, purchased semi-finished products, components purchased for production, but sold on the side without processing. c.u., (%)	6,203.65 (6.4%)	4,449.38 (4.4%)
Fuel consumption expenses. c.u., (%)	10,607.68 (10.9%)	8,804.92 (8.8%)
Energy expenses. c.u., (%)	1,952.2 (2%)	2,187.22 (2.2%)

As for the oil and gas pipeline transportation systems, there is a slightly different situation in terms of the ratio of the cost shares for "fuel and energy."

In Tyumen region, oil transportation energy consumption is 98.8% for electricity and 1.2% - for fuel (Table 4), and in gas transportation it is 70.2-70.6% for electricity and 28-29% - for fuel (Table 4) [10,11].

Table 4. Use of fuel and energy resources in Tyumen region in 2015

	Production, total	Actual expenses for production	
		Electricity, thousand kW (tons of f.e.)	Heat power, Gcal (tons of f.e.)
Oil transportation through main oil pipelines, thousand tons*km	224,822,935	3,897,130 (1,248)	43,567 (7,395)
%		98.8%	1.2%
Gas transportation through main oil pipelines, thousand tons*km	582,628,062	3,610,661 (1,248)	1,525,483 (259,318)
%		70.2%	29.8%

Table 5. Use of fuel and energy resources in Tyumen region in 2016

	Production, total	Actual expenses for production	
		Electricity, thousand kW (tons of f.e.)	Heat power, Gcal (tons of f.e.)
Oil transportation through main oil pipelines, thousand tons*km %	284,228,441	3,698,400 (1.184) 98.7%	45,961 (7,820) 1.3%
Gas transportation through main oil pipelines, thousand tons*km %	500,375,061	3,137,757 (0.992) 70.6%	1,306,560 (222,105) 29.4%

5. Conclusion

Therefore, analysis and evaluation of the data of the statistical bulletins discussed in this paper made it possible to draw a conclusion that at compressor stations of the main gas pipelines, passing through the territory of Tyumen region, the use of gas turbine engines is more expedient, since the costs for "fuel and energy" in gas transport are determined mainly by the tariff for fuel gas, which is lower than the tariff for electricity. In other regions of the country, the situation can be diametrically opposite; everything depends on the development of the infrastructure. If the region has well-developed power supply channels and its cost is not high, then the enterprises will install a GPU with an electric drive. In other cases, it will be expedient to install a GPU with a gas turbine engine. However, gas transport enterprises conduct a number of experiments on the introduction of gas distribution stations and shops equipped with electric motors, which is due to close proximity to large sources of electricity. A more comprehensive evaluation of the efficiency of an electric or gas turbine engine requires research that is more thorough. Using the technique of selecting and calculating equipment depending on the technical and economic indicators, as well as on the geographical location and resource support of each of the regions of the Russian Federation will allow obtaining optimization solutions and more efficient resource consumption taking into account regional features. All the developed measures to ensure energy efficiency and to justify the drives should take into account both the requirements of international standards, such as the international standard ISO 50001:2011, and the requirements of the Energy Strategy of the Russian Federation and facilitate the implementation of import substitution programs.

References

[1] *Official website of the Ministry of Energy of the Russian Federation. General provisions.* https://minenergo.gov.ru/node/5195
[2] *Energy Strategy of the Russian Federation for the period up to 2035* https://minenergo.gov.ru/node/1920 Access date:15.05.2017
[3] Burganov F S, Tuzhilkin V N, Shvarts G R, Shpilevoy V A 2003 *Energy and electrification of compressor stations of main gas pipelines.* (Tyumen: FSU PPE "Tyumen")
[4] Kozachenko A N, Nikishin V N, Porshakov B P 2001 *Power system of gas transmission pipelines.* (Moscow: State Unitary Enterprise "Oil and Gas")
[5] *Gas-pumping unit GPU-32 "Ladoga" with a capacity of 32 MW.* http://www.reph.ru/production/gas/211/ Access date:15.09.2017
[6] *Electric motor SDG2-12500-2R UHL3.1.* http://privod-lysva.ru/index.php?id=104 Access date:15.09.2017
[7] Shpilevoy V A, Tyrylgin I V, Zemenkov Yu D 2012 Alternative drive systems for pumping units for new main oil pipelines. News of higher educational institutions. *Oil and Gas* **5**
[8] Stepanov V S, Stepanova T B 1994 *Energy efficiency.* (Novosibirsk: VO "Nauka")
[9] Expenses for production and sales of products in the Tyumen region for 2015-2016
[10] Use of fuel and energy resources in the Tyumen region in 2015. Stat. bul. Territorial organ of the Federal State Statistics Service for the Tyumen region
[11] Use of fuel and energy resources in the Tyumen region in 2015. Stat. bul. Territorial organ of the

Federal State Statistics Service for the Tyumen region

[12] Pirogov S P, Cherentsov D A, Chuba A Y 2016 Study of elastic sensing elements for vibration-resistant pressure gauges. *IOP Conference Series: Materials Science and Engineering* 012015

[13] Pirogov S P, Cherentsov D A 2016 Theoretical foundations of the design of vibration-resistant manometers. *Measurement Techniques* 1-5

[14] Voronin K S, Ogudova E V 2016 The Effect of Dynamic Processes in the System "Pipe-Soil" on the Pipeline Deviation from Design Position. Transport and Storage of Hydrocarbons. *IOP Conf. Series: Materials Science and Engineering* **154** 012019 doi:10.1088/1757-899X/154/1/012019

IOP Conf. Series: Materials Science and Engineering **357** (2018) 012034 doi:10.1088/1757-899X/357/1/012034

Monitoring of energy efficiency of technological modes of gas transport using modern gas-turbine equipment

V V Golik[1], M Yu Zemenkova[1], A N Shipovalov[2] and K A Akulov[1]

[1] Industrial University of Tyumen, 38 Volodarskogo St., Tyumen, 625000, Russia
[2] Gazprom transgaz Yugorsk, 15 Mira Str., Yugorsk, 628260, Russia

E-mail: darkpaint@mail.ru, muzemenkova@mail.ru, shipovalov_nir@mail.ru

Abstract. The paper presents calculations and an example of energy efficiency justification of the regimes of the equipment used. The engineering design of the gas pipeline in the part of monitoring the energy efficiency of a gas compressor unit (GCU) is considered. The results of the GCU characteristics and its components evaluation are described. The evaluation results of the energy efficiency indicators of the gas pipeline are presented. As an example of the result of the analysis, it is proposed to use gas compressor unit GCU-32 "Ladoga" because of its efficiency and cost effectiveness, in comparison with analogues.

1. Introduction

One of the priorities in the Energy Strategy of Russia recently is the development of resource-saving equipment, as well as a significant increase in the volume of gas exports and at the same time ensuring the technological independence of the gas industry, which means the use of domestic highly efficient gas transport technologies. In this field, the works of professors N.A. Malyushin, A.B. Shabarov, V.A. Shpilevoy, M.N. Chekardovsky, V.M. Pisarevsky, E.I. Krapivsky and other well-known authors are known.

2. Methodology

At the Transport of Hydrocarbon Resources Department of the Industrial University of Tyumen, complex studies are carried out to monitor the reliability, safety, and energy efficiency of complex process equipment for pipeline transport facilities of hydrocarbon raw materials at various stages of the life cycle. Monitoring the energy efficiency of gas pipeline modes is an important and complex multifactor complex task, both at the design stage and during the direct operation. For example, the existing methodology for the energy efficiency of PJSC Gazprom gas transmission facilities and systems evaluation [11] involves the calculation of indicators for various cases, including such indicators as indicators of efficiency coefficient, specific fuel gas consumption, specific electricity consumption for compression, etc.

The proposed methodological complex is presented in the form of numerous algorithms and mathematical- and software provided in the works of the authors and colleagues on various aspects of reliability, efficiency, and safety evaluation [2-10]. Production experience analysis has shown that at each stage of the life cycle of the facility there is a specific control, management and specific methodological aspects. As an example, the results of the analysis of the problematic field are presented to ensure the quality of managing the properties of objects at different stages (Table 1).

IOP Conf. Series: Materials Science and Engineering **357** (2018) 012034 doi:10.1088/1757-899X/357/1/012034

Table 1. Analysis of the problematic field regarding maintenance of the hydrocarbon transport system life cycle

Life Cycle Stage	Object and technical system properties and level maintenance factors		
	Reliability	Safety	Efficiency
Designing	Quality of methods for substantiating design solutions and their application in the field of materials and equipment selection, mode optimization		
	Quality of structural calculations and technological operations	Quality of the project and security solutions, protection against tie-ins, environmental monitoring	Quality of technological calculation, equipment selection, optimization solutions
Improvement at the design stage	Evaluation of design quality and development of reasoned technical solutions, justification methodologies improvement		
Construction	Control of the production process of construction and the design documentation adherence; quality control of work and testing	Control of the production process of construction and design solutions execution; quality control of works and testings	Production process monitoring, technological innovations, modern equipment
Improvement at the construction stage	Improvement of technology, quality control and testing		
Operation	Monitoring of object reliability with the use of proactive control systems, satellite and remote technologies	Security monitoring with the use of proactive control systems, satellite and remote technologies, security equipment	Monitoring of technological modes with the use of proactive control systems
Improvement at the operation stage	Development and modernization of expert systems		
	Making decisions on technical intervention, reconstruction	Making decisions on improving the effectiveness of the security system	Making decisions on technical re-equipment, modernization

At the design stage, taking into account the implementation of the import substitution program, a competent analysis of energy consumption options and the structure of operating costs is important. The general algorithm for implementing optimization solutions is shown in Fig. 1.

At the operational stage, the development of technology for monitoring the efficiency of operating modes of technological equipment for main oil and gas pipelines on-line is of particular importance. The structure of the information flow at the operational stage is shown in Fig.2.

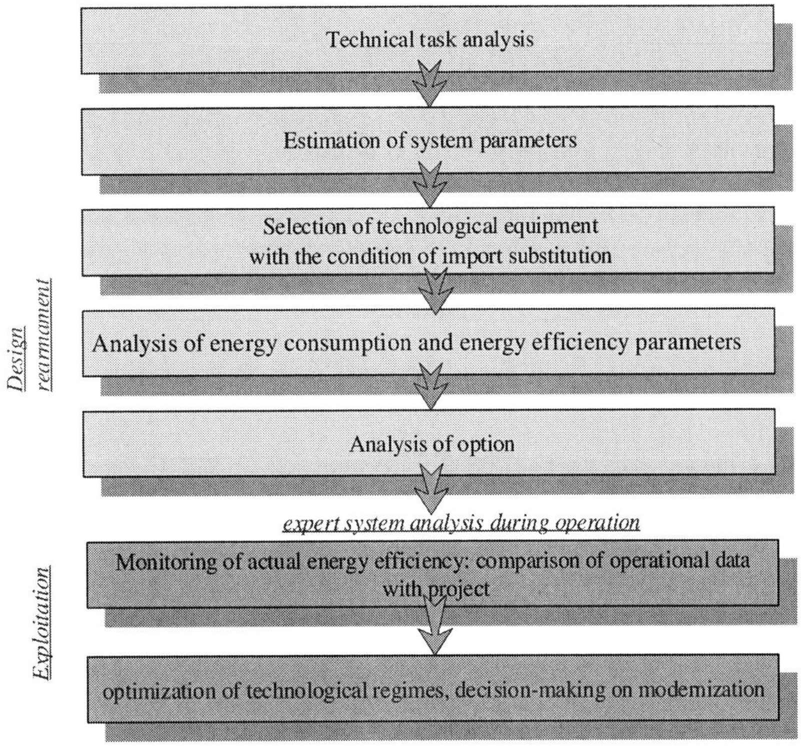

Figure 1. General algorithm for implementing optimization solutions for energy efficiency in the design

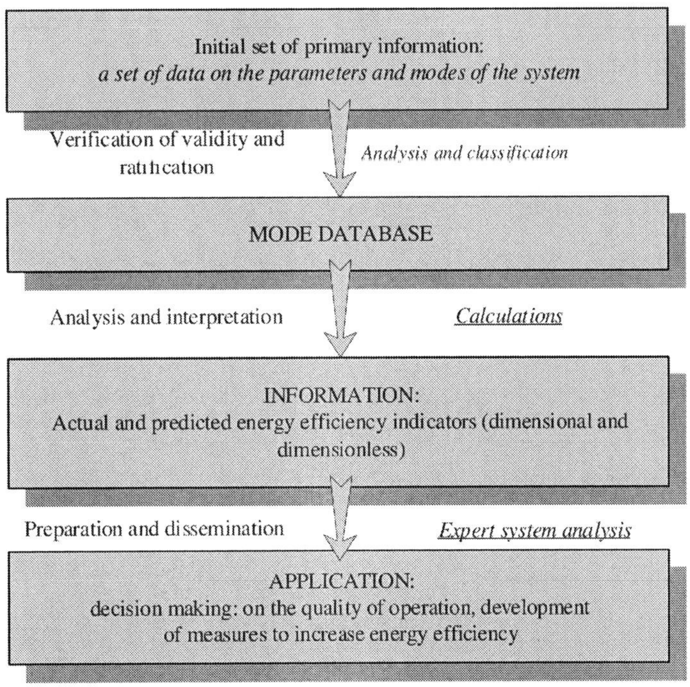

Figure 2. A structure of the information flow in the system for monitoring the energy efficiency of gas transport processes during the operation

3. Industrial testing on a technological project

Let us consider the option of using the algorithmic complex in terms of the "Power of Siberia-2" ("Altai") project, a gas pipeline between the gas fields of Western Siberia and the Xinjiang Uygur Autonomous Region in the west of China; the approximate length of the gas pipeline is about 6700 km, 2,700 km of which should pass on the territory of Russia.

The technological equipment of the "Altai" gas pipeline meets all modern requirements both from the point of view of the application of new domestic construction materials, and from the point of view of energy and resource saving. The newest gas compressor units are used, which ensure a high initial pressure in the pipeline and a high productivity when transporting gas. The "Altai" gas pipeline starts in the Urengoy field. The productivity of the pipeline with a diameter of 1420 mm, having a wall thickness of 32 mm, is 45 billion cubic meters of gas per year. Testing for strength and deformability is fully satisfied. Pipes of the gas pipeline "Altai" are electro-welded, longitudinal, with the factory insulating coating for the gas pipeline of the category III. The gas pipeline capacity is 133 million cubic meters per day [14].

When operating a gas pipeline, the criteria characterizing its efficiency and cost effectiveness are important, as well as the results of calculations in Table 2.

Table 2. Results of the analytical justification of the project

Name	Value
Throughput performance	61.639 *mln.m^3/day.*
Hydraulic resistance coefficient	0.0099099
Reynolds number	80465019
Pressure at the end of section	8.695 Mpa
Coefficient characterizing the intensity of the decrease in the gas temperature along the length of the section	0.002241
Average gas temperature	261K
Gas temperature at end of section	268.36K

Throughput performance:

$$Q_n = 1{,}33 \cdot 10^6 \frac{D^{2,5} \cdot \eta}{\Delta}; \tag{1}$$

where D - pipeline diameter, mm; η - dynamic viscosity of gas, Pa·s; Δ - relative density by air.

Hydraulic resistance coefficient:

$$\lambda_{fr} = 0{,}067 \cdot \left(\frac{158}{Re} + \frac{2 \cdot k}{D_{vn}}\right)^{0,2}, \tag{2}$$

where k=0,03 – equivalent pipe roughness, mm; $D_{\text{вн}}$ – pipe inner diameter, mm; Re – Reynolds number.

Reynolds number:

$$Re = 17{,}75 \cdot 10^3 \cdot \frac{Q\Delta}{D\eta}. \tag{3}$$

Pressure at the end of section:

$$P_k = \sqrt{p_H^2 - \frac{Q^2 \cdot \Delta \cdot \lambda \cdot Z \cdot T_{cp} \cdot L}{105{,}087^2 \cdot D_{vn}^5}}. \tag{4}$$

Coefficient characterizing the intensity of the decrease in the gas temperature along the length of the section:

$$a = \frac{k\pi d}{MC_p}. \tag{5}$$

Average gas temperature:

$$T_{av} = T_0 + \frac{T_н - T_0}{aL}\; 1 - e^{-aL} \;\; - Di\frac{P_H^2 - P_K^2}{2aLP_{sr}}\; 1 - \frac{1}{aL}\; 1 - e^{-aL} \;\; . \qquad (6)$$

After carrying out calculations on this gas pipeline, it can be concluded that with the CS capacity Q_{ks}=133.76 mln.m^3/day, which is more than 15 mln.m^3/day, superchargers are more economical than GICC; therefore, in the centrifugal blowers they are used as a type of compressor machines.

Since there is a large capacity and gas is transported under high pressure, the GCU-32 Ladoga (2 working and 1 standby unit) is used.

GCU-32 Ladoga is a new, highly efficient unit developed for the Russian market based on the basic family of GCU MS5002, which is successfully operated at many compressor stations in Russia and has a total operating time of more than 16 million hours worldwide.

GCU-32 Ladoga includes elements specially designed for operation in difficult conditions of the Russian market: separate compartments of auxiliary equipment and gas turbines (to solve the problem of very low temperatures of the ventilation air), the use of special low-temperature steels for frames production, as well as the application of GOST (all-Union State Standard) standards and Russian certification.

The supercharger 400-21-1C is full-forcing and is designed for compression and transportation of natural gas through main gas pipelines. The type of the supercharger - centrifugal, two-stage with the system of end "dry" gas seals (DGS) and oleic supporting and resistant bearings.

The supercharger is equipped with two identical in design nodes of seals of the DGS developed by JSC "John-Crane-Iskra". Seals are designed to minimize gas losses. The supercharger is driven by a gas turbine MS5002E.

Table 3. Basic parameters of the supercharger 400-21-1C

Name	Value
Productivity, referred to 20°C and 0,1013 MPa, mln.nm/day	78.9
Volume productivity, referred to the initial conditions, m^3/min	505
Final, absolute pressure of the gas when exiting the discharge nozzle, MPa	11.86
Polytropic efficiency	0.85
Pressure ratio	1.4
Power consumed by the supercharger, MW	30.4
Initial gas pressure, at the inlet to the suction pipe of the supercharger, MPa	8.45
Initial gas temperature, at the inlet to the suction pipe of the supercharger, °C	5
The density of the gas, referred to 20°C and 0.1013 MPa, kg/m	0.689
Rotor speed, rpm	5550
Operating speed range, % out of nominal	70-105

"Nevsky Zavod" is manufacturing a gas turbine unit (GTU) MS5002E under license and in cooperation with GeneralElectricOil&Gas (NuovoPignoneS.p.A.).

The first batch of GTU MS5002E (Fig.3.4) is designed for operation on the gas trunk pipelines of OAO Gazprom as a drive of the natural gas blower 400-21-1C in the GCU-32 "Ladoga" made by "Nevsky Zavod" at the frequency of rotation of the power shaft equal to 5714 rpm. However, GTU MS5002E can work in the drive mode from the generator.

IOP Conf. Series: Materials Science and Engineering **357** (2018) 012034 doi:10.1088/1757-899X/357/1/012034

Figure 3. GTU MS5002E [12]

In the variant of using GTU MS5002E as a mechanical drive, it has a capacity of 32 MW and an efficiency of 36%, in the generator version - a capacity of 31.1 MW and an efficiency of 35%.

GTU is equipped with a low-emission combustion chamber and is extremely reliable in operation. Elements of the unit are designed by methods allowing to ensure its reliable operation in difficult climatic conditions and to provide all necessary volume of on-site service. GTU is installed in an easy-to-assemble hangar-shelter.

The 11-stage axial compressor provides a compression ratio of 17. It is designed on the basis of the axial compressor of the well-known GTU GE10 (more than 80 operating units, over 700,000 operating hours). The input guiding device and the guide blades of the first and second stages of the axial compressor are adjustable; their rotation is carried out by a hydraulic oil drive. Atmospheric air entering the inlet pipe of the compressor is pre-cleaned in an air filtering and conditioning system (AFCS).

The combustion chamber of type DLN2 is countercurrent and structurally made in the form of 6 separate flame tubes, evenly distributed (each in its housing) in the circumferential direction on the outlet housing of the axial compressor. Each of the six flame tubes has 5 burners designed to work on gaseous fuels.

A two-stage high-pressure turbine provides high efficiency in a wide range of operating modes. Both stages have a developed system of air cooling of convective and film types. For cooling, air drawn from the delivery pipe of the axial compressor is used.

The two-stage power turbine is based on the well-known PGT2S + turbine (over 80 operating units, over 500,000 operating hours). Turbine bearings are easily accessible for inspection.

Structurally GTU MS5002E is located on two separate frames (one is for turbo-block, the other is for the auxiliary equipment). On the auxiliary equipment frame, which is also an "oil tank", there are:

- oil supply system with pumps and filters;
- hydraulic oil system;
- starting system with a barring device and torque converter;
- fuel gas system, etc.

IOP Conf. Series: Materials Science and Engineering **357** (2018) 012034 doi:10.1088/1757-899X/357/1/012034

The body of the gas generator is made with a horizontal connector, the power turbine module does not have it.

Figure 4. GCU-32 "Ladoga" [13]

Both frames (of turbo-block and auxiliary equipment) are equipped with acoustic enclosure and heat insulation, which allows the maintenance of GTU in field. External cooling of the turbo-block and the equipment installed under the enclosures is carried out by supplying atmospheric air to the spaces under the enclosures with the help of blowers taking air from the atmosphere through separate sections of the AFCS.

The control system is based on the MARKVI control unit supplied by GeneralElectric.

Table 4. Technical specifications of GTU MS5002E

Parameters	Generator drive	Mechanical drive
Power on the turbine shaft, kW	31100	32000
GTU efficiency, %	35	36
Compression ratio in axial compressor	17	17
Exhaust gas flow rate behind the power turbine, kg/s	101	101
Exhaust gas temperature, ^{0}C	510	510

4. Conclusion

In this paper, the features of ensuring the energy efficiency of equipment for gas transportation at the design stage were shown. As an example, a variant of an algorithm implementation with preliminary calculation and selection of the latest gas pumping equipment is shown. Herewith, the construction stage of the cycle implies that construction and installation work must be carried out in full accordance with design decisions and regulatory documents such as SNiP, administrative regulations, technical conditions and equipment passports. Thus, compliance with the requirements of efficiency, reliability, safety at all stages and their qualitative monitoring will ensure the functioning of the gas transportation system at a high technological level. Analysis results and algorithms can be used by design and expert organizations.

References

[1] "Altai" gas pipeline will pass directly to China, probably along the Ukok Plateau

[2] http://neftegaz.ru/news/view/132396-Gazoprovod-Altay-proydet-v-Kitay-napryamuyu-veroyatno-po-plato-Ukok 20.05.2017
[3] Zemenkova M, Shalay V, Zemenkov Y, Kurushina E 2016 Improving the Efficiency of Administrative Decision-Making when Monitoring Reliability and Safety of Oil and Gas Equipment *MATEC Web of Conferences* **73** DOI: 10.1051/matecconf/20167307001
[4] Vlasenko V S, Slesarenko V V, Gulkov A N and Lapshin V D 2015 Preparation of arctic oil and gas condensate deposit formation fluid for storage and transport in the form of hydrocarbon hydrate-containing dispersed system Proceedings of the International Offshore and Polar Engineering Conference 1863
[5] Kurushina V, Zemenkov Y 2014 Innovative cyclical development of the Russian pipeline system *WIT Transactions on Ecology and the Environment* **190** **2** 881-888
[6] Zemenkov Yu D, Shalay V V, Zemenkova M Yu 2015 Expert Systems of Multivariable Predictive Control of Oil and Gas Facilities Reliability. *Procedia Engineering* **113** 312-315
[7] Kurushina E V, Kurushina V A 2014 Evolution of economic development aims. Assessment of the smart growth *Life Science Journal* **11-11** 517-521
[8] Kurushina E V 2014 On regularities of economic dynamics in times of crisis *Canadian Journal of Science, Education and Culture.* **2 (6)** 378-384
[9] Mamadaliev R A, Kuskov V N, Zemenkov U D and Popova A A 2015 Influence of high-concentrated heat sources on alloying elements transition into the weld metal *Applied Mechanics and Materials* **770** 19-22
[10] Gorelik J B, Shabarov A B, Sysoyev Yu S 2011 The dynamics of frozen ground melting in the influence zone of two wells *Earth's Cryosphere* **12 (1)** 59
[11] Methodology for the energy efficiency evaluation of gas transmission facilities and systems. CS Gazprom 2-3.5-113-2007
[12] GTU T32 Access mode: free. http://www.reph.ru/production/type/713/713/
[13] Gas compressor unit GCU-32 "Ladoga" with a capacity of 32 MW. Access mode: free. http://www.reph.ru/production/gas/211/
[14] Demchenko N P, Krapivsky E I 2010 Geophysical monitoring of underground pipelines technical state 4th International Conference and Exhibition: New Discoveries through Integration of Geosciences (Saint Petersburg)
[15] Zemenkova M, Zemenkov Y, Gladenko A, Podorozhnikov S 2016 Estimation of Emissions During Monitoring of Pipelines in the Dynamic Mode of Operation. *5th International Scientific Conference Integration, Partnership and Innovation in Construction Science and Education* **86** DOI: 10.1051/matecconf/20168604053

IOP Conf. Series: Materials Science and Engineering **357** (2018) 012035 doi:10.1088/1757-899X/357/1/012035

Experimental studies of thermal preparation of internal combustion engine

N N Karnaukhov, Sh M Merdanov, Konev V V and D M Borodin

Industrial University of Tyumen, 38 Volodarskogo St., Tyumen, 625000, Russia

E-mail: karnauhovnn@tyuiu.ru, merdanovsm@tyuiu.ru, konevvv@tyuiu.ru

Abstract In conditions of autonomous functioning of road construction machines, it becomes necessary to use its internal sources. This can be done by using a heat recovery system of an internal combustion engine (ICE). For this purpose, it is proposed to use heat accumulators that accumulate heat of the internal combustion engine during the operation of the machine. Experimental studies have been carried out to evaluate the efficiency of using the proposed pre-start thermal preparation system, which combines a regular system based on liquid diesel fuel heaters and an ICE heat recovery system. As a result, the stages of operation of the preheating thermal preparation system, mathematical models and the dependence of the temperature change of the antifreeze at the exit from the internal combustion engine on the warm-up time are determined.

1. Introduction

The practice of operating the construction and road machines in the North shows that their potential for productivity, reliability, serviceability and other indicators is not fully realized [1,2,3,4].

One of the main problems of efficient operation of road construction machines in the conditions of the Russian North is to ensure the thermal state - pre-start thermal preparation of the internal combustion engine (ICE). Pre-heating of the internal combustion engine can reduce its starting wear, fuel consumption, toxicity of exhaust gases; increase the service life of the starter and battery. At the same time, it must be taken into account that the equipment is operated autonomously away from bases in the field. This makes it difficult or completely excludes the possibility of using traditional pre-start thermal preparation systems (PTPS) of ICE. Therefore, there is a need to use autonomous heat sources [5,6,7].

Rational use of fuel and energy resources can be achieved through the utilization of heat during the operation of internal combustion engines.

One of the solutions to the problem of pre-start heat preparation is the provision of mobile equipment with heat recovery systems (HRS) using heat accumulators (HA). This makes it possible to provide the need for additional heat energy necessary for pre-start thermal preparation of the ICE by using heat that is lost in the surrounding space during the operation of the machine (about 40% of the thermal energy is transferred to the cooling liquid during operation of the internal combustion engine, and up to 25% of heat is lost with exhaust gases) [1,2].

In accordance with the foregoing, the purpose of the work is to increase the efficiency of the pre-start heating of the ICE of road construction machines operated in conditions of temperatures below freezing.

Theoretical research has shown that it is advisable to combine the existing systems of thermal preparation of the machines of "Cold execution" and the heat recovery system based on thermal accumulators of different potentials [8,9]. As a result, the system of pre-start thermal preparation (PTPS) of the internal combustion engine was developed [10, 11].

At the same time, it is necessary to determine by experimental studies the stages of application of the liquid diesel preheater (LDP) and HRS by the temperature limits.

2. General methods of experimental studies

The aim of the experimental studies was to test the hypotheses put forward in theoretical studies and to determine the numerical values of the parameters of mathematical models.

During the experiment, the following tasks were solved:

1. To check the form of the mathematical model and its parameters, describing the dependence of the heating of the internal combustion engine from the developed pre-start system.

2. To compare the heating time with the help of LDP-44 and the developed PTPS at different stages of its operation at various ambient temperatures (-15 ^{0}C,-30 ^{0}C,-40 ^{0}C).

To solve the problems, an experimental setup was designed and built, containing a HRS and from LDP-44. The diagram of the system for pre-start thermal preparation of the ICE is shown in Fig. 1.

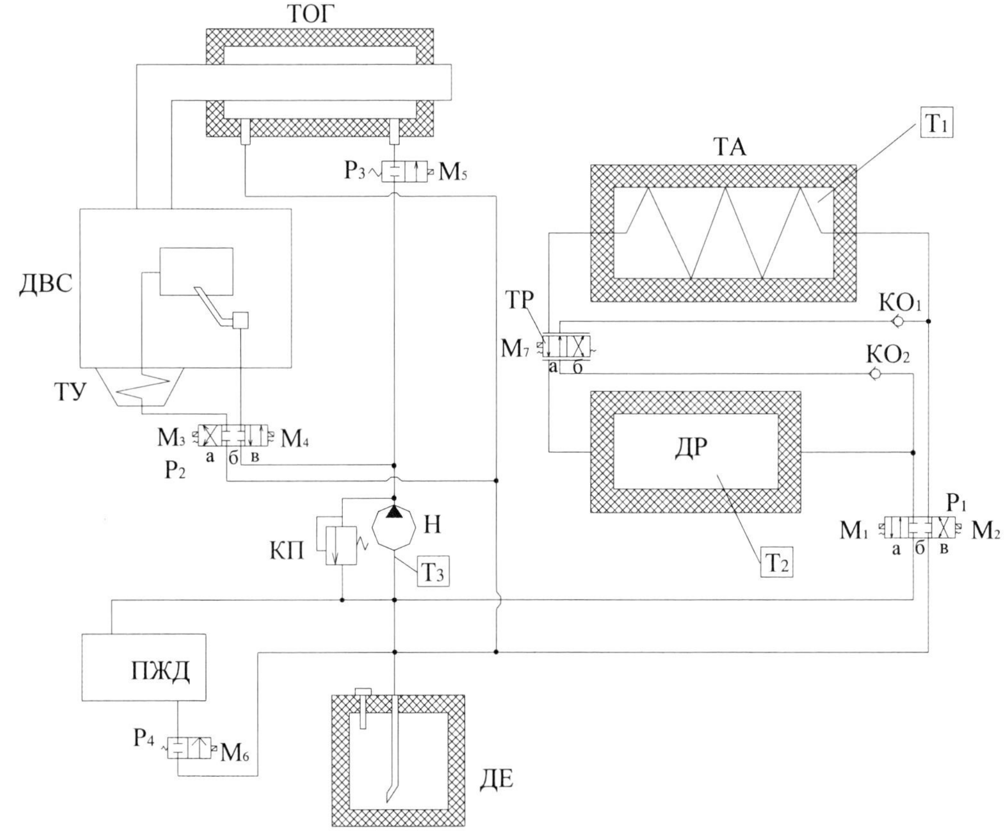

Figure 1. A scheme of the system of pre-start thermal preparation of ICE: ДВС (ICE) - internal combustion engine, ДЕ(AT) - additional tank, ДР (AR) - additional reservoir, ТА (HA) - heat accumulator, LDP)- liquid diesel preheater, ТОГ (EGHE) - exhaust gas heat exchanger, ТУ (HE) - heat exchanger, P_1-P_3 – distributor, ТР (TR) - thermoregulator, H – pump, КП – bypass valve, T_1-T_3 – temperature sensors, M_1-M_7- pressure sensors

3. Devices and equipment

To measure the temperature on the street, a liquid thermometer with a measurement range from -50°C to +50°C, an accuracy class of 1.0 and a division rate of 1°C was used. A systematic error over the whole measurement range is:

$$(50-(-50)) \times 1{,}0/100 = 1\ ^{0}\text{C}. \quad (1)$$

The temperature inside the tank was measured by a digital multimeter, working with a thermocouple of "K" type. The measurement error from -20 to +250 ^{0}C is not more than ±3 ^{0}C. In order to reduce the error in measuring the temperature with a multimeter with a thermocouple, they were subjected to a control test and selection, which made it possible to increase the accuracy of the temperature measurement up to ± 1 ^{0}C. This is determined by comparing the thermocouple readings with the reference thermometer. Multimeters with thermocouple of type "K" have high reliability in operation and allow one to observe the temperature changes visually, directly during the experiment.

Cooling chamber - Refrigerant: R134A, EK35BASIC DK 9620. Characteristics: minimum temperature is (-40 ^{0}C), volume-1m^{3}.

A manometer with a single-turn tubular spring was used to measure the pressure. The accuracy class is 2.5; the measurement range is from 0 to 10 kgf/cm^{2}. A systematic error is: (10-0)·2,5/100= 0,25 (kgf/cm^{2}). Thus, the systematic (instrumental) error is within the permissible deviation when monitoring the pressure in the line.

The LDP-44 pump with mass flow - G_a = 0,2…0,3 kg/s was chosen from the condition of optimal circulation of the coolant, which is justified in the work [5].

4. Description of the experimental unit

To test mathematical models, evaluate their adequacy and also to compare the heating time of the ICE A-01M from LDP-44 and the developed PTPS at various stages of its operation, active experiments were conducted. Preliminary experiments have shown that the prevailing error is a systematic one. Therefore, 3 parallel experiments at a point can be considered sufficient [12,13].

A pilot plant for pre-start heat preparation of the A-01M diesel engine was developed and manufactured. The pilot system was placed on a turntable next to the counterweight of the excavator, and next to the regular pre-heater LDP-44, making up one system (Fig. 2). Experimental studies were carried out on the street by selecting for the experiment the periods of the year with justified maximum and minimum ambient air temperatures [1,5]. Analysis of the methods of pre-start thermal preparation showed that the boiler of a LDP-type preheater is the closest in terms of indicators in accordance with the basic requirements for operation. This made it possible to propose the use of LDP-44 in conjunction with the HRS, and to compare the rate of ICE heating from the developed PTPS and LDP-44.

Figure 2. Pilot plant

The PTPS was developed, which was built with reference to the A-01M engine of the EO-4121 single-bucket excavator. Tests of a standard sample of such system have shown that it is possible to heat the A-01M engine for 25...30 minutes, after which it is launched at the first attempt. Unlike

IOP Conf. Series: Materials Science and Engineering 357 (2018) 012035 doi:10.1088/1757-899X/357/1/012035

systems with standard preheaters like LDP-44, in the pilot system, oil in the diesel crankcase is heated by a tubular radiator [1], structurally installed at the entrance to the "jacket" of the internal combustion engine. As a heat transfer fluid, grade 65-antifreeze is used. The tests were carried out using a single-bucket excavator EO-4121.

The diagram of the pilot unit (Fig. 3) contains: A-01M internal combustion engine in modular construction (1). Heat accumulators with different thermal potentials are made in three versions: additional tank (AT) (2), additional reservoir (AR) (3), thermal battery (TB) (4) and LDP-44 (5). All heat accumulators are insulated (as a heat insulator, foam was used over the asbestos sheet). AT (2) with a capacity of 40 liters, AR (3) with a capacity of 50 liters and TB (4) with a capacity of 70 liters. Electric heaters with a capacity of 6 kW are built into AT and TB, in addition, the following things are built in the TB: a water-antifreeze heat exchanger, reverse and safety valves (4); heat transfer device (TU) (6) for heating the diesel oil in the form of a tubular coil; pump unit (7) with a direct current motor; pipelines connecting DE-AT (2), DR-AR (3), TA-TB (4), pump unit (7); TU-HU (6) and the cooling system of the starting motor (8); valves (9,10,11,12,13,14,15,16) - to disconnect the heat accumulators DR-AR (3) and TA-TB (4) from the system and to regulate the inlet temperature of the coolant in the heated ICE. The temperatures of heat accumulators and antifreeze at the inlet and outlet of the ICE were measured by multimeters with thermocouples built into the pipeline. The liquid pressure in the line was fixed by the MP4-U manometer, and the time - by a stopwatch.

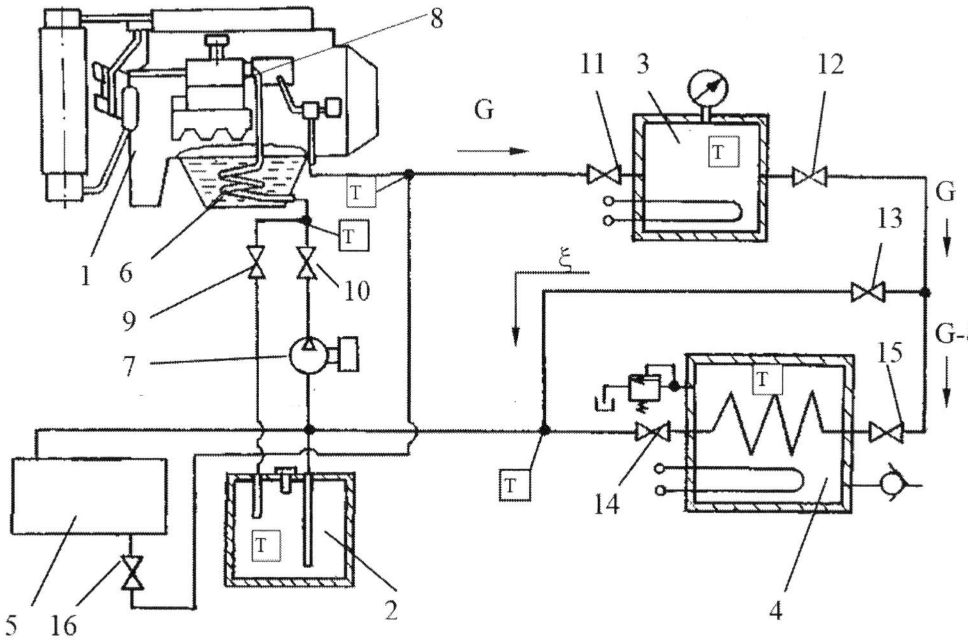

Figure 3. The scheme of the pilot unit

Lubrication and power supply systems for ICE are refueled, in accordance with the instruction manual for the ICE A-01M, with oil M-8G and diesel fuel of grade "3".

The completion of the heating process and the possibility of launching the ICE into operation is determined by the temperature of the antifreeze at the outlet of the ICE [1, 5], as noted in the work of other researchers and confirmed by experimental data on the heating of the ICE with the experimental PTPS.

The operating principle of the pilot unit is as follows (see Fig. 3). Initially, the antifreeze is drained from the "jacket" of the engine in the AT (2) through the valve (9), then the ICE cools to ambient temperature, the valves are closed. At the end of this process, electric heaters are turned on; antifreeze in AR (3) and water in TB (4) are heated to the set temperatures.

Three stages of heating the internal combustion engine were studied. At the first stage, the engine was heated from AT (2). The valve (10) is opened and the electric motor of the pump unit (7) is activated, by means of which the coolant in the AT (2), through the tubular heater (6), and the cooling system of the starting motor (8), is pumped into the "jacket" of the internal combustion engine (1).

In the second stage, the valves (11, 12) are opened on the AR and the electric motor of the pump unit (7) is activated, which ensures the circulation of the heat carrier through the ICE (1) and simultaneous flow of the heat transfer from the two heat accumulators AR (3) and TB (4) while maintaining the inlet temperature coolant +110 ^{0}C in the ICE (1) with the help of valves (13, 14, 15) [10,11,14]. The experiments were stopped when the temperatures were equalized in the heat accumulators and at the exit from the ICE. In this case, the measurement error was ±2 ^{0}C. A low error in measuring the temperature of heat accumulators is due to the high thermal conductivity of the steel and the made blind holes in the metal wall where the thermocouples were placed.

At the third stage, the A-01M engine was heated from LDP-44 (5) with the valves opened (16) and closed (9,11,12,13,14,15).

At all stages of the work of the PTPS with the HRS and LDP, the heating time of the engine was measured. The temperature changes in the heat accumulators and in the internal combustion engine were fixed in every 30 seconds at the first stage. At the second and third stages of heating the engine - every minute.

The ICE start after effective heating is carried out in accordance with the operating instructions by the operator in the excavator's cab. The idling work of ICE continued during the testing lasts no more than 15 minutes.

5. The results of the experiment, their processing and evaluation

Tests of the pilot pre-start heating system of the A-01M engine in the range of negative temperatures (from -40 ^{0}C to -15 ^{0}C) have shown its efficiency. The system in this range of negative ambient temperature has always ensured a qualitative heating of the ICE when it is subsequently launched at the first attempt.

The tables show the temperature change of the antifreeze in the A-01M ICE in the first stage of the PTPS (AT) operation and in comparison with the preheating from the LDP-44 (Table 1), at the second stage of the PTPS (AT + TAU) - (Table 2), at the third stage of work of PTPS (AT + TAU + LDP) - (see Table 3).

Table 1. Change in the temperature of antifreeze in the A-01M engine at the first stage of the PTPS (AT) operation and comparison with the LDP-44

at 263K	of PTPS						of LDP-44					
Time (min.)	Temperature at input (K)			Temperature at output (K)			Temperature at input (K)			Temperature at output (K)		
0	349	350	349				262	262	263			
0,6	353	352	351				273	272	273			
1,3	351	351	351	262	263	262	280	281	280	271	273	272
2,2	352	351	352	266	267	267	285	286	286	280	279	279

Table 2. Change in the temperature of the antifreeze in the A-01M engine at the second stage of the PTPS (AT + TAU) operation and comparison with the LDP-44

at 243K	of PTPS						of LDP-44					
Time (min.)	Temperature at input (K)			Temperature at output (K)			Temperature at input (K)			Temperature at output (K)		
0	349	349	350				242	243	243			

IOP Conf. Series: Materials Science and Engineering **357** (2018) 012035 doi:10.1088/1757-899X/357/1/012035

0,5	353	352	353				254	255	254			
1,3	351	352	351	245	246	245	269	270	270			
2,1	352	352	352	246	247	247	279	278	280	270	272	271
3	359	360	359	292	293	291	297	296	297	279	281	280
4	374	375	375	310	312	311	310	311	312	294	295	293
5	377	377	378	322	322	323	318	319	317	304	303	305
6	376	376	377	331	330	332	322	321	323	310	312	311
7	377	377	376	337	338	337	327	325	326	312	313	314
8	375	379	374	343	344	343	334	335	334	317	316	316
9	374	374	374	347	349	348	340	342	341	318	318	318
10	372	372	373	351	351	352	344	346	345	322	321	323
11	371	371	372	354	355	354	342	344	343	328	328	329
12	368	369	368	356	356	355	346	345	347	333	334	335
13	367	367	366	358	357	356	352	350	351	335	336	337
14							356	354	355	341	340	342

Table 3. Change in the temperature of antifreeze in the A-01M ICE at the third stage of the PTPS (AT+ TAU + LDP) and comparison with LDP-44

at 233K	of PTPS						of LDP -44					
Time (min.)	Temperature at input (K)			Temperature at output (K)			Temperature at input (K)			Temperature at output (K)		
0	233	234	233				234	233	233			
0,5	235	236	234				249	250	251			
1,3	235	235	235	237	235	235	267	268	267			
2,1	236	237	238	238	236	237	276	276	275	271	271	270
3	376	378	377	268	269	269	296	298	297	279	279	280
4	375	376	376	281	280	282	307	306	308	294	296	295
5	375	377	375	297	298	296	319	320	319	305	304	304
6	373	373	373	306	307	308	321	323	322	310	311	311
7	361	363	362	314	315	315	325	326	326	315	314	313
8	357	355	356	320	320	319	333	332	333	315	316	315
9	350	349	348	324	323	322	338	339	339	319	320	319
10	352	351	350	326	325	326	341	342	341	323	323	324
11	350	349	348	330	331	330	343	344	344	325	327	326
12	350	350	351	339	340	347	346	346	345	335	334	334
13	353	352	352	346	345	350	349	350	349	340	340	339
14	358	359	357	346	345	345	351	352	352	342	341	341
16	360	361	361	344	346	346	353	352	353	343	343	342
17	369	371	370	346	348	347	354	353	354	343	344	343
18	373	375	373	352	352	353	356	355	354	342	347	342
19	377	376	377	357	358	357	359	356	355	344	349	344
20	378	379	378	365	367	366	362	360	361	353	354	353
21	379	379	379	366	368	367	368	369	370	359	358	359
22							372	373	371	364	363	364
23							376	375	374	367	368	367

24	377	379	378	368	369	368
25	382	380	381	368	368	367
26	380	379	381	369	368	368

Based on the correlation-regression analysis of the experimental results, one-factor models of the antifreeze heating at the output of the ICE were obtained:

for the second stage: $$T = 336.37-310.78e^{-\tau}; \quad (2)$$

for the third stage: $$T = 329.12-439.34e^{-\tau}, \quad (3)$$

where, T- antifreeze temperature at the output of the ICE; τ- heating time.

Numerical values of the basic statistical characteristics of models, determined on the basis of experiments. The values of Fisher's dispersion ratio were 2.17-2.9. These values are greater than the table values of the F-criterion for the confidence probability of 0.9, which indicates sufficient adequacy of single-factor models for heating the antifreeze at the output of the internal combustion engine based on the results of the experiment. The average approximation error is within acceptable limits (5-7 %) [12,15].

Based on the calculation results and the experimental data, the graphs are constructed shown in Fig. 4. When heating from the AT (the first stage), the temperature of the A-01M engine rises slightly (on average - 4 ^{0}C); however, no additional energy is required to raise the temperature of the antifreeze in the "jacket" of the internal combustion engine.

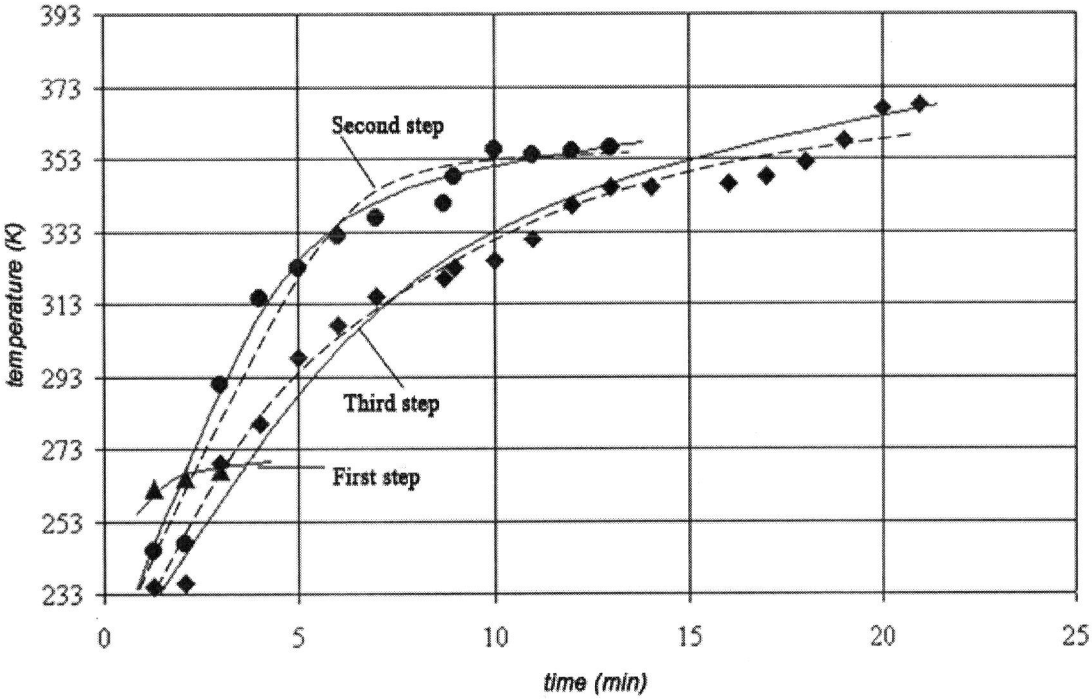

Figure 4. Dynamics of heating the antifreeze at the output of the A-01M engine: ●, ♦, ▲ - the results of the experiment; --- - calculation results

The results of the experiments were processed using Microsoft Excel, Regress software packages [15].

From the analysis of the graphs, it can be seen that the heat transfer processes during the heating of the internal combustion engine are qualitatively consistent with the calculated data, which indicates the correctness of the theoretical premises for the representation of these processes. However, the available quantitative discrepancies show that the heat transfer from the antifreeze to the massif of the internal combustion engine parts proceeds somewhat more intensively than it does in the simulation model of the diesel engine. This indicates a high efficiency of heat transfer from the antifreeze to the ICE in its course in the aisle space and inaccuracies of the model. In the middle of the stages, the errors reached 8 ... 12%, which is explained by inaccuracies in the accepted models of the heat transfer process and the inertia of the heat transfer process during the ICE heating. However, inaccuracies in the model make it possible to obtain results within acceptable limits and do not lead to appreciable errors in determining the temperature of the internal combustion engine (Fig. 4).

It is also established that as a result of the specificity of the heat transfer processes in the internal combustion engine, the equalization of the thermal fields in it proceeds more intensively than it does with the LDP-44. In the initial periods of ICE warming from the pilot system, the heat flow reached 35-45 kW, while for the regular system it did not exceed 20 kW.

6. Conclusion

As a result of the study, the following conclusions were obtained:

- good convergence of analytical studies and experimental data in the process of heating the A-01M ICE has been obtained. The temperatures of the antifreeze at the input and output of the ICE vary exponentially;
- It is determined that when the internal combustion engine is heated by the developed PTPS, the temperature of the antifreeze at the ICE output increases faster than the preheating from the regular LDP-44 system. The heating time of the engine A-01M at an air temperature of -40 °C is 21 minutes and 26 minutes, respectively.

References

[1] Garkavi N G, Bardychev O A, Teslenko N G 1980 *Ensuring the operational reliability of construction machinery in the winter* (L.: LDNTP)

[2] Merdanov Sh M 2013 *Mechanized complexes for the construction of temporary winter roads: a scientific publication* (Tyumen State Oil and Gas University. Tyumen: TSOGU)

[3] Lakhov R 2009 Exploitation of heavy equipment at low temperatures *Osnovnye sredstva* **11**

[4] Code of Regulations for the design and construction SP 12-104-2002 - "Mechanization of construction. Operation of construction machinery in winter "

[5] Vashurkin I O 2005 *Thermal preparation of construction machines in severe climate.* (SPb .: Science)

[6] Merdanov Sh M, Konev V V, Zakirzakov G G 2016 *Hydraulic drives of construction and road machines for operation at low temperatures: monograph* (Tyumen: IUT)

[7] Pavlov I 2016 Winter cares of an excavator driver. *Osnovnye sredstva* **12**

[8] Riljakin E G 2007 *Improving the performance of the tractor's hydraulic system a control fluid: abstract. dis* (Penza: PGSKHA)

[9] Vashurin O I 1993 *Substantiation of the parameters and method of constructing the heat recovery system in the drive earthmoving machines* (S-Pb)

[10] Karnaukhov N N, Konev V V, Zakirzakov G G 2003 *Pre-start thermal preparation system for the internal combustion engine: pat. 2211943 RF: IPC7 F02N 17/06.; the applicant and the patent holder- TSOGU. -2001128475 / 06*

[11] Kramskoj V F 1996 *Design and bases of calculation of thermal pre-launch preparation digger drive (for example, a bulldozer-ripper DZ-117A). Diss. for a degree of Cand.* (Voronezh)

[12] Samoilova M I 1995 *Device of the hydraulic machinery for operating conditions at low temperatures (for example, excavator EO-4121A): dis. kand. tech. Sciences* (Voronezh)

[13] Konev V V, Zakirzakov G G, Merdanov M Sh 2014 Mathematical modeling of the thermal state of road construction machines. *Modern problems of science and education* **6** 320

[14] Exner H, Freitag R, Gais H, Lang R, Oppolzer J, Schwab P, Zumpf E, Ostendorff U, Raik M 2003 *Hydraulic training course. Volume 1. Rexroth Bosch Group: Bosch Rexroth AG Service Automation Didactics*. (Erbach, Germany)

[15] Zakharov N S 1999 *The program "REGRESS". User's manual* (Tyumen: TSOGU)

IOP Conf. Series: Materials Science and Engineering **357** (2018) 012036 doi:10.1088/1757-899X/357/1/012036

On issue of substantiating availability of main pipelines in Western Siberia

B P Elkin and V A Ivanov

Industrial University of Tyumen, 38 Volodarskogo St., Tyumen, 625000, Russia

E-mail: borelpet@yandex.ru

Abstract. The urgency of providing year-round access to main pipelines (MP) constructed in wetlands is due to: the considerable length of MP strings in wetlands; ineffective and sometimes destructive use of the technology of operational and other enterprises due to the lack of year-round opportunities for carrying out preventive works; the impossibility of immediate response in the occurrence of accidents and failures in operating MPs and, as a result, material damage to operators and consumers of products; inevitable negative impact on the environment in case of emergency.

1. Introduction

The presence of a year-round operating road excludes many of the listed problems, the solution of which requires significant costs, usually unplanned, which negatively affects the financial well-being of the oil and gas industry and operational organizations. The road routes (RR) provide a high production culture, the ability to use the latest diagnostic technologies and equipment. It is very important to build such roads along the MP operating for more than 30 years, where, according to theory, the probability and the number of failures will only grow. Moreover, modern studies of the theory of the origin of oil (Larin V.N., 2005, 2016) show that oil reserves are inexhaustible, which should ensure the long life of MP.

2. Materials and methods

The purpose of the article is to show the constructive and technological solutions for the construction of RR with the use of a widespread local material - peat. Scientific novelty lies in the development of types of constructive solutions for RR with peat in the body of the mound for various territories.

The main feature that causes new problems of road construction in such conditions is a rigid binding of the route of the future road to the MP track passing through the wetlands. At the same time, it is necessary to build a road in the immediate vicinity of the operating MP.

The prerequisites for the use of peat in the body of roads along the operating MP are based on the historical experience of peat application in road construction. The normative document [1] allows the use of peat, with certain design constraints, in the lower part of the bodies of highways of the III - IV categories with a traffic intensity of 100 cars per day or more. Similarly, in Northern European countries, where the territories are covered by a multitude of marshes of different types, peat in the body of the roads is also allowed for public roads. The website [2] contains reports on solving problems of peat roads with low traffic intensity (LTI).

The oil and gas industry in the years of development of the pipeline network used peat actively and productively, located in the immediate vicinity of the pipeline route. A lot of works are devoted to studies of peat as a constructive local material and the rationale for its application [3-5]. The wide use of synthetic and rubber-fabric materials in the construction of temporary roads with peat in the body significantly increased the bearing capacity of structures [6, 7].

The technological schemes proposed in VSN 26-90 [1] are possible only in winter, and it is impossible to develop a reserve with peat from one side of the future RR because of the nearby MP.

Thus, the prerequisites for the wider use of peat for RR construction during the operational stage of the MP are:

- ultra-low traffic intensity (ULTI) of the RR technique (unlike public roads - units per day);
- proven ability of peat deposits to withstand significant loads (over 60 tons);
- the necessity of creating an all-year-round RR in wetland conditions growing due to the aging of the MP strings.

The basic design solution of the RR with the maximum replacement of mineral soil with peat is shown in Fig. 1.

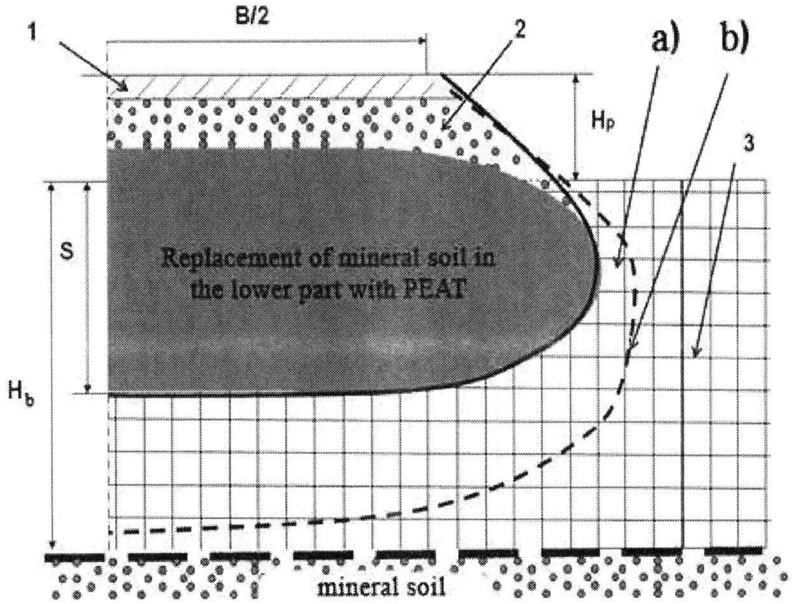

Figure 1. The basic design of the roadbed in the marshes with the use of peat in the body of the banket for RR

a – on the marsh of type I; b - on the marsh of types II and III (B – banket width, S – final sediment of the base of the banket, Hp – working height of the banket, Hb – marsh depth, 1 – road surface, 2 – sand fill, 3 – peat layer)

Fig. 1 shows the possibility of replacing mineral soil with peat for marshes of Type I; for marshes of types II-III, let us fill the marshy thickness with preformed peat (PP) along the contour b). The basis for the replacement is the calculation of the construction of road topping by [1]. Calculations are made for the road structure shown in Fig.1. The road covering of the transitional type is a slag layer of 15-20 cm. Replacement of mineral soil, according to Fig. 1, is produced by preformed peat, aged in burts before reaching the necessary humidity. The depth of the marsh varied from 3 to 4.5 m. The elevation of the bottom of the mineral soil above the surface of the marsh is 0.3 m plus the reserve for the sediment of the peat banket, calculated from the "age-old" precipitation of 2 cm/year (M.P. Bolshtyansky, 1990, 1998). The selection of the thickness of the sand backfilling between the PP layer and the slag coating was made by calculations, depending on the intensity of the working motion along the RR. Fig. 2 shows the dependence of the layer of mineral soil on the intensity of the machinery traffic intensity during the operation of the MP.

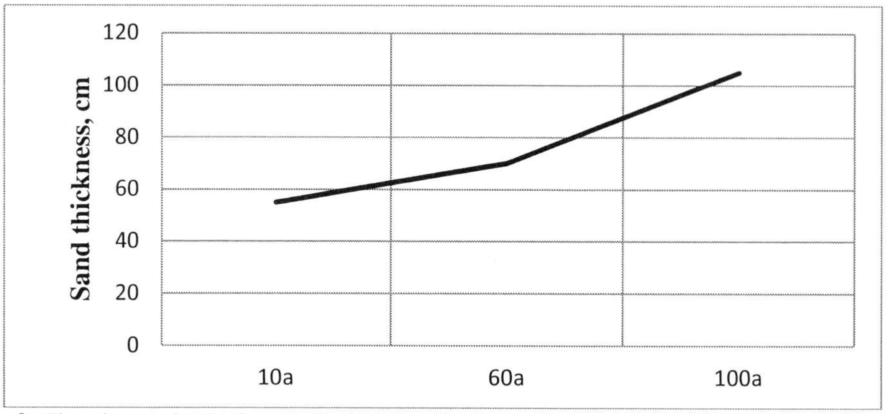

Figure 2. The change in the layer of mineral soil, depending on the traffic intensity along the RR (the abscissa axis, standard trucks)

The height of the banket at the beginning of operation depends on the planned service life of the RR, so for a service life of 30 years, the height Hp will be 1.1-1.2 m, while the "age-old" peat sediment with a low intensity of impact on the peat thickness is unknown.

The typification of organizational constructive and technological solutions for the construction of a RR is carried out, including the name of the structure (peat banket-TH), the type of marsh, lake (marsh of the 2nd type - B2, Oz) and the performance time (L, Z). Fig. 3 shows an example of the use of different design and technological solutions for the construction of RR.

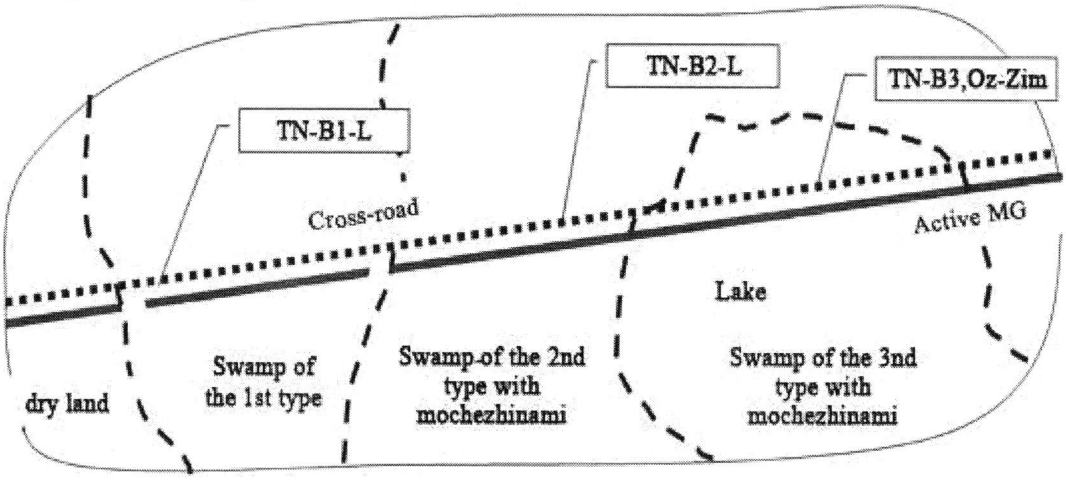

Figure 3. An example of the implementation of organizational design and technological solutions of laying MP in the wetlands
(rigid binding of the route of the future road to the highway MP)

Peat bankets (PB) are fairly peaty, 10-50 thousand m^3 are required for 1 km of RR, depending on the thickness of the marshy deposit. For the implementation of technological processes for the erection of a PB, preformed peat is required with a moisture content not exceeding the limit of its moisture capacity (300-900%). Fig. 4 shows a schematic diagram of the peat bulging. The formation of lakes at places of peat deposits is ecologically expedient, since in this case, according to biologists, biological diversity of wildlife is increasing.

IOP Conf. Series: Materials Science and Engineering 357 (2018) 012036 doi:10.1088/1757-899X/357/1/012036

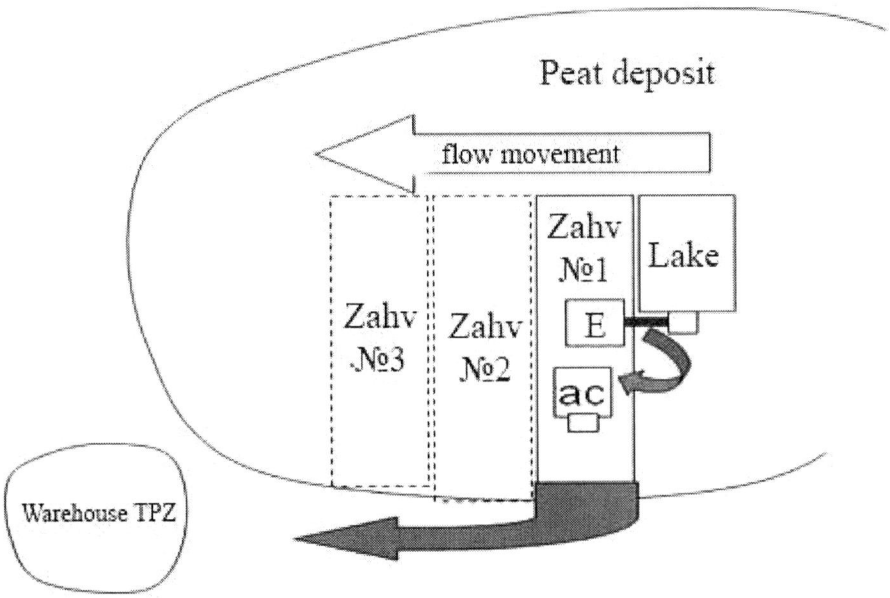

Figure 4. A schematic diagram of the peat deposit development

The economic efficiency of peat application in the body of RR bankets is related to the elimination of problems arising in the absence of such roads. Moreover, the continued availability of MP leads to a range of economic benefits [8]:

- the elimination of peak seasonal resource requirements (machinery and people);
- maximum utilization of the production capacity of operational organizations;
- reduction in the share of contingent expenses as part of the annual program expenditure;
- an increase in the annual program of works, therefore, an increase in the volume of repair and restoration works in the MP system of Russia;
- increase in profit due to more efficient use of funds for depreciation of fixed assets.

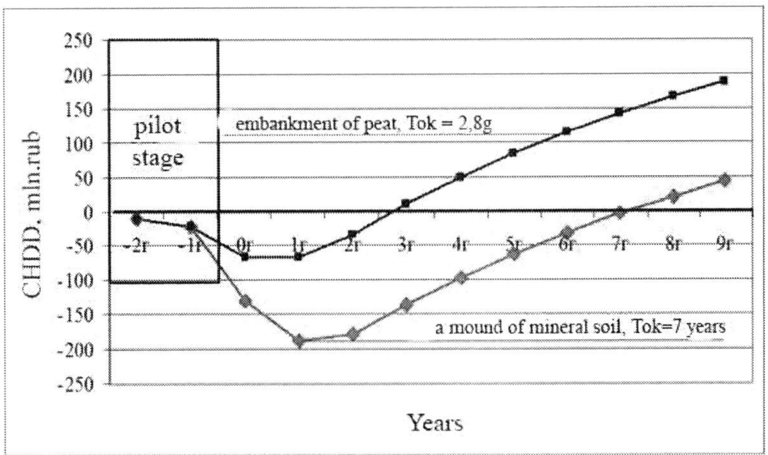

Figure 5. Comparative economic efficiency and payback period of roads with peat in the body of banket and traditional roads

The calculation of the efficiency of investment in RR using peat soils is based on [9]. Fig. 5 shows the comparative data for the calculation of net present value for a section of MP of 100 km length laid in wetland conditions. Initial data: "the project" takes into account the costs of performing the work of the experimental stage, then within 3 years, it will be possible to construct a RR using peat; "without a project", the possible costs for the construction of levies, long-term products cutting-off, etc. were taken into account.

At the experimental stage, the following problems need to be solved:

- To work out new ways of erecting PBs on the basis of the construction of experimental sites in marshes of different types and lake-moss complexes.
- To work out the technology of preliminary peat formation.
- To check the stability of the PB with low traffic intensity and the impact of long-standing construction machines in the pilot areas.
- To assess the consequences of peat extraction for the needs of the RR.
- To conduct a study of the economic parameters of organizations that perform routine and emergency work on pipeline transport to obtain a more reliable assessment of the integrated economic effect.

3. Conclusions

1. Construction of permanent roads along the MP is technically feasible and cost-effective.
2. The use of peat in the body of the banket, which gives a multiple economy of investment, is the most effective one;
3. To develop recommendations for the construction of roads along the MP, it is necessary to carry out Scientific Research and Experimental Design Work, the tasks for which have already been set.

4. Acknowledgement

The authors are grateful to Bolshtyansky M.P., a PhD in Technical Sciences, one of the developers of VSN 26-90, who once walked tens of kilometers in the marshes of Western Siberia, who observed the "life" of peat roads in real conditions for many years, for valuable advice in the preparation of this article.

References

[1] VSN 26-90. 1990 *The instruction on designing and construction of highways of oil and gas fields of Western Siberia.* (Moscow, Ministry of Transport and Communications)

[2] Roadex: roads on weak grounds http://www.ador.ru/data/files/static/kolarctic-5-07.pdf

[3] VSN 2-105-78. 1978 *Instructions for the construction of temporary roads for pipeline construction in difficult conditions (on watered and marshy areas).* (Minneftegazstroy. Moscow)

[4] P 491-83. 1983 *Recommendations on technology and organization of year-round construction of pipelines in marshes and watered areas.* (Minneftegazstroy. Moscow)

[5] Chirskov V G, Berezin V L, Telegin L G 1991 *Construction of main pipelines.* (M.: Nedra)

[6] Kazarnovsky V D, Polunovsky A G, Ruvinsky V I *1984 Synthetic textile materials in transport construction.* (Moscow: Transport)

[7] Recommendations for the use of geosynthetic materials in the construction of field roads. St. Petersburg. First edition. www/miakoming.ru/infaTech/promislovie_dorogi/

[8] Bobrova T V 2000 *A feasibility study for the production of road construction in the winter. Textbook* (Omsk: publishing house of SibADI)

[9] CS Gazprom 2004 *Intracorporate rules for assessing the effectiveness of SR&EDW / RD 1.12-096-2004*

IOP Conf. Series: Materials Science and Engineering **357** (2018) 012037 doi:10.1088/1757-899X/357/1/012037

Mathematical Model of Stress-Strain State of Curved Tube of Non-Circular Cross-Section with Account of Technological Wall Thickness Variation

S P Pirogov[1], **N N Ustinov**[2] **and N I Smolin**[2]

[1]Industrial University of Tyumen, 38 Volodarskogo Str, Tyumen, 625000, Russia
[2]Northern Trans-Ural State Agricultural University, 6 Respubliki Str, Tyumen, 625003, Russia

E-mail: piro-gow@yandex.ru, ustinovniknik@mail.ru, tlt_dir@mail.ru

Abstract. A mathematical model of the stress-strain state of a curved tube of a non-circular cross-section is presented, taking into account the technological wall thickness variation. On the basis of the semi-membrane shell theory, a system of linear differential equations describing the deformation of a tube under the effect of pressure is obtained. To solve the boundary value problem, the method of shooting is applied. The adequacy of the proposed mathematical model is verified by comparison with the experimental data and the results of the calculation of tubes by the energy method.

1. Introduction

Curved tubes of a non-circular cross-section are widely used as structural elements of pipelines, elastic sensitive elements of devices, power elements of machines in various industries [1-3]. As a result of tube bending during the manufacturing process, the fibers at the maximum bend radius are stretched, while at the minimum radius they are compressed. This leads to a change in the calculated wall thickness of the tube along the perimeter of the cross section [7-10].

2. Materials and methods

Let us consider the possibility of constructing a mathematical model of the stress-strain state of a curved tube of a non-circular cross-section with allowance for technological wall thickness variations based on the semi-membrane shell theory [4].

The stress-strain state of a curved tube of a non-circular cross-section (Fig. 1 a) is described by differential equations [3, 4]:

$$\begin{cases} \left(\dfrac{1}{t}\psi'\right)' - \mu_0 \vartheta \cos\alpha_0 = -m\sin\alpha_0 \\ \left(t^3\vartheta'\right)' + \mu_0 \psi \cos\alpha_0 = -qf_0 \end{cases} \quad (1)$$

In these equations, the unknown functions are ψ and ϑ of dimensionless coordinate η:

$$\eta = s / r ; \; r = P / 2\pi , \tag{2}$$

where s – the length of the meridian measured from the selected origin; r – the "reduced radius" of the section; П– the perimeter of the cross section of the middle surface of the tube.

Function ϑ is the angle of rotation at the point of section η (Figure 1 b), that is, the change in the initial angle of inclination of tangent $\alpha_0(\eta)$:

$$\vartheta(\eta) = \alpha_0(\eta) - \alpha(\eta) . \tag{3}$$

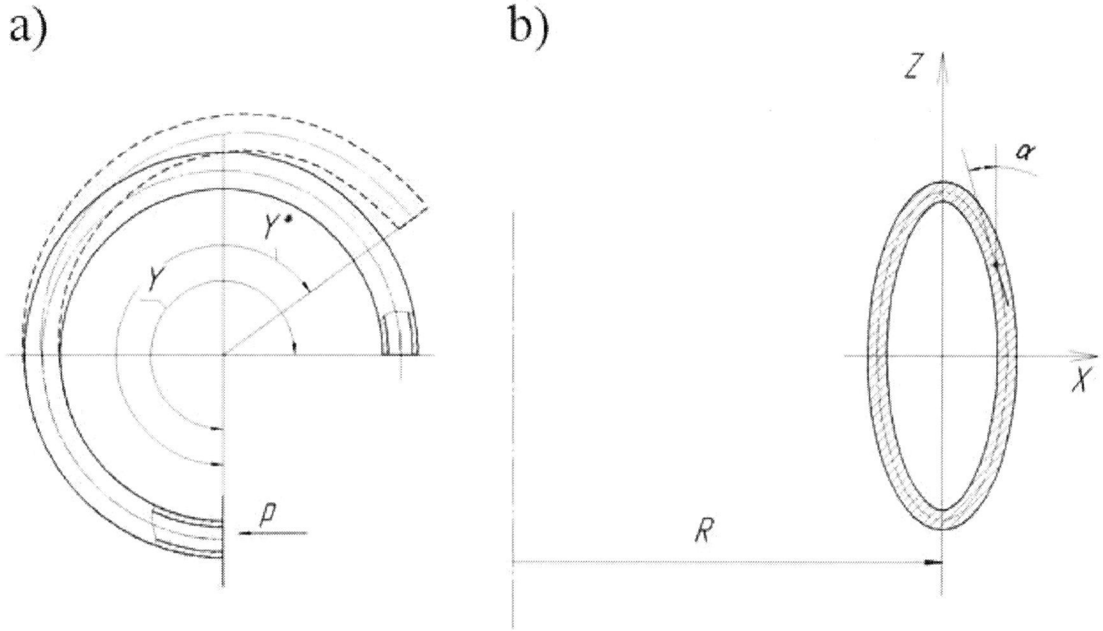

Figure 1. A curved tube of a non-circular cross-section
a – deformation under pressure; b – meridional section

Function ψ(η) characterizes the stress state in the rod and has the form:

$$\psi(\eta) = \sqrt{12(1 - \nu^2)} \frac{r}{Eh_m^2} \int_0^\eta T_1(\eta) d\eta , \tag{4}$$

where ν – Poisson's ratio; E – Young's modulus of the tube material; h_m– the thickness of the spring wall at the point with coordinate η=π/2; T_1 – the normal force per unit length of the cross-section. Values with index 0 refer to the state of the tube without deformation.

Function t(η) depends on the law of variation of the wall thickness and is defined as follows:

$$t(\eta) = h(\eta) / h_m , \tag{5}$$

where *h(η)* – the cross-section wall thickness.

Parameters q, μ and m take into account normal pressure p, the curvature of the axis and its variation:

$$q=12(1-\nu^2)\frac{pr^3}{Eh_m^3},\quad \mu=\sqrt{12(1-\nu^2)}\,\frac{r^2}{Rh_m},\quad m=\mu-\mu_0, \tag{6}$$

where R – the radius of the central axis of the tube.

The shear force in the section with r=1 from the unit load is represented in (2) by the function:

$$f_0=-\cos\alpha_0\int_0^{\eta}\cos\alpha_0 d\eta-\sin\alpha_0\int_{\pi}^{\eta}\sin\alpha_0 d\eta. \tag{7}$$

Equations (1) are valid if the material of the tube is elastic and homogeneous, the rotation angle is small compared to the unity, and dimensions of the cross section are small compared to R.

In equations (1), the shape of the midline of the cross-section is given by functions cosα and sinα. Due to the symmetry of the mean line of the section with respect to the x and z axes (Fig. 2), it is sufficient to determine the given functions on interval $0\le\eta\le\pi/2$, so expressing the values of angle α(η) through the geometric parameters of the "universal" section [1], one gets:

$$\alpha=\begin{cases}\alpha_1=arctg\left(\dfrac{B-B_1}{A-B_1}\right), & 0\le\eta<\eta_1;\\[2mm] \alpha_2=\dfrac{r(\eta-\eta_1)}{B_1}, & \eta_1\le\eta<\dfrac{\pi}{2}.\end{cases} \tag{8}$$

In (8):

$$r=\frac{2}{\pi}\sqrt{A^2+2B_1^2+B^2-2B_1(A+B)+B_1},\ \eta_1=\frac{\sqrt{(B-B_1)^2-(A-B_1)^2}}{r}.$$

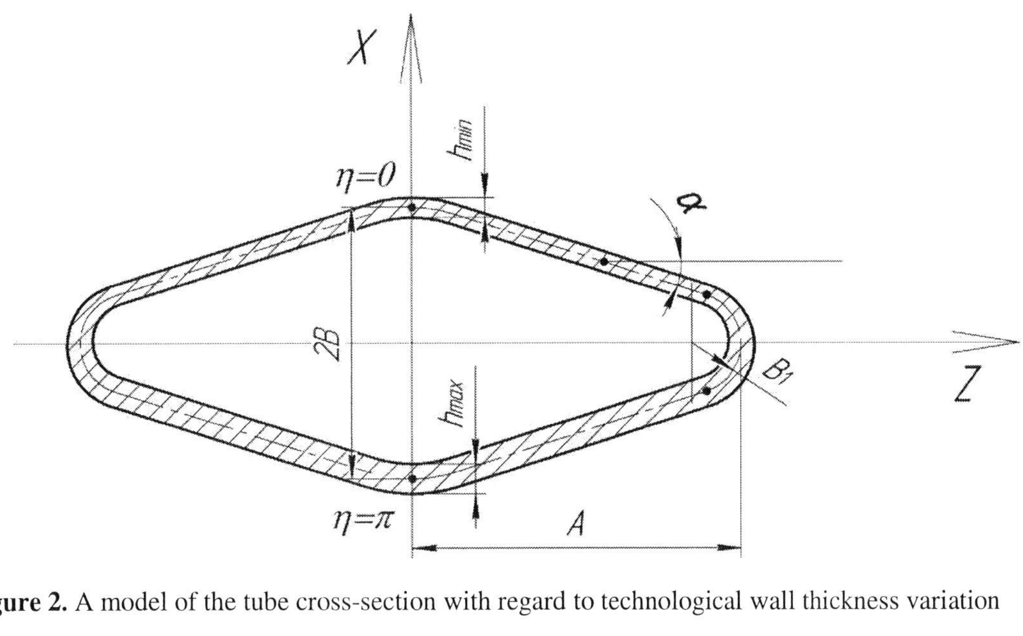

Figure 2. A model of the tube cross-section with regard to technological wall thickness variation

To model the technological wall thickness variation, the law of variation of the wall thickness along the perimeter of the section is represented in the form of a piecewise function:

$$h(\eta)=\begin{cases} h_{\max} & 0\le\eta\le\eta_1 \\ h_{\min}+(h_{\max}-h_{\min})(\eta_2-\eta)/(\eta_2-\eta_1) & \eta_1<\eta\le\eta_2 \\ h_{\min} & \eta_2<\eta\le\eta_3 \\ h_{\min}+(h_{\max}-h_{\min})(\eta-\eta_3)/(\eta_4-\eta_3) & \eta_3<\eta\le\eta_4 \\ h_{\max} & \eta_4<\eta\le\eta_5 \end{cases}, \tag{9}$$

where h_{max} – maximum section wall thickness; h_{min} – minimum section wall thickness; η_1, η_2, η_3, η_4 - coordinates that determine the law of variation of the wall thickness along the perimeter of the section (Fig. 1).

Let us introduce notations: $\frac{1}{t}\psi'=\varphi$,$t^3\vartheta'=u$. Then the system of equations (1) in a normal form can be written as:

$$\begin{cases} \varphi'=\mu_0\cos\alpha_0\vartheta-m\sin\alpha_0 \\ u'=-\mu_0\cos\alpha_0\psi-qf_0 \\ \psi'=t\varphi \\ \vartheta'=\frac{1}{t^3}u \end{cases}. \tag{10}$$

The system of equations is supplemented by boundary conditions:

$$\begin{cases} \psi(0)=0 \\ \vartheta(0)=0 \end{cases}, \begin{cases} \psi'(\pi)=0 \\ \vartheta(\pi)=0 \end{cases}. \tag{11}$$

In the linear approximation, the required functions can be represented as the sum of two parts proportional to the parameter of the change in curvature *m* and to the parameter of the normal pressure:

$$\text{q}: \psi=m\psi_m+q\psi_q ,\vartheta=m\vartheta_m+q\vartheta_q . \tag{12}$$

As a result of substituting expressions (12) into system (1), let us obtain two systems:

$$\begin{cases} \left(\frac{1}{t}\psi'_m\right)'-\mu_0\vartheta_m\cos\alpha_0=-\sin\alpha_0 \\ \left(t^3\vartheta'_m\right)'+\mu_0\psi_m\cos\alpha_0=0 \end{cases}, \begin{cases} \left(\frac{1}{t}\psi'_q\right)'-\mu_0\vartheta_q\cos\alpha_0=0 \\ \left(t^3\vartheta'_q\right)'+\mu_0\psi_q\cos\alpha_0=-f_0 \end{cases}. \tag{13}$$

Using the algorithm for solving system (1), let us determine from systems (13) functions ψ_m,ψ_λ ,$\vartheta_m,\vartheta_\lambda$.

To solve the boundary value problem, the shooting method was used [4], also a program was written in the MATLAB environment, the integration of the Cauchy problems was carried out using the Runge-Kutta method of the 4th order of accuracy using the standard solver ode45.

In the case of fixing one of the ends of the tube, its free end under the action of normal pressure is able to move; the system of resolving equations is supplemented by the condition that the moment in the cross-section be equal to zero:

$$M = \frac{Eh_m^2 r}{\sqrt{12(1-\nu^2)}} \oint \psi \sin\alpha_0 d\eta = 0. \tag{14}$$

From (12), taking into account (14), let us determine the relative opening angle of the tube:

$$\frac{\Delta Y}{Yp} = \frac{12(1-\nu^2)}{\mu_0 E}\left(\frac{r}{h_m}\right)^3 \frac{m}{q}, \text{ where } \frac{m}{q} = \frac{\oint \psi_q \sin\alpha_0 d\eta}{\oint \psi_m \sin\alpha_0 d\eta}. \tag{15}$$

3. Results and Discussion

It is of interest to assess the adequacy of the proposed model by comparing the calculation results with the experimental data and the results of calculating the tubes by the energy method. Table 1 shows the parameters of a number of manometric springs, borrowed from the work of M.P. Shumsky [6].

Table 1. Parameters of manometric springs of flat-oval cross-section with technological wall thickness variation

No. of sample	Measured pressure, [Mpa]	Semimajor axis of section A, [mm]	Semiminor axis of section B, [mm]	Radius of rounding at the ends of the major axis of section B1, [mm]	Maximum wall thickness h_{max}, [mm]	Minimum wall thickness h_{min}, [mm]	Radius of the central axis of the spring R, [mm]	Modulus of elasticity of material E, [Mpa]
1	0.4	10.20	2.00	1.72	0.65	0.63	52.5	$1.2 \cdot 10^6$
2	0.4	10.21	1.82	1.18	0.64	0.62	54.0	$1.2 \cdot 10^6$
3	0.6	10.14	1.70	1.11	0.83	0.76	54.5	$1.2 \cdot 10^6$
4	1	9.85	2.28	1.59	0.93	0.85	54.0	$1.2 \cdot 10^6$

The study revealed a deviation in the value of the wall thickness from the calculated along the contour of the cross-section within wide limits.

Comparison of the results of calculating the relative opening angle with experimental and calculated data by M.P. Shumsky [6] is given in Table 2. In view of the lack of information on the law of variation in the wall thickness along the perimeter of the cross-section, the wall thickness at the outer and inner sections of the cross-section was assumed to be constant, equal to the minimum and maximum values, respectively, and in the rounding sections a linear law of variation in wall thickness was taken.

Table 2. Comparison of results of calculating the sensitivity of springs taking into account technological wall thickness variation

No. of sample	Relative opening angle $\Delta Y/(Y)\cdot 10^{-2}$, [1/ MPa]				
	Experiment	Calculation by the Ritz method in the second approximation	Error Δ, [%]	Calculation by the proposed method	Error Δ, [%]
1	0.152	0. 142	6.57	0.143	5.92
2	0.165	0.144	12.72	0.209	-26.66
3	0.155	0.172	-10.96	0.186	-20.0
4	0.149	0.150	-0.67	0.162	-8.72

As can be seen from Table 2, deviations of the calculation results using the proposed method and the Ritz method in the second approximation, with allowance for deviations of the real geometry of the tube from the theoretical one, should be considered satisfactory.

4. Conclusion

Thus, it can be concluded that the proposed model with sufficient accuracy allows us to evaluate the stress-strain state of a curved tube with allowance for technological wall thickness variations and can be used in solving problems of calculation and design.

References

[1] Pirogov S P, Cherentsov D A. 2016 Theoretical Foundations of the Design of Vibration-Resistant Manometers, *J Measurement Techniques* **59** 845–849

[2] Ustinov N, Maratkanov A, Martynenko 2017 A Experimental study of the parameters of the active tool of a cultivator with a frame in form a flexible tubular element. *MATEC* Web of *Conferences* **106**

[3] Kokoshin S, Ustinov N, Kirgincev 2016 B The use of flexible tubular elements of the overhaul and tunnels reconstruction, *Procedia Engineering* **165** 817-828

[4] Akselrad E L 1976 *Flexible Shell* (Moscow, Nauka)

[5] Shumsky M N, 1966 *Calculation and optimal design of manometric springs*: dissertation for the degree of Candidate of Technical Sciences (Tomsk)

[6] Voronin K S, Ogudova E V 2016 The Effect of Dynamic Processes in the System "Pipe-Soil" on the Pipeline Deviation from Design Position *Transport and Storage of Hydrocarbons. IOP Conf. Series: Materials Science and Engineering* **154**

[7] Dudin S, Voronin K, Yakubovskaya S, Mutavaliev S 2016 Modeling Hydrodynamic State of Oil and Gas Condensate Mixture in a Pipeline *MATEC Web of Conferences. XV International conference topical problems of architecture, civil engineering, energy efficiency and ecology*

[8] Dudin S, Bahmat G, Zemenkov Y, Voronin K, Shipovalov A 2017 Strategy for monitoring and ensuring safe operation of Russian gas transportation systems *MATEC Web of Conferences* **106**

[9] Shalay V, Zemenkova M, Zemenkov Y 2016 Modeling Parameters of Reliability of Technological Processes of Hydrocarbon Pipeline Transportation *XV International Conference Topical Problems of Architecture, Civil Engineering, Energy Efficiency and Ecology MATEC Web of Conferences* **73**

[10] Toropov E S,Vorobieva T I, Toropov S Y, Toropov V S 2016 Finding the Displacement of Elastically Bent Pipeline Disposed on Roller Supports under the Action of the Axial Shear Force *IOP Conference Series: Materials Science and Engineering* **154**

Study of torsion oscillations of pumping unit shafts

V M Loginovskikh[1], D A Cherentsov[1], K S Voronin[1] and S P Pirogov[2]

[1] Industrial University of Tyumen, 38 Volodarskogo St., Tyumen, 625000, Russia
[2]State Agrarian University of the Northern Trans-Urals, 5 Repablic St., Tyumen, 625003, Russia

E-mail: cherentsovda@bk.ru, voronin_tsogu@mail.ru

Abstract. The method of detuning from resonant frequencies is an effective method of reducing the vibration activity of the HA at the design stage, which allows selecting the necessary operating mode for the HA, which will provide the necessary reliability of operation. In this connection, works in the field of increasing the vibration protection of NA are relevant. All calculations are based on the dynamic model of free torsional vibrations of the pumping unit, for which the oscillation equation was derived on the basis of Lagrange equations of the second kind, where the angle of rotation of the rotor was taken as the generalized coordinate. Next, expressions were obtained for the determination of the potential and kinetic energies that take into account all the characteristics of the shaft and the disk of the rotor HA; in particular, for the first time the mass of the shaft is taken into account. Then an expression was obtained for determining the frequencies of free torsional oscillations and the equation of motion of the rotor HA. Knowing the parameters of the oscillatory motion makes it possible to estimate the stresses that arise when the shaft rotates. The following is a case of a sudden stop of the shaft, for example, when the support is destroyed, and a graph is obtained from which the main conclusions are drawn.

1. Introduction

In modern conditions of operation and management of oil and gas transportation systems, the relevance of control of the technological process is of great importance. One of the main parameters of technological processes in the oil and gas industry, which requires a high level of control, is frequency of oscillations [1-5].

As operating conditions become more severe, the intensity and frequency of vibrations increase; pressure pulsations have a negative effect leading to the breakdown of pumping units and, as a consequence, disruption of the technological process [6-8].

A method of vibration protection of pumping units (PU) by means of tuning from resonant frequencies implies knowledge of natural oscillation frequencies of PU and frequencies of the disturbing force.

2. Materials and methods

In [1], free torsional vibrations of a cylindrical shaft with n disks fixed to it are determined, neglecting the mass of the shaft as compared to the mass of the disk.

Below, a procedure for determining natural frequencies of torsional oscillations of PU is presented, obtained with the help of the Lagrange equation of the second kind taking into account the mass of the shaft [9, 10].

Fig. 1 shows the design scheme of the construction.

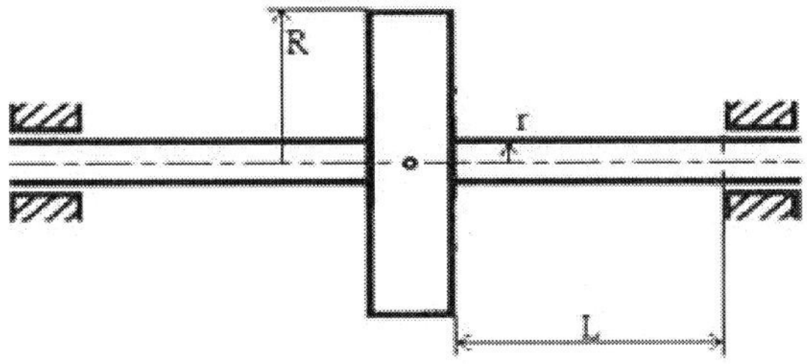

Figure 1. System - shaft and rigidly fixed disk.

The angle of rotation of the shaft - φ is taken as the generalized coordinate.

The Lagrange equation of the second kind for natural torsional oscillations of PU has the form:

$$\frac{d}{dt}\left(\frac{\partial T}{\partial \dot{\varphi}}\right) + \frac{\partial \Pi}{\partial \varphi} = \frac{\partial T}{\partial \varphi} \quad (1)$$

where t – time, s; T – kinetic energy; Π – potential energy.

The expression for the kinetic energy of the system - shaft and rigidly fixed disk, will be combined from the kinetic energy of rotation of the shaft and the disk:

$$T = \frac{\left(M(R^2 + r^2) \ + mr^2\right)}{4}\dot{\varphi}^2 \quad (2)$$

where M – disc weight, kg; R – outer radius of disk, m; r – shaft radius, m; m – shaft weight, kg.

The potential energy of the system is determined as:

$$\Pi = \frac{GI_p}{L}\varphi^2 \quad (3)$$

where G – modulus of elasticity of the second kind, Pa; I_p – polar moment of inertia, m^4; L – distance from bearing to disc, m.

Substituting the expressions for the partial derivatives in the La Grange equation, let us obtain a second-order differential equation with constant coefficients:

$$\frac{\left(M(R^2+r^2)+mr^2\right)}{2}\ddot{\varphi} + \frac{GI_p}{2L}\varphi = 0.$$
$$\ddot{\varphi} + k^2\varphi = 0 \quad (4)$$

This equation will be a mathematical model of this object.

The value of the frequencies of natural torsional oscillations:

$$k = \sqrt{\frac{GI_p}{(M(R^2 + r^2) + mr^2)L}}. \quad (5)$$

Thus, an expression was obtained for determining the frequency of natural torsional vibrations, allowing taking into account the characteristics of the shaft and the disk.

Knowing the parameters of the oscillatory motion makes it possible to estimate the stresses that arise when the shaft rotates.

The tangential stresses arising in the cross sections of the shaft are equal:

$$\tau = \frac{M_k}{W_p} \leq [\tau], \quad (6)$$

where M_k – torque in section; W_p – polar moment of circular section resistance.

Let us imagine the case of a sudden stop of the shaft, for example, when the support is broken. The solution of (4) is sought in the form:

$$\varphi = A\sin(kt + \alpha), \tag{7}$$

whence:

$$\dot{\varphi} = kA\cos(kt + \alpha). \tag{8}$$

The initial conditions for this problem can be formulated in the form: $\varphi_0 = 0$, $\dot{\varphi}_0 = \omega$. Substituting it into the solution, one obtains:

$$\dot{\varphi} = kA\cos(kt + \alpha), \qquad 0 = A\sin(\alpha). \tag{9}$$

From here $= 0$, $\omega = kA$, $\varphi = \frac{\omega}{k}\sin(\alpha)$. The maximum value will be reached when $\sin(\alpha) = 1$ whence $\varphi = \frac{\omega}{k}$.

The angle of rotation of a shaft of length L will be:

$$\varphi = \frac{M_k L}{G I_p}, \tag{10}$$

whence:

$$M_k = \frac{G I_p \varphi}{L}, \tag{11}$$

Then the tangential stresses will be:

$$\tau = \frac{G I_p \varphi}{W_p L}. \tag{12}$$

For a rod of circular cross section, the polar moment of section resistance and the polar moment of inertia can be determined from the formulas:

$$W_p = \frac{\pi d^3}{16}, I_p = \frac{\pi d^4}{32}. \tag{13}$$

Then the tangential stresses will be:

$$\tau = \frac{G d \omega}{2kL}. \tag{14}$$

3. Results and Discussion

Fig.2 shows the dependence of the tangential stresses on the angular velocity.

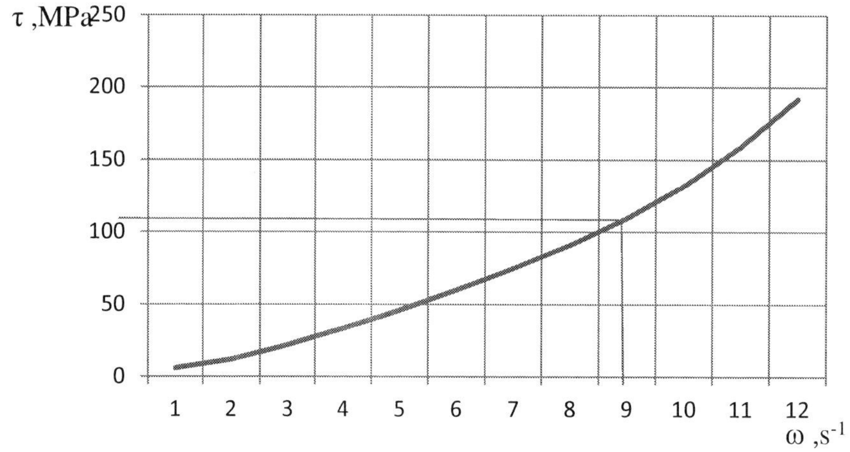

Figure 2. Dependence of tangential stresses on angular velocity.

It can be seen from the graph that even at a speed of about 9-10 rad/s, stresses will exceed the permissible values. Therefore, to prevent accidents, special shock-absorbing devices should be provided.

Thus, to improve the efficiency of the equipment, the resulting expressions can be used to develop instructions for the design and operation of PU.

4. Conclusion

As a result, a mathematical model based on the Lagrange equations of the second kind was developed for free torsional oscillations of the rotor HA, allowing taking into account the characteristics of the shaft and rotor disk HA. The stresses resulting from the rotation of the shaft are evaluated, and the conditions for the emergence of an emergency situation for a specific task are determined.

References

[1] Dubnischev Yu N 2011 *Oscillations and waves* (Moscow: Lan)

[2] Mirzadjanzadeh A H 2005 *Theory of oscillations in the oilfield engineering* (Institute of Computer Studies; Izhevsk)

[3] Chelomey V N 1980 Vibrations in technology *Oscillations of machines, structures and their elements* **3** 544

[4] Cherentsov D A, Pirogov S P 2014 Investigation of damped oscillations of a manometric tubular spring in a viscous medium. *News of higher educational institutions. Oil and gas* **5** 66-72

[5] Yablonsky A A 2003 *Course of the theory of oscillations* (Krasnodar: Lan)

[6] Toropov V S, Tamer O S, Toropov S Y, Nikiforov N A 2016 Dynamics of Pipeline Pulling Process by Horizontal Directional Drilling *IOP Conference Series: Materials Science and Engineering* **154**

[7] Dudin S M, Zemenkov Y D, Shabarov A B 2016 Modeling the Phase Composition of Gas Condensate in Pipelines *IOP Conference Series: Materials Science and Engineering* **154**

[8] Shalay V, Zemenkova, M, Zemenkov Y 2016 Modeling Parameters of Reliability of Technological Processes of Hydrocarbon Pipeline Transportation *MATEC Web of Conferences* **73**

[9] Pirogov S P, Cherentsov D A, Gulyaev B A 2016 Prospects of applying vibration-resistant pressure gauges in the oil and gas industry *IOP Conference Series: "Transport and Storage of Hydrocarbons"*

[10] Pirogov S P, Cherentsov D A, Chuba A Y 2016 Study of elastic sensing elements for vibration-resistant pressure gauges *IOP Conference Series: "Transport and Storage of Hydrocarbons"*

IOP Conf. Series: Materials Science and Engineering **357** (2018) 012039 doi:10.1088/1757-899X/357/1/012039

Modeling flows of heterogeneous media in pipelines when substantiating operating conditions of hydrocarbon field transportation systems

S M Dudin and D V Novitskiy

Industrial University of Tyumen, 38 Volodarskogo St., Tyumen, 625000, Russia

E-mail: srghome@mail.ru

Abstract. The works of researchers at VNIIgaz, Giprovostokneft, Kuibyshev NIINP, Grozny Petroleum Institute, etc., are devoted to modeling heterogeneous medium flows in pipelines under laboratory conditions. In objective consideration, the empirical relationships obtained and the calculation procedures for pipelines transporting multiphase products are a bank of experimental data on the problem of pipeline transportation of multiphase systems. Based on the analysis of the published works, the main design requirements for experimental installations designed to study the flow regimes of gas-liquid flows in pipelines were formulated, which were taken into account by the authors when creating the experimental stand. The article describes the results of experimental studies of the flow regimes of a gas-liquid mixture in a pipeline, and also gives a methodological description of the experimental installation. Also the article describes the software of the experimental scientific and educational stand developed with the participation of the authors.

1. Introduction

At various times, many researchers have been engaged in modeling the flow of hydrocarbon raw materials in pipelines, among them are N.A. Slezkin, S.G. Teletov, S.S. Kutateladze, M.A. Styricovich, H.A. Rakhmatullin, A.N. Kraiko, L.E. Sternin, A.K. Dyunin, Yu.T. Borshchevsky, A.I. Guzhov, V.F. Medvedev, R.I. Nigmatulin, N.A. Yakovlev, M.A. Huseynzadeh, A.B. Shabarov, Yu.S. Danielyan, V.A. Yufin, V.N. Antipiev, V.A. Zysin, E.L. Kitanin, A.K. Gallyamov, SE Kutukov, G. Wallis, and others.

By now, many authors have developed a number of models of the flow of hydrocarbon liquids in pipelines using the classical provisions of the theory of continuum mechanics. Differences are in the quantity and quality of the conditions and features of pipeline transportation of hydrocarbons that are taken into account.

2. Materials and methods

Common preconditions for modeling heterogeneous media flows through pipelines are the following [1, 2, 3, 4, 5 et al].

1. When hydrocarbon media move through pipelines, homogeneous media, homogeneous mixtures and multiphase (heterogeneous) media are distinguished. A model of a homogeneous medium is used when

the proportion of impurities is negligible, compared with the proportion of the main carrier liquid or gas. In homogeneous mixtures, the constituents are mixed at the molecular level. In heterogeneous mixtures there are isolated heterogeneities: drops, gas bubbles, solid particles, etc. The most studied are homogeneous media, homogeneous mixtures and two-phase (dispersed) mixtures. Two-phase flows are a special case of a heterogeneous medium, the mathematical description of which can be produced on the basis of multi-rate continua. In this case, each phase, filling part of the total volume of the mixture, is characterized at a given point by the true phase density, velocity, temperature and enthalpy.

2. In the mathematical modeling of heterogeneous mixtures, there are usually three main assumptions. The first assumption is that dimensions of inhomogeneities (diameters of particles, droplets or their conglomerates, etc.) are much larger than the characteristic distances between molecules. This assumption makes it possible to use the approaches adopted in the mechanics of continuous media in describing processes near the surface of inhomogeneous inclusions. The second assumption is that the dimensions of the inhomogeneities are many times smaller than the diameter and linear dimension of the pipeline. This assumption allows correctly averaging the parameters and using the equations of continuous medium mechanics for averaged quantities. When the problem is formalized, the third assumption about the continuity of the medium is adopted, which makes it possible to simplify considerably the problem and to represent the flow in the form of a continuous liquid and gas phase. Due to this, it is possible to use the apparatus of continuous functions and differential calculus for studies.

3. The presence of phase transformations in the flow and the use of a two-velocity model leads to the need to take into account the change in mass and momentum of each phase by using the equations of conservation of mass and momentum for each of the phases separately, and also take into account the kinetics of the formation of a two-phase flow, i.e., two-phase transformations.

In Tyumen region, a generalized model of the quasi-one-dimensional flow of multiphase hydrocarbons in pipelines, developed by Professor A.B. Shabarov, serves as the basis for the development of questions of the physico-mathematical modeling of technological processes in the oil and gas industry - a physical and mathematical model of the flow of multiphase hydrocarbon media in pipeline systems, the essence of which is the following [6, 7].

The basis of the model is the principle of constructing a quasi-one-dimensional model, i.e. a one-dimensional model of the flow of the medium, in which, in one way or another, the properties of the real three-dimensional flow are approximately taken into account. With this approach to the description of fluid motion, the main flow parameters, the cross-sectional variables of the channel, are replaced by some constants throughout the section at a fixed time. When averaging, an irregular flow in an arbitrary section is replaced by a canonical flow, some of whose parameters may differ from the actual flow. The canonical flow is characterized by a certain set of constant averaged parameters, under which all properties of a real non-uniform flow are preserved [10, 11].

Most publications on the pipeline transport of multiphase hydrocarbons reveal the issues of transport of two-phase hydrocarbon systems, where special attention is paid to the flow regimes of gas-saturated and gas-liquid mixtures with the separation of gas and liquid phases in the flow structure.

It is known that under given specific operating conditions of pipelines, hydrocarbon media can be in a liquid, gaseous or two-phase gas-liquid state. For hydrocarbons taken in their pure form, the boundary pressure between the gaseous and liquid states at a given temperature is their vapor pressure. At the vapor pressure, the complete transition of the hydrocarbon in question from one phase to another occurs [2].

In the pipeline transport of hydrocarbon media, the transition from one state to another occurs when the pressure and temperature change, which causes evaporation or condensation. In practice, the movement of hydrocarbon media through pipelines is usually accompanied by evaporation of hydrocarbons, which occurs under the influence of pressure reduction at a relatively constant temperature.

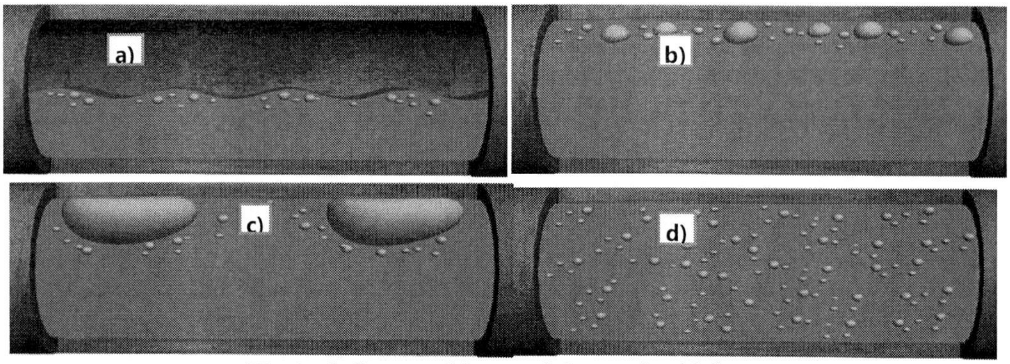

Figure 1. Structural forms of the gas-liquid flow:
a – separated; *b* – bubbly; *c* – slug; *d* - emulsion

In commercial pipelines the most common structural forms of the flow are slug and emulsion.

In contrast to the analytical method of studying regularities and describing the flow regimes of gas-liquid mixtures through pipelines, few works have been devoted to experimental studies. First of all, this is due to the fact that the movement of two-phase gas-liquid systems in pipes, complicated by mass-exchange processes and other factors, is quite difficult to model in the laboratory [1, 2]. However, the research of VNIIgaz, Giprovostokneft, Kuibyshev NIINP, Grozny Petroleum Institute, etc. is known, which, based on the data obtained, has developed methods for calculating pipelines transporting multiphase products. The developed methods distinguish the number of fixed and varying values, and in objective consideration they represent a bank of experimental data on the problem of pipeline transport of multiphase systems [8, 9].

At the department of "Transport of Hydrocarbon Resources", IUT, the research and experimental stand has been developed, the general layout of which is shown in Fig. 2. This stand allows simulating the movement of a gas-liquid flow in the pipeline and is a closed hydraulic circuit consisting of three series-connected sections with a glass tube-insert for visualization of structural flow forms.

Experimental studies were carried out in accordance with the technique of preliminary planning and carrying out the experiment, according to which the experiments simulated conditions characterized by different regimes, the magnitude of the hydraulic resistance, and the physico-chemical properties of the pumped product.

At the first stage of the study, parameters of a single-phase flow were measured, and at the second stage - of a gas-saturated liquid. As a working fluid, water and a water-glycerine mixture saturated with carbon dioxide were used. In accordance with the experimental procedure, a total of 51 series of measurements were performed, an average of about 40 measurements in each, in which four gas-liquid systems with a liquid component viscosity of 1.13 cSt to 3.38 cSt and a density of 998 kg/m^3 to 1098 kg/m^3. The experiments were carried out in the range of gas saturation from 0.54 to 0.97.

3. Results

Based on the results of the research, experimental graphs of the change in the basic hydrodynamic parameters of gas-liquid mixtures are constructed and empirical relationships are obtained. Fig. 3 shows experimental data on the investigation of flow regimes of a two-phase two-component system. Fig. 4 presents experimental data on the structural forms of a gas-liquid flow.

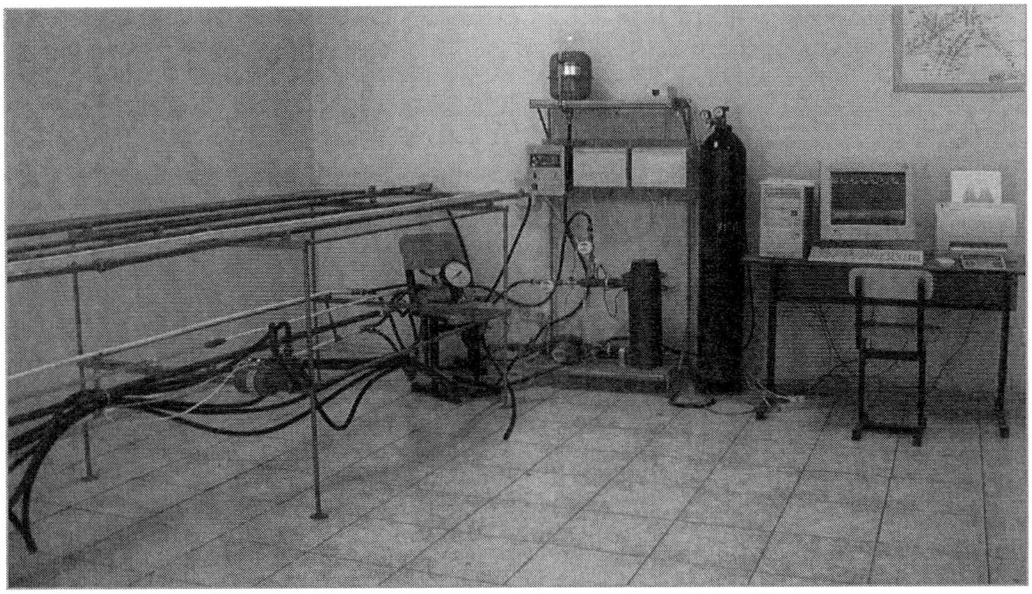

Figure 2. A general layout of the experimental installation

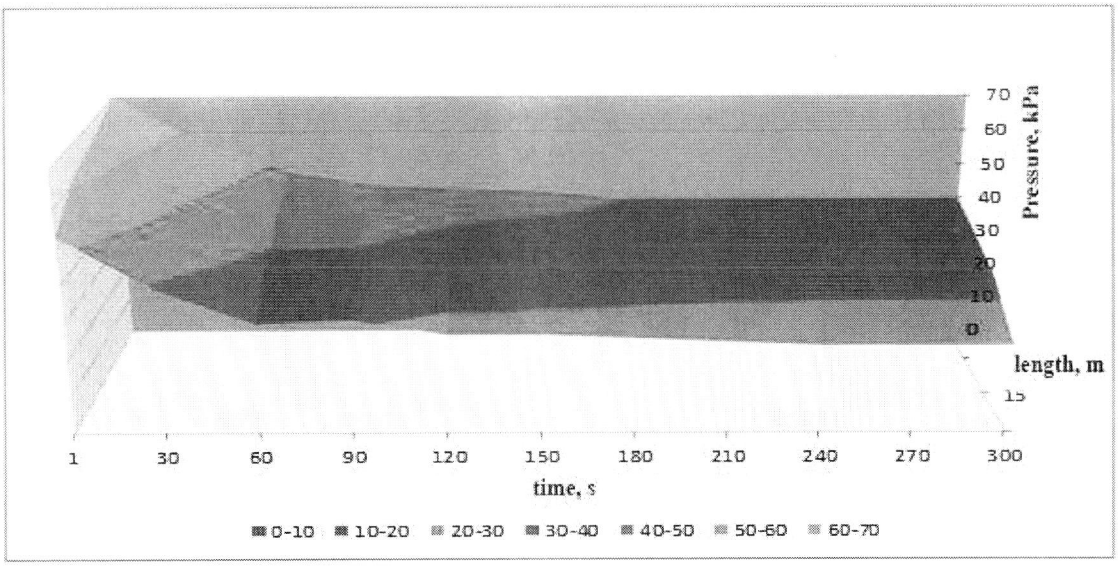

Figure 3. Changes in the pressure of the gas-saturated flow in the slug flow regime

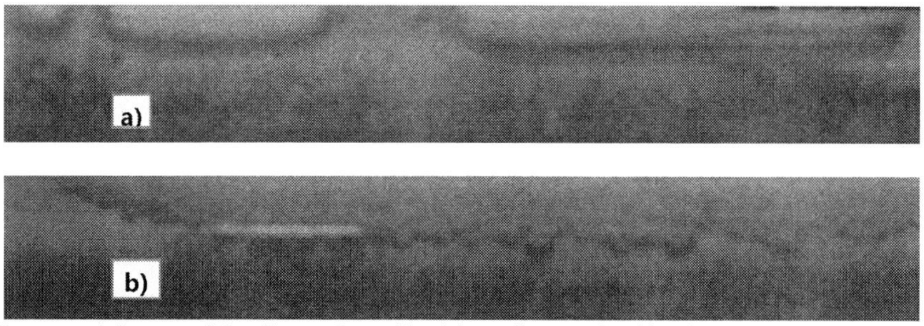

Figure 4. Structural forms of the flow of gas-liquid media, marked in the experimental study of a two-phase flow in a pipeline:
a – slug; *b* – emulsion (foamed)

4. Conclusions

1. A research and development stand has been designed that allows simulating and studying the flow regimes of heterogeneous media in horizontal pipelines.

2. The results of the experimental studies and the obtained empirical dependencies of the hydrodynamic characteristics of two-phase flows can be used in the analysis of heterogeneous medium flows in pipelines, in particular, in the study of slug and emulsion flow regimes.

3. The created research laboratory stand and the experimental data obtained with its use make it possible to predict and carry out the selection of rational operating modes of hydrocarbon transportation systems.

References

[1] Guzhov A I 1973 *Joint collection and transport of oil and gas* (M.: Nedra)

[2] Mamaev V A, Odishariya G E, Semenov N I, Tochilin A A 1969 *Hydrodynamics of gas-liquid mixtures in pipes* (M.: Nedra)

[3] Zemenkov Yu D 2002 *Diagnosis of damage and leakage in the pipeline transportation of multiphase hydrocarbons* (Vector Buk, Tyumen)

[4] Dudin S, Voronin K, Yakubovskaya S, Mutavaliev S 2016 Modeling hydrodynamic state of oil and gas condensate mixture in a pipeline *MATEC Web of Conferences* **73**

[5] Dudin S, Zemenkov Yu D, Shabarov A B 2016 Modeling the Phase Composition of Gas Condensate in Pipelines *IOP Conf. Series: Materials Science and Engineering* **154**

[6] Antipyev V N, Zemenkov Yu D, Shabarov A B 2002 *Technical and parametric diagnostics in pipeline systems* (Vector Buk, Tyumen)

[7] Shabarov A B, Saranchin N V 2009 Methods of calculating equilibrium parameters of liquid hydrocarbon mixture in pipelines *Bulletin of Tyumen State University* **6** 112-19

[8] Dudin S M 2015 Experimental studies of gas-liquid flows in pipelines *Modern problems of science and education* **2**

[9] Guzhov A I, Titov V G, Medvedev V F, Vasilyev V A 1978 *Collection, transportation and storage of natural hydrocarbons* 223

[10] Zemenkova M, Zemenkov Y, Gladenko A, Podorozhnikov S 2016 Estimation of Emissions During Monitoring of Pipelines in the Dynamic Mode of Operation. *5th International Scientific Conference Integration, Partnership and Innovation in Construction Science and Education* **86**

[11] Zemenkova M, Shalay V, Zemenkov Y, Kurushina E 2016 Improving the Efficiency of Administrative Decision-Making when Monitoring Reliability and Safety of Oil and Gas Equipment *MATEC Web of Conferences* **73**

[12] Kurushina V, Kurushina E, Shabarov A 2016 Technology Audit: Assessment of Innovative Portfolio *MATEC Web of ConferencesVolume* **73**

[13] Nelyubov D, Shabarov A, Podorozhnikov S 2016 Complex-action reagent for oil-infield and oil-trunk pipelines *Solid State Phenomena* **871** 185

IOP Conf. Series: Materials Science and Engineering **357** (2018) 012040 doi:10.1088/1757-899X/357/1/012040

Numerical simulation of fire vortex

D D Barannikova[1], V E Borzykh[2] and A.G.Obukhov[2]

[1]Tyumen State University, 6 Volodarskogo St., Tyumen, 625003, Russia
[2] Industrial University of Tyumen, 38 Volodarskogo St., Tyumen, 625000, Russia

E-mail: lusy_and_jam@mail.ru, borzykh@mail.ru, aobukhov@tsogu.ru

Abstract. The article considers the numerical simulation of the swirling flow of air around the smoothly heated vertical cylindrical domain in the conditions of gravity and Coriolis forces action. The solutions of the complete system of Navie-Stocks equations are numerically solved at constant viscosity and heat conductivity factors. Along with the proposed initial and boundary conditions, these solutions describe the complex non-stationary 3D flows of viscous compressible heat conducting gas. For various instants of time of the initial flow formation stage using the explicit finite-difference scheme the calculations of all gas dynamics parameters, that is density, temperature, pressure and three velocity components of gas particles, have been run. The current instant lines corresponding to the trajectories of the particles movement in the emerging flow have been constructed. A negative direction of the air flow swirling occurred in the vertical cylindrical domain heating has been defined.

1. Introduction

In the experimental studies [1] with thermal ascending currents of air simultaneously with the air vertical movement upward, there is its swirling: in the positive direction for the Northern hemisphere and in the negative direction for the Southern hemisphere. A similar air behavior is also observed in nature tornadoes and hurricanes [2]. In the research [3-6], it was proved using the constructed solutions of gas dynamics equations set that with existence of gas discharge on the vertical cylinder of nonzero radius in the external relative to the cylinder current, an air vortex generates in the respective direction. These directions correspond to the directions of the swirling of atmospheric ascending currents of tornado or hurricane type. In a number of studies, it was demonstrated that for generation of swirling in the radial flow of gas no matter what method is used to obtain a discharge, either a local heating, or a vertical blowing [7-9]. It is the method of initiating the upward swirling currents of air that has been proved experimentally [10].

2. Materials and methods

In nature, one also observed [11] so called fire vortexes (fire tornadoes), which swirling has a different direction: negative in the Northern hemisphere and positive in the Southern hemisphere. In some experiments the researchers managed to create currents close to fire vortexes observed in nature, and without using the forced swirling at that [12]. In the article, a possibility in principle has been demonstrated of physical simulation of free concentrated fire vortexes in the laboratory conditions without using the swirling devices. It gives a description in detail of the experimental unit designed for generation of free fire vortexes and presents a detailed analysis of these vortexes origin specifics. A frame-by-frame recording enabled to obtain the information about the main parameters of the fire vortexes generation process and their integral characteristics.

During the study [13] on numerical simulation of thermal ascending swirling currents there was defined a specific feature of the moving gas stream behavior at the initial time of its formation during the local heating of the calculation domain lower plane. On the heating domain boundary the counter currents occur which have the opposite direction of the swirling. This inevitably leads to a generation of several local vortex formations with opposite orientation of rotation.

This fact, as well as the fire vortexes observed in nature, gave impetus to the formulation and solution of the specific initial-boundary problem. In the study [14], the solution of the gas dynamics equations system reproducing the interesting property of the ideal gas flows was rigorously mathematically described. It is shown, that if in the gas, being initially at rest in the gravity field beginning from some time moment, the surface of the vertical non-zero radius cylinder in the gas smoothly heats, then along with the radial movement of the gas due to the Coriolis force action its swirling also generates. The direction of this swirling is negative in the Northern hemisphere and positive in the Southern hemisphere, i.e. opposite to the direction of tornado and tropical hurricanes swirling.

In the present work the numerical simulation of the air swirling flow is run around the smoothly heated vertical cylinder in the conditions of gravity and Coriolis forces action.

For description of complex flows of compressible viscous heat-conducting gas, possessing the dissipative properties of viscosity and heat conductivity, a complete system of Navie-Stocks equations is used, which, being written in dimensionless variables taking into account the action of gravity and Coriolis forces in the vector form, has the following form [15]:

$$\begin{cases} \rho_t + \vec{V}\cdot\nabla\rho + \rho\, div\, \vec{V} = 0, \\ \vec{V}_t + (\vec{V}\cdot\nabla)\vec{V} + \dfrac{T}{\gamma\rho}\nabla\rho + \dfrac{1}{\gamma}\nabla T = \vec{g} - 2\vec{\Omega}\times\vec{V} + \dfrac{\mu_0}{\rho}\left[\dfrac{1}{4}\nabla(div\,\vec{V}) + \dfrac{3}{4}\Delta\vec{V}\right], \\ T_t + \vec{V}\cdot\nabla T + (\gamma-1)T\,div\,\vec{V} = \dfrac{\kappa_0}{\rho}\Delta T + \dfrac{\mu_0\gamma(\gamma-1)}{2\rho}\left\{\left[(u_x - v_y)^2 + \right.\right. \\ \left.\left. + (u_x - w_z)^2 + (v_y - w_z)^2\right] + \dfrac{3}{2}\left[(u_y + v_x)^2 + (u_z + w_x)^2 + (v_z + w_y)^2\right]\right\}, \end{cases} \quad (1)$$

where the constant values of dimensionless coefficients of viscosity and heat conductivity are as follows: $\mu_0 = 0.001$, $\kappa_0 \approx 1.46\mu_0$.

In system (1): t – time; x, y, z – Cartesian coordinates; ρ – gas density; $\vec{V} = (u, v, w)$ – gas velocity vector with projections on the respective Cartesian axes; T – gas temperature; $\vec{g} = (0, 0, -g)$ – gravitational vector; $\gamma = 1.4$ – polytropic index for air; $-2\vec{\Omega}\times\vec{V} = (av - bw, -au, bu)$ – acceleration vector of Coriolis forces, where $a = 2\Omega\sin\psi$, $b = 2\Omega\cos\psi$, $\Omega = |\vec{\Omega}|$; $\vec{\Omega}$ – angular velocity vector of the Earth rotation; ψ – width of point O – origin of Cartesian coordinate system $xyzO$, rotating together with the Earth.

As the initial conditions, at describing the given flow of compressible viscous heat-conducting gas in case of constant values of viscosity and heat conductivity coefficients, the functions are taken which set an exact solution of the system (1):

$$u = 0,\ v = 0,\ w = 0,\ T_0(z) = 1 - kz,\ k = \frac{l x_{00}}{T_{00}},\ l = 0.0065 \text{ К/м},\ x_{00} = 50 \text{ м},\ T_{00} = 288^o K \quad (2)$$

and

$$\rho_0(z) = (1 - kz)^{\nu - 1}; \quad \nu = \frac{\gamma g}{k} = const > 0. \quad (3)$$

The calculation domain presents a rectangular parallelepiped with sides lengths $x^0 = 1$, $y^0 = 1$ and $z^0 = 0.2$ along axes Ox, Oy and Oz, respectively.

On the lower and upper edges of the calculation domain there were "the conditions of non-percolation" (the vertical velocity component equals zero), and for the remaining gas dynamics parameters on all sides – "the continuity conditions". A gradual heating up to temperature of 300°C in the center of the vertical portion of the calculation domain for each discrete value of the height was simulated by the following functional relationship temperature vs. heating radius:

$$T(x, y, t) = 1 + T^*(1 - exp(-10t))\, exp\left(-\frac{(x-0.5)^2 + (y-0.5)^2}{r_0^2}\right), \qquad (4)$$

where T^* is an excess of the maximum dimensionless value of temperature over the scale unit value, r_0 is a dimensionless value of the effective radius of the heating domain.

The calculation domain is filled by three-dimensional grid of nodes of three families of planes crossing $x = x_i$, $y = y_j$, $z = z_k$, где $x_i = i \cdot \Delta x$, $y_j = j \cdot \Delta y$, $z_k = k \cdot \Delta z$, $0 \le i \le L$, $0 \le j \le M$, $0 \le k \le N$. The difference steps by three spatial variables $\Delta x = x^0 / L$, $\Delta y = y^0 / M$, $\Delta z = z^0 / N$. A calculation of three-dimensional non-stationary flow is run by the explicit finite-difference scheme by moving from one n–st time layer to the next $n+1$ time layer with a constant set increment in time Δt.

The calculations were run with the following input parameters: scale dimensional values of density, velocity, distance and time are equal respectively to:

$$\rho_{00} = 1.2928 \text{ kg/m}^3,\ u_{00} = 333 \text{ m/c},\ x_{00} = 50 \text{ m},\ t_{00} = x_{00} / u_{00} = 0.15 \text{ c}.$$

There are difference steps by three spatial variables $\Delta x = \Delta y = 0.01$, $\Delta z = 0.04$, and the increment in time $\Delta t = 0.001$. The numerical construction of solutions of the complete system of Navie-Stocks equations with the set initial and boundary conditions enables one to find by the explicit finite-difference scheme the values of the desired five functions in all nodes of the calculation domain at the arbitrary calculation step in time.

3. Results of calculations

In figures 1-2 the results are shown of calculation of thermal dynamics characteristics of the air flow, generated during heating the vertical cylinder domain, for the average value of height at given fixed moment of time. On axes Ox and Oy, the numbers of the calculation grid nodes are plotted. The temperature in the calculation process varies in agreement with the specified correlation (4). The peripheral dimensionless values remain equal to the scale unit value (the dimensional value of 15 degrees of centigrade). In the center, the temperature smoothly changes up to the maximum dimensionless value of 1.99 (300 degrees of centigrade). A similar distribution of temperature is assumed for all discrete values of height, thereby the vertical heating of the entire calculation domain is simulated.

Fig. 2 presents the distribution of gas density at a given fixed moment of time. The behavior of the density in the process of the heated gas flow formation is characterized by fluctuations of peripheral dimensionless values near the unit scale value. Moreover, its decline down to the minimum value of 0.65 (dimensional 0.84 kg/m^3) is observed in the central vertical domain of heating.

Fig.3,4 present the results of calculation of two components of velocity of the air flow occurred at heating for the central portion of the calculation domain at the same specific fixed moment of time.

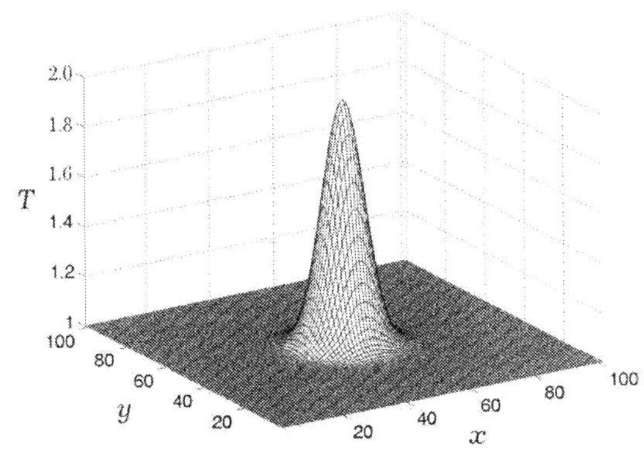

Figure 1. Temperature distribution

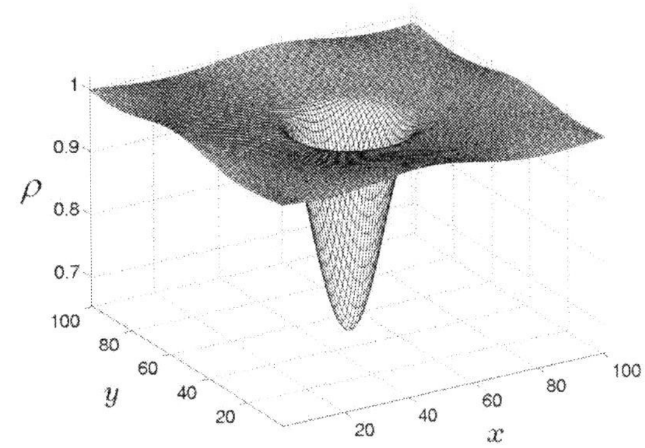

Figure 2. Density distribution

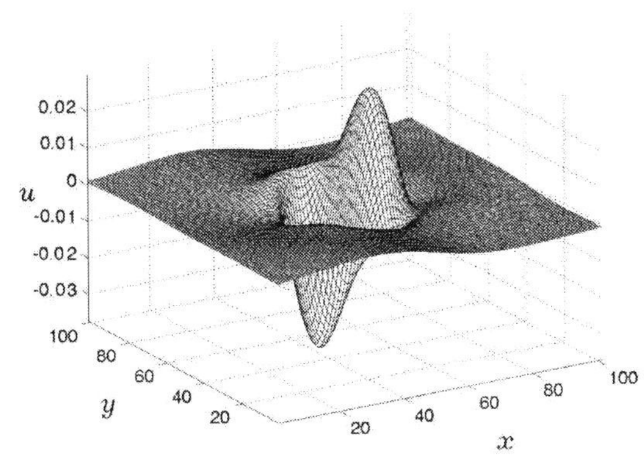

Figure 3. x - component of velocity

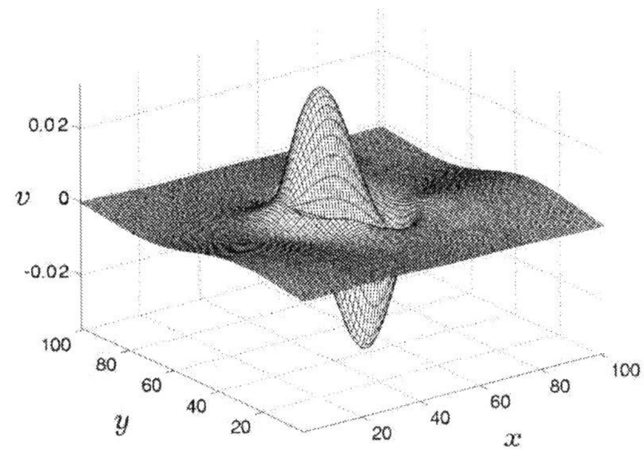

Figure 4. y - component of velocity

The behavior of the x - component of velocity of the gas particles can be characterized as follows. At the initial moment of time this velocity component, as well as the two other ones, are equal to zero. With the course of time of calculation in the heating domain two basic flows oriented in the opposite directions from the flow center generate, which are characterized by velocities growing in value and opposite by the sign. The modules of this velocity component increase up to the values of 0.03 (dimensional 9.99 m/s). Next to the basic local extrema of the x- component of the gas particles velocity with time there generate two domains with less and opposite by sign values of velocities which clearly point to a presence of the swirling movement of gas around the heating center in the negative direction (clockwise, if to look at the flow from above). The similar behavior is observed also for the second component of velocity presented in Fig.4. One can clearly see two domains of positive and negative velocities on the left and on the right of the basic extrema, which correspond to the presence of the swirling movement in the flow in the negative direction.

The complex character of the gas movement in such flow is reflected by lines of the moving gas particles current in Fig. 5, 6. These figures show several lines of the flow which are let out at the middle height of $z = 0.1$ from the central portion of the calculation domain. In this very domain, the intensive heating occurs. In Fig. 5, the side view is shown and in Fig. 6 the top view of the same lines of flow is illustrated. In the last two figures along axes Ox and Oy, the dimensionless distances are plotted.

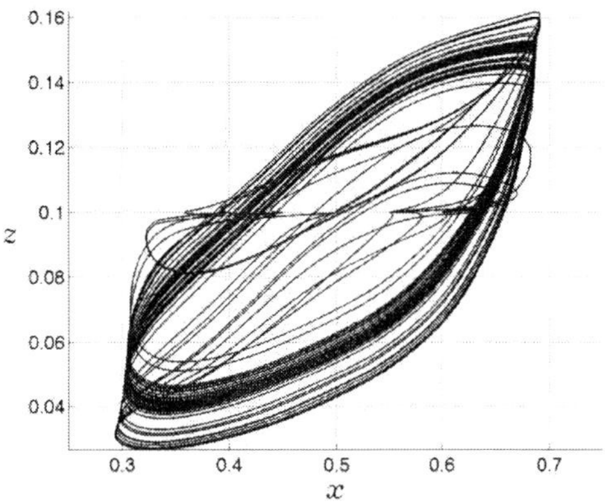

Figure 5. Lines of flow (side view)

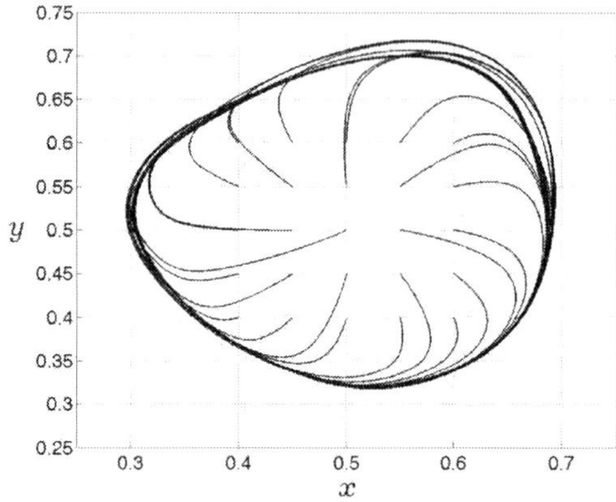

Figure 6. Lines of flow (top view)

In Fig. 5, one can see that the current lines gradually drift from the center in the vertical direction and the inclination of separate loops planes is clearly visible. The left-hand portion of the current lines corresponds to the negative values of vertical velocity, and the right-hand portion to the positives ones.

The top view of these current lines permits to assume that the current swirled in the negative direction, forms a certain cylindrical surface, which has an asymmetric form, intensively varying with time. This again points to the fact that in principle, a non-stationary and three-dimensional process of fire vortex formation is calculated. The average value of the cylindrical surface diameter coincides with the diameter of the vertical domain effective heating. The important point here is the calculated free character of the heat vortex. In the process of calculation, it freely moves along the horizontal plane, in some moments of time it disappears, then occurs anew.

4. Conclusion

The results of calculations run reveal a complex structure of the current generated during heating, its pronounced nonstationarity. The calculations enabled one to prove a generation of the negatively directed swilling of the air flow during heating of the vertical cylindrical domain, which agrees with the conclusions made in the studies referenced above.

References

[1] Varaksin A Yu, Romash M E and Kopeitsev V N 2011 *Tornado* (Moscow: Physmatlit)

[2] Nalivkin D V 1969 *Hurricanes, storms and tornados Geographical peculiarities and geological activities* (L: Nauka)

[3] Bautin S P 2008 *Tornado and Coriolis force* (Novosibirsk: Nauka)

[4] Bautin S P and Obukhov A G 2012 *Mathematical modeling of destructive atmospheric vortexes* (Novosibirsk: Nauka)

[5] Bautin S P, Krutova I Yu, Obukhov AG and Bautin K V 2013 *Destructive atmospheric vortexes: theorems, calculations, experiments* (Novosibirsk: Nauka)

[6] Bautin S P and Krutova I Yu 2012 Swirling of gas at smooth discharge in the conditions of gravity and Coriolis forces action *Thermal physics of high temperatures* **50** 473–475

[7] Abdubakova L V and Obukhov A G 2014 Numerical prediction of velocity characteristics of three-dimensional upward swirling flow of gas *Proceedings of Higher Education Institutions. Oil and Gas* **3** 88–94

[8] Obukhov A G and Abdubakova L V 2014 Numerical prediction of thermodynamic characteristics of three-dimensional upward swirling flow of gas *Proceedings of Higher Education Institutions. Oil and Gas* **7** 157–165

[9] Abdubakova L V and Obukhov A G 2014 Numerical prediction of thermodynamic parameters of the swirling gas flow, initiated by cold vertical blowing *Proceedings of Higher Education Institutions. Oil and Gas* **5** 57–62

[10] Bautin K V, Bautin S P and Makarov V N 2013 Experimental verification of possibility of creation of the air current swirled by the Coriolis force effect *Bulletin of UrGUPS* **2(18)** 27–33

[11] Nalivkin D V 1984 *Hurricanes* (Moscow: Nauka)

[12] Varaksin A Yu, Romash M E, Kopeitsev V N and Gorbachev M A 2014 On possible generation of fire vortexes without using forced swirling *Proceedings of Academy of Sciences* **456** 159–161

[13] Obukhov A G, Barannikova D D 2014 Specific features of gas flow at the initial stage of generation of heat ascending swirling current *Proceedings of Higher Education Institutions. Oil and Gas* **6** 65–70

[14] Bautin S P, Krutova I Yu and Obukhov A G 2015 Swirling of fire vortex when considering the gravity and Coriolis forces *Thermal physics of high temperatures* **53** 961–964

[15] Bautin S P 2009 *Characteristic Cauchy problem and its applications in gas dynamics* (Novosibirsk: Nauka)

IOP Conf. Series: Materials Science and Engineering **357** (2018) 012041 doi:10.1088/1757-899X/357/1/012041

Study of a method for reducing fuel consumption and the amount of specific emissions of harmful substances with exhaust gases of passenger cars when using the "climate control" system

L N Burakova, I A Anisimov, A D Burakova and O D Burakova

Industrial University of Tyumen, 38 Volodarskogo str., Tyumen, 625000 Russia

E-mail: tkcc@list.ru, burakova.1992@mail.ru, burakova.96@bk.ru

Abstract. The article deals with the issue of improving the fuel economy and environmental friendliness of motor vehicles which serve the administrative and management personnel of the oil and gas industry. It is established that fuel consumption and the amount of specific emissions of harmful substances with exhaust gases of cars when using the "climate control" system depend on the effective ambient temperature, the color of the opaque car body elements, the power of the car engine and the interior volume. However, the simplest controlled factor is the color of the opaque car body elements, which is characterized by the coefficient of light reflection. In the course of experimental studies, we established the dependences of a change in fuel consumption and a share of reducing emissions of harmful substances with exhaust gases of passenger cars with the "climate control" system on the coefficient of light reflection. A method has been developed to reduce fuel consumption and the amount of specific emissions of harmful substances with the exhaust gases of passenger cars using the "climate control" system, which involves painting the vehicle roof white and allows reducing fuel consumption by 5.5-10.3%, and the amount of specific emissions of harmful substances by 0.8-2.3%.

1. Introduction

At the present time in the field of transport services of administrative and management personnel of oil and gas industry the number of motor vehicles is increasing. Limited availability of natural energy sources causes an increase in prices for all types of fuel. This leads to the need to improve the fuel economy of vehicles [1-4]. However reducing fuel consumption increases the environmental friendliness of vehicles. According to the Transport Strategy of the Russian Federation until 2030, the ecological nature of the transport system plays an important role in the socio-economic development of the country, so reducing the harmful impact of transport on the environment is one of its main tasks [5]. GOST R 52033-2003 "Cars with gasoline engines. Emissions of pollutants with exhaust gases. Norms and methods of control in assessing the technical condition (Amend. No. 1 of 01.07.2012)" indicates values of the content of carbon monoxide and hydrocarbons in exhaust gases.

Thus, passenger cars serviced by the administrative and management personnel of the oil and gas complex should have minimum fuel consumption and specific emissions of harmful substances with exhaust gases [6-12].

Currently, there are some specific features of the passenger car operation: low density of the street and road network, rapid growth in the number of vehicles, separation of pedestrian and traffic flows at intersections. This reduces the throughput and increases the average idle time of vehicles. As a result of the analysis of previous studies, it has been established that for the last 30 years the idle time of cars with a running engine has increased by 18% and in modern conditions is 30 ... 35% [13,14,15,16].

When identifying features of the passenger car operation, one should also take into account the effect of high temperatures [17,18]. In the Russian Federation, the number of car-days in the operation of cars at temperatures above +20 ° C is more than 51%. Deviation of operating conditions from standard affects the motor vehicles by changing the thermal state of the elements of their design. In particular, in the warm period of the year, there is an increased temperature of the ambient air (above + 20 ° C) and an increase in the amount of solar radiation entering the Earth's surface. This causes an increase in the amount of heat entering the interior of the car, hence increasing the air temperature inside the vehicle. According to the norms of POT RM-008-99 "Inter-industry rules for labor protection" in the operation of industrial vehicles, the air temperature at the driver's workplace during the warm season should be 20-25°C or no more than 3°C above the average outside air temperature at 13.00 of the hottest month, but not more than 28°C. To create the required microclimate in the car, "climate control" systems are applied, which are the auxiliary equipment of the car. The inclusion of this in the work leads to an increase in the proportion of the car engine power spent on the work of the auxiliary equipment, which causes an increase in the consumption of fuel by the vehicle. The greatest impact is observed during the idle time of the car with the running engine; it is caused by the absence of streams of air cooling the interior during movement, and the minimum value of the effective capacity of the car engine.

The increase in fuel consumption of cars with the use of "climate control" during the idle time with the engine running increases the amount of exhaust gases [19-22]. Their composition is 95 ... 98% products of complete combustion, unused oxygen and nitrogen of the air, and 2 ... 5% - several hundred components that determine the degree of a harmful impact of the vehicle on the environment. However, in the products of complete combustion, there are chemical compounds that have a negative effect on the ecological state of the atmosphere, for example, carbon dioxide, which contributes to the development of the greenhouse effect. Reducing the concentration of this compound in the exhaust gas can be achieved by reducing the fuel consumption of cars.

Thus, the purpose of this study is to improve the fuel economy and environmental friendliness of passenger cars when using the "climate control" system during idle time with the engine running.

2. Methodology

The ecological nature of the transport system can be characterized as the ability of cars to carry out transport work or transportation of passengers with the minimum possible emissions of harmful substances in strictly defined conditions. Vedenyapin G.V., Govorushchenko N.Ya., Gutarevich Yu.F., Evtin P.V., Gelagin O.I., Zvono V.A., Magaril E.R. Reznik L.G. studied the problems of reducing the toxicity of exhaust gases and increasing fuel economy. Fuel consumption and the content of harmful substances in the exhaust gases of cars depend on many factors, in particular, on incorporating the auxiliary equipment and operating conditions.

An increase in the ambient air temperature leads to an increase in the number of heat gain to the car interior, therefore, the microclimate inside the vehicle changes [23]. Assessment of the impact of operating conditions on the microclimate in the vehicles was carried out by Palutin Yu.I., Kormin A.M., Semikin S.N., Shelyakin A.S. and others [24]. The given works proposed a set of natural and climatic factors of the conditions of vehicle operation to evaluate the effective temperature of the ambient air with the help of a complex index, the numerical value of which corresponds to the true temperature of the still and saturated air. However, the effective temperature of the ambient air is the determining factor only when heat is supplied to the car interior by heat transfer.

When assessing the heat balance of the vehicle interior, it should be taken into account that heat gain from external sources is realized not only by heat transfer, but also by thermal radiation. The amount of heat entering the passenger car interior through the opaque car body elements with the help of thermal radiation depends on the reflectivity of the paintwork. The color characterizes the reflectivity of the opaque body elements. For example, an absolutely white body has the coefficient of light reflection equal to 1, and absolutely black (soot) -0. However, it is impossible to achieve absolutely white or black colors when painting, therefore, as a rule, cars with a white color of the opaque body elements have the coefficient of light reflection equal to 0.95, and with a black color - 0.05. Thus, the amount of heat entering the interior of the car depends on the effective ambient temperature and the coefficient of light reflection of the opaque body elements and influences the microclimate inside the vehicle. An increase in the amount of heat input causes an increase in the temperature of the air inside the car; in order to reduce and maintain it, within the limits corresponding to comfort conditions, a "climate control" system is used. The main characteristics of its operation affecting the fuel consumption of a car are the cooling capacity and power consumption which depend not only on the amount of heat entering the interior of the car, but also on the interior volume. An increase in the interior volume, as well as the amount of incoming heat lead to an increase in the power consumption of the "climate control" system. It is established that the increase in the fuel consumption of a car when the "climate control" system is switched on is determined by an increase in the proportion of the power take-off of the automobile engine for the operation of this auxiliary equipment, which is characterized by the ratio of the cooled volume of the interior to the car engine power. This value is characterized by the specific climatic power parameter, which determines the part of the power of the automobile engine spent for cooling the unit volume of the car interior.

The fuel consumption of a passenger car when using the "climate control" system that affects the amount of specific emissions of harmful substances with exhaust gases depends on the effective ambient temperature, the coefficient of light reflection of the opaque car body elements and the specific climatic power parameter (the ratio of the passenger car interior volume to its engine power). However, the effective ambient temperature is an uncontrolled factor, and the coefficient of light reflection of the opaque car body elements is a controlled input indicator that can be quite simply changed by giving the car the color characterized by the maximum coefficient of light reflection. Thus, it is necessary to develop a method for reducing the increase in fuel consumption of a passenger car using the "climate control" system.

2.1. Equipment and devices used in studies

The experiment involved cars of three makes - Ford Focus, Toyota Corolla, and Mitsubishi Lacer. The light reflection coefficient was measured with a thermal imager. Illumination of the site was assessed using a luxmeter. The effective ambient temperature was taken into account by measuring the three factors: ambient air temperature and humidity - by the Assmann psychrometer and wind velocity - by the Fuss anemometer. The air temperature in the car was measured with mercury and digital thermometers. Fuel consumption was measured using a BT-ECU CAN adapter connected to the diagnostic socket of the car with an adapter and a computer program Check-Engine. To monitor the volume content of harmful substances in the exhaust gases of engines, a gas analyzer GIAM-29M-1 was used. The resulting volume concentrations were converted into specific CO and CH emissions.

2.2. Methodology of research

Passenger cars participating in the experiment consisted of a series of pairs of cars (black, silver, gray and white), one of which had an original roof color and the other was given a white color with the help of a vinyl film. Initially, the area of the opaque car body elements, in particular the roof, was determined.

Then, the vehicles were installed on a uniformly illuminated platform without shading. At the time of the maximum warm-up of the car with the "climate control" system switched off, the air temperature in the car was measured.

Then, the "climate control" system was switched on. When the constant thermal conditions were achieved, the air temperature in the car was measured and checked for compliance with its set value in the "climate control" system. The coefficient of light reflection for opaque body elements of different

IOP Conf. Series: Materials Science and Engineering **357** (2018) 012041 doi:10.1088/1757-899X/357/1/012041

colors was determined with an illumination of 60,000 lux (illumination at noon on a cloudless summer day) using a thermal imager.

Fuel consumption data of a passenger car equipped with a "climate control" system were measured during the idle time at a steady coolant temperature in the operating range of 80-100°C from the vehicle ECU using the BT-ECU CAN adapter and the Check- Engine computer program. Together with the measurement of fuel consumption, carbon dioxide and hydrocarbons in the exhaust gases were measured with a gas analyzer.

3. Results and discussion

Analyzing the state of the matter, it has been revealed that the factors that have the greatest effect on the fuel consumption of a passenger car using the "climate control" system are the effective ambient temperature, the power of the car engine, the interior volume and the color of the opaque car body elements (the light reflection coefficient). Of the presented factors only the power of the car engine, the interior volume and the color of the opaque car body elements are controllable, but the change in the first two factors is difficult. Thus, one of the ways to reduce fuel consumption of a car with the use of the "climate control" system is to impart a color that is characterized by the largest coefficient of light reflection (white color). A change in the increase in fuel consumption of a Ford Focus passenger car using the "climate control" system during the idle time with the engine running from the coefficient of light reflection of the opaque body elements is shown in Fig. 1.

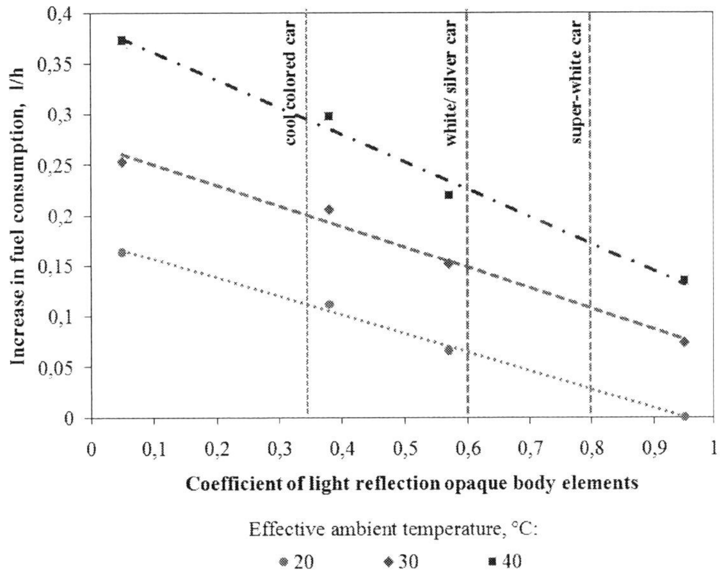

Figure 1. A change in the increase in fuel consumption of a Ford Focus passenger car using the "climate control" system during the idle time with the engine running from the coefficient of light reflection of the opaque body elements.

An increase in the fuel consumption of a car with the use of the "climate control" system during the idle time with the engine running becomes greater with an increase in the effective ambient temperature and a change in the color of the opaque body elements from light to dark characterized by a smaller coefficient of light reflection. This causes an increase in specific emissions of harmful substances with exhaust gases into the atmosphere, which is shown in Fig. 2 and 3.

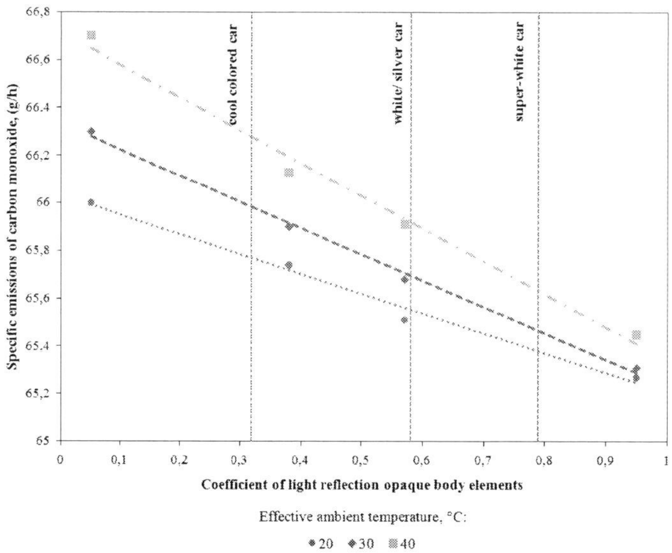

Figure 2. Reduction in specific emissions of carbon monoxide with exhaust gases of a passenger car when using the "climate control" system during the idle time with the engine running from the coefficient of light reflection of the opaque body elements.

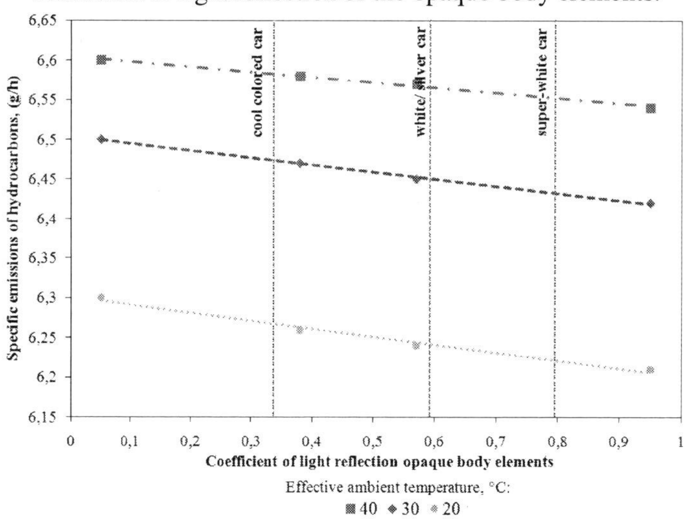

Figure 3. Reduction in specific emissions of hydrocarbons with exhaust gases of a passenger car when using the "climate control" system during the idle time with the engine running from the coefficient of light reflection of the opaque elements.

An increase in the coefficient of light reflection leads to a decrease in specific emissions of carbon monoxide and hydrocarbons with exhaust gases. Thus, in order to reduce fuel consumption and specific emissions of harmful substances with exhaust gases, opaque body elements must be painted in colors characterized by a high coefficient of light reflection. However, the amount of heat entering the interior of the car is also affected by the internal arrangement of the opaque body elements. The diagram of the internal arrangement of the opaque body elements is shown in Fig. 4.

Sequence of layers inside the roof structure: steel, noise and vibration insulating film, air, plastic. In the doors, there are additional layers of expanded polystyrene and air. Each of them is characterised by the coefficient of thermal conductivity which is the ability of a substance to conduct heat. Depending on this indicator, some of the heat will be absorbed by air interlayers and noise, vibration and thermal insulation. The absence of additional layers inside the roof causes an increased amount of heat entering the interior through this opaque body element.

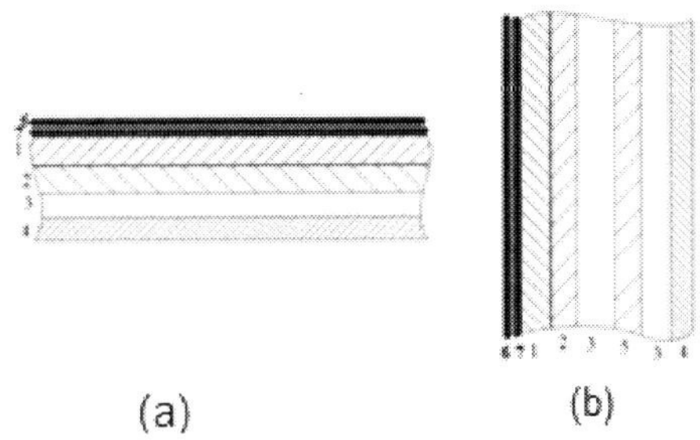

1 steel; 2 - noise, vibration and thermal insulation; 3 - air;
4 - plastic; 5 - foam polystyrene; 6 - paint; 7 - primer

Figure 4. Diagram of the internal arrangement of the (a) roof and (b) doors.

When studying the peculiarities of heat gain inside the car, it is established that the area of this structural element of a passenger car is between 45 and 52%. The results of measurements of the area of vehicle structural components are presented in Table 1.

Table 1. Areas of structural elements of cars.

	Ford Focus	Mitsubishi Lancer	Toyota Corolla
Roof area, m^2	1.69	1.66	1.56
Door area, m^2	1.50	1.57	1.70
Area of opaque body elements, m^2	3.19	3.23	3.26

The greatest amount of heat enters the interior through the roof, due to its internal features, which consists in the absence of two additional layers (foam polystyrene and air) in comparison with the structural design of doors, a large area, and a perpendicular location relative to solar radiation at the warmest time of day (from 10:00 to 16:00). Thus, a reduction in fuel consumption and specific emissions of harmful substances with exhaust gases of a passenger car using the "climate control" system can be achieved by painting the roof of a car white.

To prove the proposed assumption, calculation of the amount of heat entering the passenger car interior through a roof painted in various colors was initially performed. The results of the calculation are presented in Table 2.

Table 2. The amount of heat coming into the car interior, at an ambient temperature of +30°C and a temperature inside the car +24°C (due to heat transfer and heat radiation).

Car	Roof area, m^2	Amount of heat entering the passenger car interior, kW/day			
		white (ρ=0.95)	silver (ρ=0.57)	silver-gray (ρ=0.38)	black (ρ=0.05)
Ford Focus	1.69	0.51	3.41	6.04	7.39
Toyota Corolla	1.56	0.48	3.16	5.59	6.83
Mitsubishi Lancer	1.66	0.51	3.36	5.94	7.26

As a result of the analysis of the received data it is established that the roof of the car equipped with the "climate control" system should be white to reduce fuel consumption and the amount of specific emissions of harmful substances with the exhaust gases. In the course of experimental studies, results are obtained that attest to the feasibility of introducing the developed method. Experimental data on the effect of the white color of the roof on the change in fuel consumption and the amount of specific emissions of harmful substances with the exhaust gases are presented in Fig. 5 and 6.

Application of the developed method allowed reducing the fuel consumption of a passenger car using the "climate control" system by 5.5-10.3%. Making the roof of a silver car white led to the achievement of fuel consumption corresponding to that of a white car.

Reducing the fuel consumption of a car with the introduction of the developed method allowed reducing the amount of specific carbon monoxide emissions by 0.37-1.13% and hydrocarbons by 0.47-1.08% with the exhaust gases.

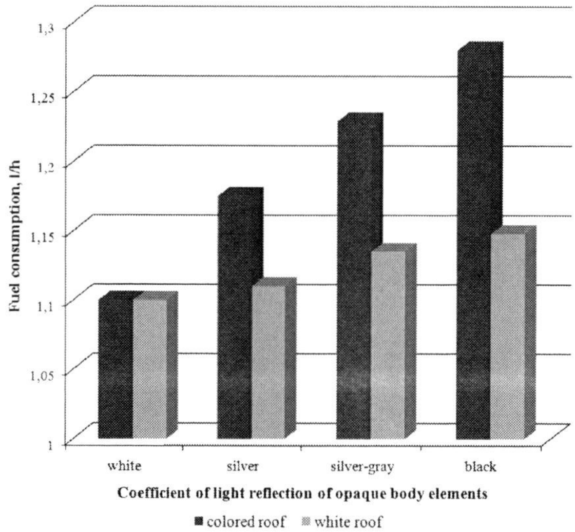

Figure 5. Change in the fuel consumption of a car equipped with the "climate control" system with the introduction of the developed method.

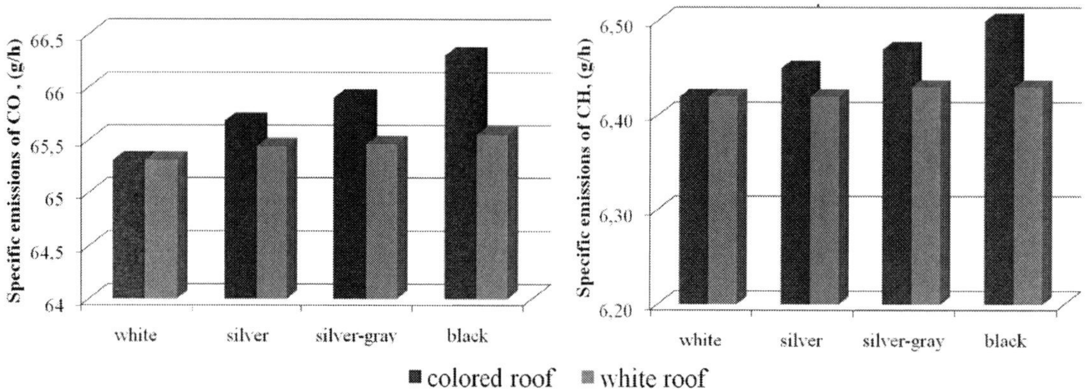

Figure 6. Decrease in specific emissions of carbon monoxide and hydrocarbons with exhaust gases of cars equipped with the "climate control" system.

Thus, painting the roof of a vehicle white will reduce the fuel consumption of a car by 5.5-10.3% and the total amount of specific emissions of harmful substances with the exhaust gases by 0.8-2.3%.

4. Conclusions

In the course of experimental studies, it has been established that when the color of the opaque body elements changes from light to dark (reducing the coefficient of light reflection), fuel consumption and concentration of carbon monoxide and hydrocarbons in the exhaust gases of passenger cars increase when using the "climate control" system.

The area of the opaque body elements and their internal arrangement are analyzed. This made it possible to reveal that the greatest amount of heat enters the interior through the roof, due to its internal features, which consists in the absence of two additional layers (foam polystyrene and air) in comparison with the structural design of doors, a large area, and a perpendicular location relative to solar radiation at the warmest time of day (from 10:00 to 16:00). Based on the results of the preliminary stage of the experiment, a method has been developed for reducing the fuel consumption of a passenger car using the "climate control" system. It is to give a white color to the roof of the vehicle.

According to the obtained results of the experiment, it is established that the application of the developed method allows reducing the fuel consumption of a car by 5.5-10.3% and the total amount of specific emissions of harmful substances with the exhaust gases by 0.8-2.3%.

References

[1] Anisimov I, Magaril E, Magaril R, Chikishev E, Chainikov D, Gavaev A, ErtmanS, Ertman Yu and Ivanov A 2016 Improving vehicle adaptability to the operating conditions of "smart" cities in the northern regions *E3S Web of Conf. Rev.* **6**

[2] Panuccio P, Amodeo L, D'agostino P, Lamari D and Scattarreggia T 2015 Urban Regeneration and smart sity according to EU strategies: an urban distribution center in city ogistics *WIT Transactions on The Built Environment* **146**

[3] Monfaredzadeh T and Berardi U 2014 How can cities lead the way towards a sustainable, competitve and smart future? *WIT Transactions on Ecology and the Environment* **191** 1063-74

[4] Russo F, Rindone C and Panuccio P 2014 The process of smart city definition at an EU level *WIT Transactions on Ecology and the Environment* **191** 979-989

[5] Government of the Russian Federation 2014 *Transport Strategy of th Russian Federation until 2030* (Moscow: Government of the Russian Federation)

[6] Chikishev E, Chainikov D and Anisimov I 2017 Increasing the use of natural gas on motor transport by an efficient location of the NGV RS *IOP Conf. Ser.: Earth Environ. Sci. Rev.* **50**

[7] Chikishev E, Chikisheva A, Anisimov I and Chainikov D 2017 Natural gas use on mimbuses, engaged in the carriage of passengers and baggage on the regular routes, as a measure for decrease I harmful environment effects *IOP Conf. Ser.: Earth Environ. Sci. Rev.* **50**

[8] Chainikov D, Chikishev E, Anisimov I and Gavaev A 2016 Influence temperature on the CO_2 emitted with exhaust gases of gasoline vehicles *IOP Conf. Ser.: Mater. Sci. Eng. Rev.* **142**

[9] Chikishev E, Ivanov A, Anisimov I and Chainikov D 2016 Prospects of and problems in using natural gas for motor transport in russia *IOP Conf. Ser.: Mater. Sci. Eng. Rev.* **142**

[10] Ertman J, Ertman S, Anisimov I, Chainikov D and Chikishev Y 2015 Complex characteristic estimation of vehicle adaptability for low temperature operation conditions *Research Journal of Pharmaceutical, Biological and Chemical Sciences* **6** 1761-70

[11] Anisimov I, Ivanov A, Chekishev E, Chainikov D and Reznik L 2014 Assesment of gas cylinder vehicles adaptabilty for operation at low ambient temperature conditions *WIT Transactions on Ecology and the Enviroment* **190** 685-695

[12] Ertman S, Ertman J and Zaharov D 2016 Raising of operating a motor vehicle effects on enviroment in winter *Innovative Technologies in Engineering Rev.* **142**

[13] Cloke J and Layfield R 1996 The environmental impacts of traffic management schemes *WIT Transactions on The Built Environment* **26** 11

[14] Ntzeremes P, Kirytopoulos K and Tatsiopoulos I 2016 Management of infrastructure's safety under the influence of normative provisions *WIT Transactions on The Built Environment* **164**

51-59
[15] Langeland A 2015 Sustainable transport in cities: learnin from best and worst practice? *WIT Transactions on The Built Environment* **146** 93-103
[16] Langer D and Mattrish G Future trends and driving forces in transport and logistics *WIT Transactions on the Built Environment* **75** 10
[17] Galderisi A 2014 Adapting cities for a chenging climate: an integrated approach for sustainable urban development *WIT Transactions on Ecology and the Environment* **191** 549-560
[18] Alisov B and Poltaraus B 1974 *Climatology* (Moscow: Publishing House of Moscow University)
[19] Jacyna M and Merkisz J 2015 Estimations in real urban traffic conditions s a deteminant of shaping sustainable urban development *WIT Transactions on The Built Environment* **146** 219-231
[20] Gong R and Waring P 1996 Investigation of emission behaviour in an urban driving cycle with a remote sensing system *WIT Transactions on The Built Environment* **26** 10
[21] Tong H and Hung W 1999 Vehicular modal emission and fuel consumption factors in Hong Kong *WIT Transactions on Ecology and the Environment* **37** 8
[22] Kazopoulo M, El-Fadel M and Kaysi I 2004 Emissions Standards Development for an I/M program *WIT Transactions on Ecology and the Environment* **74** 10
[23] Bogoslovski V 1985 *Air conditioning and refrigeration* (Moscow: Stroiizdat)
[24] Ertman S, Ertman J and Zakharov D 2016 Raising of operating a motor vehicle effects on environment in winter *Innovative Technologies in Engineering Rev.* **142**

IOP Conf. Series: Materials Science and Engineering **357** (2018) 012042 doi:10.1088/1757-899X/357/1/012042

Analysis of technical university information system

N A Savelyev and M A Boyarkin

Industrial University of Tyumen, 38 Volodarskogo St., Tyumen, 625000, Russia

E-mail: bojarkinma@tyuiu.ru.

Abstract. The paper covers a set and interaction of the existing higher education institution automated control systems in ф state budgetary educational institution of higher professional education "Industrial University of Tyumen ". A structural interaction of the existing systems and their functions has been analyzed which has become a basis for identification of a number of system-related and local (related to separate modules) drawbacks of the university activities automation. The authors suggested a new structure of the automated control system, consisting of three major subsystems: management support; training and methodology support; distance and supplementary education support. Functionality for each subsystem has been defined in accordance with the educational institution automation requirements. The suggested structure of the ACS will solve the challenges facing the university during reorganization and optimization of the processes of management of the institution activities as a whole.

1. Introduction

The first two stages of integrational transformation at the level of the RF Ministry of Education had been accomplished by the beginning of 2016, which resulted in the establishment of 10 Federal and 29 national research universities. The third, decisive stage of the reorganization of higher education institutions has come: an establishment of regional base universities when a pool of the state leading higher education institutions will be finally formed. They will assume major responsibility for training highly qualified specialists for various branches of national economy of the Russian Federation [1].

IT penetration into basic management processes and modernization of the university management system are among the key directions of the base regional technical universities development.

One of the factors affecting the support of educational, administrative and financial activities of a technical university is a creation and development of common information space by means of modern information technologies. Because of the requirements imposed on base regional institutions in terms of educational process management and control, the old methods of record keeping do not ensure a possibility of effective managerial decision-making, which would be aimed primarily at improving the quality of specialists training [2].

Thus, in today's dynamic environment managing a technical university is impossible without an open and distributed automated control system [3].

In this regard, the task of development of the university automated control system is urgent and will allow solving the following tasks:

- record of all the university internal processes;
- analysis of key performance indicators of the institution;
- optimization of management decisions at all management levels;
- improvement of the learning service quality as the educational institution target function.

2. Materials and methods

Currently, in the Industrial University of Tyumen (IUT), there are a number of systems to automate the workflows of the institution. The automated control system of the university is presented in Fig. 1:

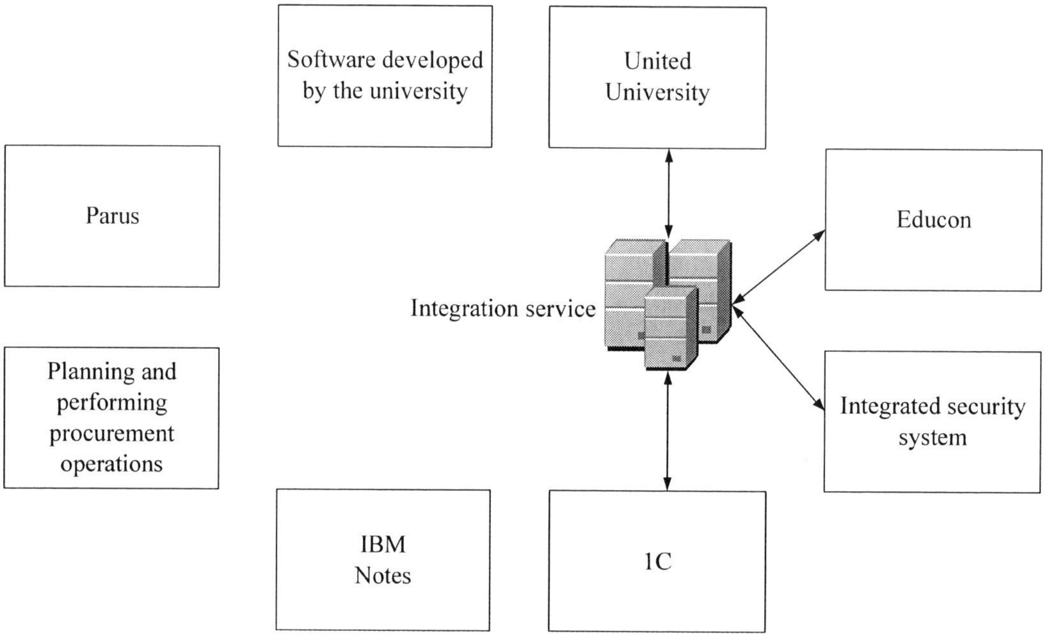

Figure 1. Current automation of the university

Thus, the university automated information system has eight distributed systems, four of which are combined by means of the integration service and four ones are not interrelated. In general, the presented system solves the following tasks.

Currently the UnitedUniversity system allows solving a number of problems related to the automation of academic activities, namely:

a) user authentication to information about students, depending on access rights and its backup [4];

b) recording and keeping personal files of applicants, students and faculty [5, 6];

c) registering training areas / specialties of higher education with the identification of education levels (masters, bachelors, specialists, graduate students), secondary vocational training specialties throughout the university in general and structural subdivisions in particular, as well as in the context of generations of state educational standards. The register includes 160 training areas / specialties (including: 103 of higher education and 57 of secondary vocational training), structured in accordance with 3 generations of educational standards, by 8 basic structural units of the university and 5 branches, as well as by 35 graduate departments and 8 secondary vocational departments;

d) keeping a directory of state education standards (32 SES, 13 FSES of higher education, 58 FSES of higher professional education and 57 FSES of secondary vocational training are in use), which includes:

- the structure of the state educational standard (components, indices, and disciplines);
- the complexity (in credit units and hours);
- a list of competencies and knowledge, skills and abilities;
- conditions of implementation of the state educational standard;
- restrictions and criteria for checking the compliance of the curricula to the standards;

e) curricula import and export in software format GosInsp (Goins) (Federal State Budgetary Institution "Information and Methodology Centre for Assessment of Educational Organizations");

IOP Conf. Series: Materials Science and Engineering **357** (2018) 012042 doi:10.1088/1757-899X/357/1/012042

f) making statistical reports on applicants and students;

g) making analytical reports through integration with Pentaho Analysis, allowing to obtain an operational analysis based on multidimensional arrays of data (OLAP cubes);

h) automation of the university admission board;

i) keeping curricula and calculation of academic and study loads;

j) automated formation of academic calendars for all training areas/ specialties for the entire academic year based on the academic calendars in the curricula;

k) monitoring the students'progress [6].

The Educon system solves the following major tasks:

a) objectified assessment of students' knowledge;

b) distance learning support;

c) students' portfolio formation and maintenance.

The university integrated security system solves the following tasks:

a) protection of personal details by means of specialized software in accordance with the requirements of the Ministry of Education and State Commission for Academic Degrees and Titles;

b) protection of information systems and the university PC users from outside threats;

c) arrangement of the access control and management system.

A set of 1C software products includes the following subsystems:

a) "1C: Budgetary institution wages and HR, edition 1.0";

b) "1C: State institution accounting".

Block "1C: Budgetary institution wages and HR" is intended for HR record keeping and the university staff wages accounting. The system also allows making analytical and statistical reports on the university staff.

Block "1C: State institution accounting" is used for accounting automation, taking into account the budgetary organization specificity.

Software product IBM Notes allows solving the following tasks:

a) arrangement of electronic document flow within the university;

b) setting of execution of instructions;

c) arrangement and control of filling university service requests in electronic form;

d) keeping the university official documents in electronic form.

The university procurement activities planning and conducting include 2 subsystems:

a) the subsystem of procurement planning in accordance with the RF legislation. It allows automating the entire procurement activity from organizations registration and accreditation to the procedure summing up and conclusion of contracts.

b) the subsystem of keeping records of the contracts is used for agreements and state contracts records concluded by the university and their execution control,.

The system "Parus" is designed for students' scholarships calculation.

The software developed by the university is intended for calculation of income from teaching.

After analyzing the university ACS structure and functions, a number of drawbacks of both the system-based (general) and local nature have been identified, namely:

- lack of common information space as a basis of effective automated control system of a complex socio-economic facility (like a base technical university);
- lack of monitoring the system key performance indicators, and as a consequence, the impossibility of assessing the ACS effectiveness in general;
- lack of a decision support system (DSS) based on actual (operational) and objectified data from common information space in real time;
- lack of automated university property accounting, including the warehouse inventory;
- lack of synchronization between the UnitedUniversity system, the students scholarships accounting module and the scholarships calculation system;
- lack of a hostel vacancies distribution module synchronized with 1C and UnitedUniversity;

IOP Conf. Series: Materials Science and Engineering **357** (2018) 012042 doi:10.1088/1757-899X/357/1/012042

– lack of a modern electronic document flow system with the possibility of using electronic digital signature;
– lack of supplementary education activity automation;
– lack of a system of automated technical support, including the one by means of the class "helpdesk"system;
– lack of automated administration of user working environment (Active Directory).

It also requires developing the university network infrastructure through hardware and software modernization.

3. Results and Discussion

After the existing management system modernization in accordance with the identified drawbacks, it is proposed to divide the ACS into three components:

1. Training and research support system – TRSS.
2. Management support system – MSS.
3. Distance and supplementary education support system – DSESS.

The structure of the automated control system is presented in Fig.2.

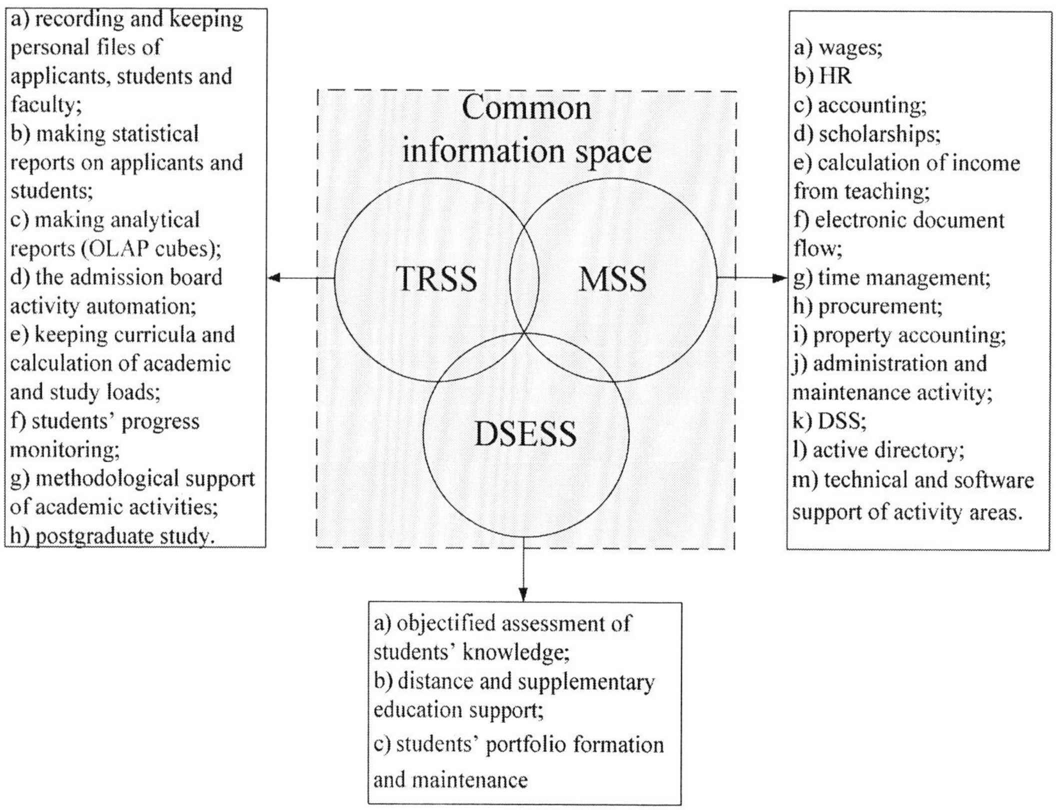

Figure 2. A structure of the automated control system

Thus, such arrangement of the automated control system will allow solving the task of support of all the activity areas characteristic of educational institutions. Information technology environment is presented in Fig.3:

IOP Conf. Series: Materials Science and Engineering **357** (2018) 012042 doi:10.1088/1757-899X/357/1/012042

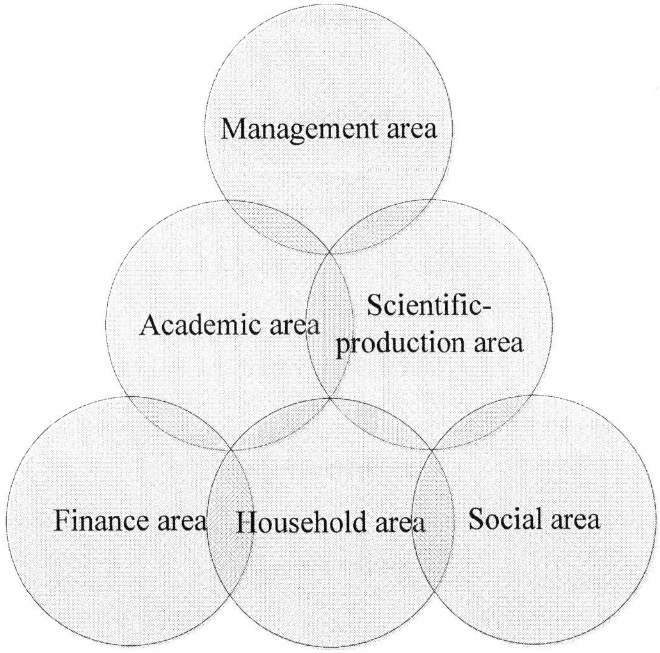

Figure 3. The university information technology environment

There are three basic approaches, within which automation and computerization of higher education institutions major activities are accomplished [7].

The first approach is based on the introduction of commercial ERP systems which are created by third-party firms that undertake the development, introduction and maintenance of automation systems by the results of inspection and optimization of customers' business processes.

The second approach presupposes a formation of automated control systems using universities' own resources. This creates an evolving system, adapted to the peculiarities of a university and providing automation of all its major departments. In this case, the university acts as both a customer and a contractor.

The third approach is based on a mixed use of customized commercial systems and software solutions developed by the university IT services. The scope of third-party firms includes mainly financial and maintenance departments. They use the tools that have approved themselves in the field of automation, for example, "1C", "Parus" and several others. Internal solutions typically relate to academic process management.

The choice of approach is an important task, which largely determines the effectiveness of a system under development.

4. Conclusion

The proposed ACS structure (Fig.2) will cover all the aspects of the university activity, which is a complex hierarchical system with a significant number of structural elements and a large number of connections between them. The proposed subsystems are directly or indirectly interlinked and form common information space, based on databases which accumulate the information about all the areas of university activity and a complex of managerial decision-making support models.

References

[1] RF State Programme «Development of Education» for the period of 2013-2020. http://минобрнауки.рф

[2] RF Ministry of Education and Science project «Base Universities». Access mode: http://опорныйуниверситет.рф
[3] Men'shikova E. Khazanov O, Verkhovskaya M 2015 Development of the Information System Structure for Continuous Improvement of QMS in an Educational Institution *SHS Web of Conferences* **28**
[4] Kiparisova A I, Azhmuhamedov I M 2016 Providing access to information systems of higher education in the case of loss of key information *International Research Journal* 23-25
[5] Kuzyakov O N, Akhmadulin R K 2015 Information systems in the academic process of the Department of Cybernetic *Proceedings of the VI All-Russian scientific and technical conference with international participation* 209-214
[6] Akhmadulin R K 2012 Automation of monitoring of classes attendance by students and of making up academic deficiencies in the Institute of Geology and Oil-Gas Production *Proceedings of the Eighth all-Russian scientific and technical conference* 257-259
[7] Glukhikh I N, Pryakhina E N 2015 Innovative educational environment of universities and competencies of graduates-innovators *Historical and social-educational ideas* **7(4)** 133-136

IOP Conf. Series: Materials Science and Engineering **357** (2018) 012043 doi:10.1088/1757-899X/357/1/012043

Formation flow rate control method in multi-layer production

H N Muzipov[1], **R K Akhmadulin**[1] **and L N Bakanovskaya**[1]

[1] Industrial University of Tyumen, 38 Volodarskogo St., Tyumen, 625000, Russia

E-mail: halim46@mail.ru

Abstract: The article describes a method of flow rate control of separate formations in multi-layer production by noises frequency response (FR). The noise FR is converted into electrical signals scaled in proportion to the flow rates using secondary facilities. The pump noise is suggested to be reduced with the quarter-wave acoustic resonator working as an acoustic filter.

1. Introduction

To control oil production from separate formations, the following instrumentation is used: flow meters, pressure gauges, thermometers [1-5].

The disadvantage of this method is that tubes, packers, downhole cameras, regulating choke restricting devices, measuring instruments, conductive cables, disconnectors, telescopic joints and other equipment are run into the hole. As a result, because of this set of equipment, the well flow rate measurement unit has low reliability and control efficiency.

2. Research

A more reliable method is frequency response (FR) measurement of noise generated by turbulent oil flows through perforations with further conversion into electrical signals proportional to formation flow rate [6].

Because the geometry of the perforations may not be identical, so the noise frequency response will also differ. This is similar to unique fingerprints of people. A change in formation flow rate causes a change in the noise frequency response.

For each oil formation, there is perfomed a measurement of amplitude-frequency spectra of the noises generated by every formation, turbulent vortexes of oil-water-gas mixtures coming from the perforations in a production well. Noise amplitude-frequency spectra measurement is implemented with recording instruments, including a hydrophone (run on a cable into a hole to the level of an oil formation), a frequency meter and an electronic unit for the information processing (installed at the wellhead).

Next, a program for the electronic unit is developed enabling processing of information about the flow rates according to the amplitude-frequency spectrra of the noises generated by a separate oil formation. This method allows a concurrent flow rate control in multi-layer oil production using a single lift.

It should be noted that the accuracy of flow rate control by means of frequency spectra is affected by a pump noise. To reduce the impact of the pump noise, it is necessary to suppress it by means of quarter-wave resonators, i.e. acoustic filters.

An acoustic filter is an acoustic resonator made in the form of a vessel communicating with the external environment through a small hole or tube. Its characteristic feature is an ability to perform low-frequency self-oscillation, with wavelength much greater than the size of the resonator. According to Helmholtz and Rayleigh theory, an acoustic resonator is regarded as an oscillation system with one degree of freedom.

The kinetic energy is considered to be concentrated in a layer, moving in the tube, called the throat of an acoustic resonator, like a hard rod; and the potential energy is associated with elastic deformation of the medium contained in the volume.

The acoustic resonator natural frequency does not depend on the shape of the vessel and the pipe cross section and is expressed by the formula [7]:

$$f = c/2\pi\times(F/Vh)^{1/2},$$

where:

- c is the speed of sound in the medium (drilling mud), m/s;
- V is the resonator volume, m^3;
- F and h are the cross-sectional area, m^2, and the tube length, m, respectively.

This property of an acoustic quarter-wave resonator, i.e. reflection and absorption of energy at a given frequency from the sound vibration spectrum can be used, first, to reduce noises if their frequency is in the information signal frequency range, thereby to increase the signal/noise ratio; for instance using an acoustic resonator for absorbing the noises created by electrical submersible pumps (ESPs).

Knowing the pump noise frequency, which is to be measured at the stand of the manufacturer, it is possible to determine the size of an acoustic quarter-wave resonator.

The property of a quarter-wave acoustic resonator to suppress low-frequency noises and convert them into ultrasound is implemented in papers [8-11].

Quarter-wave acoustic resonators are placed in the tubing body above an electric submersible pump (ESP). This arrangement allows the resonators to suppress the pump noise.

In Fig. 1 there is a diagram of a production oil well, with a hydrophone and a quarter wave resonator in the tubing above the ESP.

From figure 1 it can be concluded that the ESP operation in a well is accompanied by the noises of productive formations and the very ESP. In case of marginal wells, occasional pumping is possible, i.e., the ESP will be periodically switched off and, accordingly, noise $S_{12}(f)$ will disappear for this period. At the same time, noises $S_{14}(f)$ and $S_{15}(f)$ will propagate in the well, creating the effect of well operation, i.e. in the absence of flow, the secondary equipment will register the flow. To prevent this phenomenon, it is proposed to use an acoustic quarter-wave resonator, designed to suppress the ESP noise to provide an accurate control of flow rates, and to create start and stop commands for the flow rate control. During the ESP operation there is an acoustic mark in the form of the frequency absorbed by an acoustic quarter-wave resonator; this mark will serve as a start signal for the secondary equipment to control the flow rate. When the ESP is switched off, in the noise spectrum there will be an acoustic mark in the form of the frequency recovered by an acoustic quarter-wave resonator, which will serve as a stop signal for the secondary equipment to control the flow rate.

The second variant to start and stop the flow-rate control can be the ESP operation control signals. The on/off control signals for the ESP may serve as on/off control signals for the flow rate control by the secondary equipment.

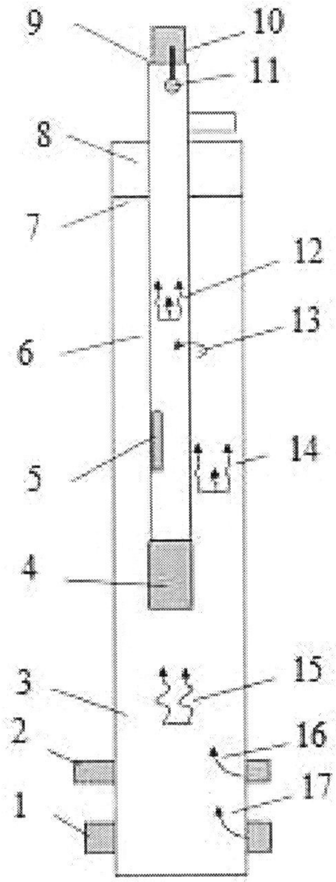

Figure 1. Production oil well diagram.
1 – oil formation P1; 2 –oil formation P2; 3 – oil-water-gas mixture; 4 – ESP; 5 – quarter-wave resonators; 6 – tubing; 7 – dynamic level; 8 – gas; 9 – casing; 10 – lubricator with a hydrophone; 11 – hydrophone; 12 – noise $S_{18}(f)$ propagating in the tubing and making a sum of noise $S_9(f)$ and $S_{17}(f)$; 13 – noise transition from the well into the liquid filling the tubing; 14 – noise $S_{17}(f)$ propagating in the production well and consisting of four amplitude-frequency spectra of noises: noise $S_{14}(f)$ is proportional to the formation P1 oil-water-gas mixture flow rate, noise $S_{15}(f)$ is proportional to the P2 oil-water-gas mixture flow rate, noise $S_{11}(f)$ generated into the well from the tubing and noise $S_{12}(f)$ generated by the ESP with quarter-wave resonators above the ESP; 15 – noise $S_16(f)$ propagating in the production well and comprising two amplitude-frequency spectra of noises $S_{14}(f)$ and $S_{15}(f)$; 16 –oil-water-gas mixture turbulent flows with varying characteristics when P2 oil formation flow rate varies; 17 – oil-water-gas mixture turbulent flows with varying characteristics when P1 oil formation flow rate varies

Figure 2 shows the amplitude-frequency spectra of noises $S_{13}(f)$ and $S_{14}(f)$ propagating from two oil formations to the recording instruments. There is an additional ESP noise source on the propagation path.

IOP Conf. Series: Materials Science and Engineering **357** (2018) 012043 doi:10.1088/1757-899X/357/1/012043

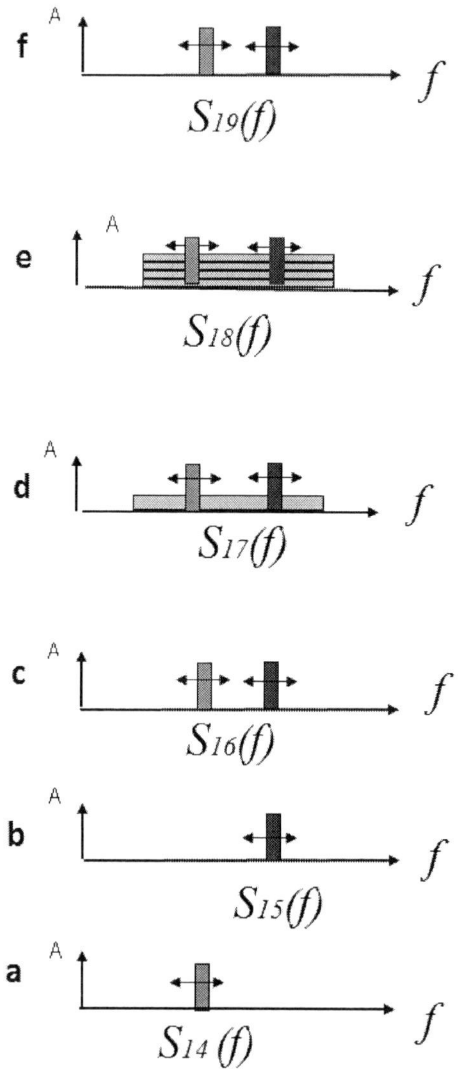

Figure 2. Amplitude-frequency spectra of noises S_{13}(f) and S_{14}(f) propagating from two oil formations: a) shows the frequency spectrum of noise S_{14}(f) generated by perforations of oil formation P1; b) frequency spectrum of noise S_{15}(f) generated by perforations of oil formation P2; c) frequency spectrum of noise S_{16}(f) propagating in the well and consisting of two frequency spectra of noises S_{14}(f) and S_{15}(f); d) the total frequency spectrum of noise S_{17}(f), consisting of noises S_{14}(f) and S_{15}(f) generated by perforations of oil formations P1 and P2, with the noise in the form of noise S_{13}(f) generated by tubing and ESP; e) the total frequency spectrum of noise S_{18}(f), consisting of noises S_9(f) and S_{17}(f), which in turn consist of noises S_{11}(f) S_{12}(f) S_{14}(f) and S_{15}(f); f) frequency spectrum of noise S_{19}(f), received by the hydrophone and filtered by the recording instument (frequency spectra of noises S_{14}(f) and S_{15}(f) of the waves are proportional to the flow rates of the respective formations, with an appropriate calibration).

Amplitude-frequency spectra of noises $S_{13}(f)$ and $S_{14}(f)$ propagating from oil formations associated with fluid flow heterogeneities and elastic oscillations of the walls of the channel, in which the flow moves. Heterogenities of the filtration flow occur due to vortex formation on the rough walls of the reservoir rock capillaries. The hydrodynamic noise acoustic properties for these conditions are also related to the channels configuration and walls apart fom the flow heterogrinities velocity and structure.

The noise amplitude-frequency spectra are also affected by additional flows that are associated with fluid motion in perforations $S_{15}(f)$ and damaged tubes at the annulus in cross-flows and in cracks in the cement stone and cavities. Acoustic noises in the well for these additional situations have intermediate characteristics between the radiation parameters of turbulent flows in pipes and filtration flows in reservoirs.

The given experimental studies form a basis for a method of downhole spectral noise logging [12]. Based on the experimental studies of hydrodynamic sound formation in the well, a flow noise spectrum can be divided into three frequency bands. Fluid flow in pipes (casing, tubing) generates noise in the frequency band up to 100 Hz. With the flow in cavernous and fissured environments, the noise spectrum has maximum in the range of 100 to 2000 Hz. The filtration flow noise spectrum in reservoirs is in the band of 2 to 20 kHz.

The noise in the form of ESP noise $S_{13}(f)$ consists of mechanical, electromagnetic and even aerodynamic noises and vibrations, which are closely related to each other.

3. Conclusion

The role of acoustic methods of study, control and diagnostics are well known. These methods allow one to obtain huge amounts of information about the state of materials and structures that are difficult to perform with sensors.

Acoustics also provides wide opportunities for study of the properties of materials, substances, and structures.

The method of well layers flow rate control in multilayer production under study is practically analoguos to spectral noise logging. In general case, the spectral noise logging is aimed at determining the technical condition of wells and their operational parameters.

The method of flow rate control of separate formations in multi-layer production by formation noise frequency response (FR) is a promising energy-saving technology, which allows one to determine the flow rate of several productive formations in oil wells and thereby to cut the costs of oil companies without major capital investment. Using this method, it is possible to determine the efficient formation zones, as well as to determine the optimal modes of wells operation that will increase the efficiency of production.

References

[1] Garipov O M 2010 *Patent RU № 2309246. Garipov well unit to study multi-layer wells in dual operation*

[2] Garipov O M 2011 Express method of flowrate calculation by variable-pressure drop in wells with dual water injection into several production zones *Neftyanoye khozyaistvo* **5** 108-110

[3] Garipov O M 2009 General trends in the development of high-tech services during the development, installation and maintenance of multi-packer systems for dual operation *Neftyanoye khozyaistvo* **9** 58-61

[4] Baryshnikov A V, Kofanov O A, Galeev D R, Garipov O M and Mustafin E L 2010 Experience of application and prospects of development of the dual injection technology in the Yuzhno-Priobskaya licensed territory *Neftyanoye khozyaistvo* **12** 66-68

[5] Davitashvili G I and Garipof O M 2006 Introduction of dual injection at Priobskoye field *Neftyanoye khozyaistvo* **12** 73-75
[6] Savinykh Yu A 2010 *Patent №2382196. Method of permanent monitoring separate oil formations flow rate in multi-layer production*
[7] Ed M Heckl and H A Muller 1980 *Technical Acoustic Guide (transl. from Germ.)* (Leningrad: Sudostroyeniye) 440
[8] Musipov H N and Savinykh Yu A 2004 New technology to enhance the production wells performance using ultrasound *Neftyanoye khozyaistvo* **12** 53-54
[9] Musipov H N, Savinykh Yu A and Dunaev S A 2005 Acoustic technology to reduce vibration of tubing fitted with centrifugal pumps *Neftyanoye khozyaistvo* **11** 82-83
[10] Savinykh Yu A, Grachev S I and Musipov H N 2011 *Methods of Production Stimulation* (Tyumen: Slovo)
[11] Savinykh Yu A 2004 Patent RUS 2263765. A method of prevention of wax accumulation in the oil well
[12] Nikolaev S A and Ovchinnikov M N 1992 Sound generation by filtration flow in porous structures *Akusticheskiy zhurnal* **1** 114-118